This book is about ways of understanding contingency and necessity in the world and how these ideas influenced the development of the mechanical philosophy in the seventeenth century. It examines the transformation of medieval ideas about God's relationship to the creation into seventeenth-century ideas about matter and method as embodied in early articulations of the mechanical philosophy. Medieval thinkers were primarily concerned with the theological problem of God's relationship to the world he created. They discussed questions about necessity and contingency as related to divine power. By the seventeenth century, the focus had shifted to natural philosophy and the extent and certitude of human knowledge. Underlying theological assumptions continued to be reflected in the epistemological and metaphysical orientations incorporated into different versions of the mechanical philosophy.

The differences between Pierre Gassendi's (1592–1655) and René Descartes' (1596–1650) versions of the mechanical philosophy directly reflected the differences in their theological presuppositions. Gassendi described a world utterly contingent on divine will. This contingency expressed itself in his conviction that empirical methods are the only way to acquire knowledge about the natural world and that the matter of which all physical things are composed possesses some properties that can be known only empirically. Descartes, on the contrary, described a world in which God had embedded necessary relations, some of which enable us to have a priori knowledge of substantial parts of the natural world. The capacity for a priori knowledge extends to the nature of matter, which, Descartes claimed to demonstrate, possesses only geometrical properties. Gassendi's views can be traced back to the ideas of the fourteenth-century nominalists, while Descartes' can be linked to the Thomist tradition he imbibed at La Flèche. Refracted through the prism of the mechanical philosophy, these theological conceptualizations of contingency and necessity in the world were mirrored in different styles of science that emerged in the second half of the seventeenth century.

T0275673

Divine will and the mechanical philosophy

Divine will and the mechanical philosophy

Gassendi and Descartes on contingency and necessity
in the created world

MARGARET J. OSLER
THE UNIVERSITY OF CALGARY

CAMBRIDGE
UNIVERSITY PRESS

PUBLISHED BY THE PRESS SYNDICATE OF THE UNIVERSITY OF CAMBRIDGE
The Pitt Building, Trumpington Street, Cambridge, United Kingdom

CAMBRIDGE UNIVERSITY PRESS
The Edinburgh Building, Cambridge CB2 2RU, UK
40 West 20th Street, New York NY 10011–4211, USA
477 Williamstown Road, Port Melbourne, VIC 3207, Australia
Ruiz de Alarcón 13, 28014 Madrid, Spain
Dock House, The Waterfront, Cape Town 8001, South Africa

http://www.cambridge.org

First published 1994
Reprinted 1998
First paperback edition 2004

Typeface Sabon.

A catalogue record for this book is available from the British Library

ISBN 0 521 46104 9 hardback
ISBN 0 521 52492 X paperback

For J'nan, who knows why

Contents

Acknowledgments   *page* ix
Introduction   1

Part I. Theology and the philosophy of nature

1. Divine power and divine will in the Middle Ages: 15
    Historical and conceptual background

2. Baptizing Epicurean philosophy: Gassendi on divine   36
    will and the philosophy of nature
        Gassendi's Epicurean project   36
        Gassendi's voluntarist theology   48
        Gassendi's providential worldview   56
        The limits of mechanization: Gassendi on the   59
        immortality of the soul
        Appendix: Gassendi's 1631 outline of his Epicurean   78
        project

3. Providence and human freedom in Christian   80
    Epicureanism: Gassendi on fortune, fate, and divination

4. Theology, metaphysics, and epistemology: Gassendi's   102
    "science of appearances"

5. Eternal truths and the laws of nature: The theological   118
    foundations of Descartes' philosophy of nature
        The creation the eternal truths   123
        The status of the laws of nature   136
        Was Descartes really an intellectualist?   146

6. Gassendi and Descartes in conflict   153

Part II. The mechanical philosophy and the formation of scientific
styles

7. Introduction: Theories of matter and their   171
    epistemological roots

8. Gassendi's atomism: An empirical theory of matter   180
    Atoms and the void   182
    Atoms and qualities: An empiricist mechanical   194
    philosophy

9. Mathematizing nature: Descartes' geometrical theory   201
    of matter

10. Conclusion: Theology transformed – the emergence of   222
    styles of science

Bibliography   237
Index   271

Acknowledgments

This book has been a long time coming. In the process, I have accrued many debts. I began working on it during my years as a young graduate student in the Department of History and Philosophy of Science at Indiana University. I vividly remember one of Richard S. Westfall's eloquent lectures in his course on the Scientific Revolution in the spring of 1964, one in which he compared the mechanical philosophies of Gassendi and Descartes. That lecture, coupled with my reading of E. A. Burtt's *Metaphysical Foundations of Modern Science,* stimulated me to explore the history of the mechanical philosophy more deeply, leading in due course to a Master's thesis, a doctoral dissertation, and eventually – so many years later – this book. Progress has not been steady, and the road has not been straight. At times I felt that I was in a trackless wilderness, and sometimes I shared Voltaire's exasperation when he described his plans for writing the history of the century of Louis XIV: "God preserve me from devoting 300 pages to the story of Gassendi! Life is too short, time too precious, to spend it speaking of useless things."[1] That I did not succumb to despair and that I continue to find meaning in this work owes as much to the support of my good friends and colleagues as to my own determination to persevere and bring to fruition a project I undertook with youthful innocence.

Since adequate thanks would require me to write an autobiography, which would extend far beyond the bounds of prefatory propriety, abbreviated acknowledgments will have to suffice.

To begin at the beginning, I am grateful to my parents who raised me to be a scholar and have always supported my academic endeavors. They will not be mistaken if they detect some resonances of our many, sometimes heated, dinner table discussions in this book.

1. Voltaire to Jean Baptiste Dubos, 30 October 1738, in *The Complete Works of Voltaire,* edited by Theodore Besterman et al., 135 vols. (Geneva: Institut et Musée Voltaire, and Toronto: University of Toronto Press, 1969), vol. 89, p. 345; my translation.

Richard S. Westfall supervised my graduate studies and thesis research. He prepared me well for teaching and writing the history of science and has always been a powerful model for me. I am grateful for his continuing friendship and support.

Several people have taken the time to read all or part of earlier drafts of this book: H. Floris Cohen, David Duncan, Paul Farber, and Lynn Joy. Their comments and criticisms – as varied as their interests and approaches – have vastly improved its quality. Lisa T. Sarasohn has been a close friend and companion in Gassendi studies since we first met – by fate, chance, or providence – in 1978. She has read and criticized my work at every stage of this project, and she has generously shared drafts of her own work.

Betty Jo Teeter Dobbs deserves special thanks for reading not just one but three complete drafts of this book. Her criticisms have always been honest and her encouragement genuine. The book has improved and I have grown in response to her comments. I have been fortunate to have her as a friend and mentor. We share memories of great days in the Canadian Rockies when she reinforced my understanding of the limits of mechanism and of the importance of the vegetable spirit.

An anonymous reader for Cambridge University Press read two successive drafts of the book, making detailed and extremely thoughtful comments that guided me through extensive revisions and saved me from some egregious errors. That the reader and I do not agree on all matters of interpretation in no way lessens the usefulness of the readings.

Muriel Blaisdell, Roberta Beeson, Catherine Gerbich Marcus, and Karen Torjesen many years ago formed a community of friendship that was personally sustaining and helped me form my point of view. Bernard Cooke, Maria Eriksen, Jean Hall, Marsha P. Hanen, Francine Michaud, Barbara J. Shapiro, Martin S. Staum, and Robert G. Weyant have, in one way or another, been excellent friends and colleagues. Jack MacIntosh has read chapters and discussed ideas with me at length. He is the only colleague I know whom I can phone in desperation at 11:30 at night for yet one more lucid and gracious explanation of future contingents. I learned a lot about nothing during many stimulating conversations with Jane Jenkins, who wrote her Master's thesis on the concept of the void under my supervision. She also ferreted out obscure journal articles for me in the University of Toronto library. Edward B. Davis, Philip Holtrop, and Jan W. Wojcik generously shared their dissertations with me.

Time for sustained work was provided by three sabbatical leaves granted by the University of Calgary, for the academic year 1980–1, for the fall term of 1984, and for the 1991 calendar year. Life was made more pleasant by Leave Fellowships from the Social Sciences and Humanities

Research Council of Canada granted for the first two of these sabbaticals. More time was provided for research by the Calgary Institute for the Humanities, which awarded me an Annual Fellowship for the academic year 1987–8 and by an Isaac Walton Killam Resident Fellowship, which I received for the fall term of 1989.

Other friends, not directly connected to this work, are a deep source of stimulation and inspiration.

Paula R. Backscheider's energy and determination stand as a model that has more than once pulled me out of the slough of despond.

Blythe McVicker Clinchy constantly reminds me, in both word and deed, that the personal and the intellectual are inseparably bound and that humor and e-mail are essential to maintaining the connection.

My friend Betty Flagler has provided close personal support for many years and has always been patient with the strange demands that this project has placed on my time and energy.

J'nan Morse Sellery, to whom this book is dedicated, has been a close friend ever since we met in 1970 as colleagues at Harvey Mudd College in Claremont, California. She has always posed challenging questions, provided persistent goading, and offered insight and comfort. Her friendship has made it possible for me to sustain this work to completion.

Introduction

Just as the poets suppose that the Fates were originally established by Jupiter, but that after they were established he bound himself to abide by them, so I do not think that the essences of things, and the mathematical truths which we can know concerning them, are independent of God. Nevertheless I do think that they are immutable and eternal, since the will and decree of God willed and decreed that they should be so.

René Descartes, "Replies, V"[1]

If some of the natures [of things] are immutable and eternal and could not be otherwise than they are, God would not have existed before them. Otherwise such things would not be natures . . . The thrice great God is not, as Jupiter of the poets is to the fates, bound by things created by him, but can in virtue of his absolute power destroy anything that he has established.

Pierre Gassendi, Disquisitio metaphysica[2]

This book is about ways of understanding contingency and necessity in the world and how those ideas influenced the development of philosophies of nature in the seventeenth century. Is the world contingent on forces beyond the possibility of human understanding and control? Or

1. René Descartes, "Fifth Set of Replies," in *The Philosophical Writings of Descartes* (hereafter *PWD*), translated by John Cottingham, Robert Stoothoff, Dugald Murdoch, and Anthony Kenny, 3 vols. (Cambridge University Press, 1984, 1985, 1991), vol. 2, p. 261, *Oeuvres de Descartes*, edited by Charles Adam and Paul Tannery, 11 vols. (hereafter AT) (Paris: J. Vrin, 1897–1983), vol. 7, p. 380.
2. Pierre Gassendi, *Disquisitio metaphysica seu dubitationes et instantiae adversus Renati Cartesii metaphysicam et responsa*, edited and translated into French by Bernard Rochot (Paris: J. Vrin, 1962), pp. 480–1; in Pierre Gassendi, *Opera omnia*, 6 vols. (Lyons, 1658; facsimile reprint, Stuttgart-Bad Cannstatt: Friedrich Frommann Verlag, 1964), vol. 3, p. 377. All translations are mine unless otherwise noted.

does the world necessarily conform to rationally intelligible principles? The interplay between these conceptions goes back to both the Greek and the Hebrew sources of Western thought, forming an important strand in the long history of the relationship between Athens and Jerusalem. The themes of contingency and necessity appear in various guises throughout the intellectual history of the West. In this study, I discuss the role they played in the seventeenth-century debates about the choice of a new philosophy of nature to replace the Aristotelianism that had dominated natural philosophy for many centuries.

Classical Greek ideas about chance and reason, about the intransigence of matter and the intelligibility of Pythagorean harmonies and Platonic forms expressed these themes.[3] To the extent that the gods were subject to the fates, reason was limited by the irrational, logical necessity by contingent fact. The same themes of contingency and necessity arose within the Hebrew tradition. In his argument with God about why piety does not guarantee prosperity and well-being, Job sought an explanation of his misfortunes, assuming that the world is a rational and just place in which the good are rewarded and the wicked are punished. Instead of receiving a satisfying answer, he found himself confronted with Jahweh's demand for uncomprehending obedience. In this way, Job discovered the absolute contingency of the world in terms of its dependence upon an omnipotent God whose actions do not necessarily conform to human standards of rationality and justice. "I know you can do all things and nothing you wish is impossible" (Job 42:1).[4]

Ideas about contingency and necessity – cast in terms of the possibilities and limits of human understanding – are closely tied to ideas about causality. Is the course of natural and human events ineluctably determined? or do some events happen by chance or coincidence, apparently free from the shackles of causal necessity? Such considerations are incorporated into scientific discourse in underlying assumptions about causality and the epistemological status of our knowledge of the natural world. These assumptions comprise the conceptual framework within

3. For a discussion of these ideas in an ethical context, see Martha C. Nussbaum, *The Fragility of Goodness: Luck and Ethics in Greek Tragedy and Philosophy* (Cambridge University Press, 1986). See also Richard Sorabji, *Necessity, Cause, and Blame: Perspectives on Aristotle's Theory* (Ithaca, N.Y.: Cornell University Press, 1980).

4. Stephen Mitchell, *The Book of Job* (San Francisco: North Point, 1987), p. 88. The more familiar King James version of this passage reads, "I know that thou canst do every thing, and that no thought can be withholden from thee."

which science is constructed. During the early seventeenth century, many natural philosophers were explicitly engaged in formulating a new conceptual framework to replace Aristotelianism, which they considered intellectually bankrupt in light of the Renaissance revival of ancient philosophies of nature, the Reformation, the skeptical crisis, and the Copernican revolution.[5] Although philosophers debated the merits of many alternative philosophies of nature, eventually the mechanical philosophy – the view that all natural phenomena can be explained in terms of matter and motion alone – was adopted as the conceptual framework for natural philosophy.

Aristotelianism had served as the conceptual framework for science since the translation of Aristotle's works from Arabic into Latin in the thirteenth century.[6] Because the various comp nents of Aristotelianism were so closely knit, the heliocentrism of Copernicanism threatened the entire philosophy, not just its geocentric cosmology. This challenge was intensified by the realist interpretation that Copernicus, Kepler, and Galileo gave to astronomical theory, meaning that for the first time since Greek antiquity, the conclusions of physics and astronomy were relevant to each other.[7] By removing the Earth from the center of the universe, the Copernicans undermined Aristotelian physics to which the assumption of

---

5. As a matter of historical fact, a viable Aristotelian tradition continued to develop well into the seventeenth century. See Charles B. Schmitt, *Aristotle and the Renaissance* (Cambridge, Mass.: Harvard University Press, 1983), p. 7. See also Edward Grant, "Celestial Perfection from the Middle Ages to the Late Seventeenth Century," in *Religion, Science, and Worldview: Essays in Honor of Richard S. Westfall* (Cambridge University Press, 1985), pp. 137–62; and Edward Grant, "Ways to Interpret the Terms 'Aristotelian' and 'Aristotelianism' in Medieval and Renaissance Natural Philosophy," *History of Science, 25* (1987): 335–58.
6. See Edward Grant, "Aristotelianism and the Longevity of the Medieval World View," *History of Science, 16* (1978): 93–106.
7. For a discussion of realism and instrumentalism in the history of astronomy, see Pierre Duhem, *To Save the Phenomena: An Essay on the Idea of Physical Theory from Plato to Galileo,* translated by Edmund Doland and Chaninah Maschler (Chicago: University of Chicago Press, 1969; first published, 1908). Not all Copernicans were realists. See Robert S. Westman, "Three Responses to the Copernican Theory: Johannes Praetorius, Tycho Brahe, and Michael Maestlin," in *The Copernican Achievement,* edited by Robert S. Westman (Berkeley: University of California Press, 1975), pp. 285–345; and Robert S. Westman, "The Melancthon Circle, Rheticus, and the Wittenberg Interpretation of the Copernican Theory," *Isis, 66* (1975): 165–93.

a geocentric cosmos was intrinsic. Galileo's new science of motion further eroded the essentialism of Aristotelian matter theory by making it impossible to infer the nature of a body from its motions.[8] Because the motions or other perceived qualities of bodies no longer revealed the essences of the bodies in question, observation alone could not lead to knowledge of forms. In this way, the foundations of Aristotle's theory of scientific knowledge were undermined.

The Renaissance recovery of classical texts made alternative ancient philosophies of nature available to humanistically inclined natural philosophers. Of particular significance were Plato's dialogues, Lucretius' exposition of ancient atomism in *De rerum natura,* and the Hermetic corpus, all of which had been little known before the fifteenth century.[9] The availability of these alternatives made it easier for natural philosophers to contemplate the abandonment of Aristotelianism in an age when many thinkers considered an ancient model to be a prerequisite for working out a philosophical position.[10] Another factor undermining the authority of Aristotelianism was the skeptical crisis of the sixteenth and early seventeenth centuries.[11] The epistemological issues raised by the Reformation debates over the criterion for a rule of faith, the recovery and translation of the writings of the ancient skeptics, and the psychological impact of the discovery of the New World gave skepticism prominence in the intellectual world.

The mechanical philosophy appealed to the practitioners of the new

8. See Margaret J. Osler, "Galileo, Motion, and Essences," *Isis,* 64 (1973): 504–9.
9. On the recovery of ancient texts, see Anthony Grafton, "The Availability of Ancient Works," in *The Cambridge History of Renaissance Philosophy,* edited by Charles B. Schmitt, Quentin Skinner, and Eckhard Kessler (Cambridge University Press, 1988), pp. 767–91.
10. Although Schmitt wisely emphasized the persistence of Aristotelianism in the Renaissance, he acknowledged that "several of the major philosophers and scientists of the generation of Bacon and Galilei on to that of Hobbes and Descartes sealed the fate of Aristotelianism as a coherent philosophy, at least from an intellectual if not wholly from a historical point of view." Charles B. Schmitt, *Aristotle and the Renaissance,* p. 5.
11. The key study here is Richard H. Popkin, *The History of Scepticism from Erasmus to Spinoza,* revised edition (Berkeley: University of California Press, 1979). For the origins of the skeptical crisis, see esp. chaps. 1 and 2. On the revival of Academic skepticism, see Charles B. Schmitt, *Cicero Scepticus: A Study of the Influence of the* Academica *in the Renaissance* (The Hague: Martinus Nijhoff, 1972).

natural philosophy. Galileo held such ideas,[12] as did the very influential, though unpublished, Isaac Beeckman.[13] So did Gassendi and Descartes, both of whom created systematic mechanical philosophies.[14] One reason for its popularity was its apparent compatibility with recent developments in astronomy, the science of motion, and physiology.[15] These developments appeared to Hobbes and other seventeenth-century natural philosophers as a new beginning:

> I know . . . that the hypothesis of the earth's diurnal motion was the invention of the ancients; but that both it, and astronomy, that is celestial physics, springing up together with it, were by succeeding philosophers strangled with the snares of words. And therefore the beginning of astronomy, except observations, I think is not to be derived from farther time than from Nicolaus Copernicus; who in the age next preceding the present revived the opinion of Pythagoras, Aristarchus, and Philolaus. After him, the doctrine of the motion of the earth being now received, and a difficult question thereupon arising concerning the descent of heavy bodies, Galileus in our time, striving with that difficulty, was the first that opened to us the gate of natural philosophy universal, which is the knowledge of the nature of *motion*. So that neither can the age of natural philosophy be reckoned higher than to him. Lastly, the science of *man's body*, the most profitable part of natural science, was first discovered with admirable sagacity by our countryman, Doctor Harvey . . . Before these, there

12. Galileo, *The Assayer*, in *Discoveries and Opinions of Galileo*, edited and translated by Stillman Drake (Garden City, N.Y.: Doubleday Anchor, 1957), pp. 275–8. See also Pietro Redondi, *Galileo Heretic*, translated by Raymond Rosenthal (Princeton, N.J.: Princeton University Press, 1987). Redondi clarifies the relationship between an endorsement of atomism and its bearing on the explanation of the Eucharist, esp. in chaps. 7 and 9.

13. Although Beeckman's journal was not published in toto before the twentieth century, extracts from it were published in 1644 in *D. Isaaci Beeckmanni medici et rectoris apud Dordracenos mathematico-physicarum meditationum, questionum solutionum centuria* (Utrecht, 1644). See R. Hooykaas, "Science and Religion in the Seventeenth Century: Isaac Beeckman (1588–1637)," *Free University Quarterly*, 1 (1951): 169; and R. Hooykaas, "Beeckman, Isaac," in *Dictionary of Scientific Biography*, edited by Charles Coulton Gillispie, 16 vols. (New York: Scribner, 1972), vol. 1, p. 566.

14. E. J. Dijksterhuis, *The Mechanization of the World Picture*, translated by C. Dikshoorn (Oxford University Press, 1961), pp. 70–6.

15. For example, Descartes considered Harvey's proof of the circulation of the blood as "much easier to conceive" if understood in mechanical terms. See René Descartes, *Discourse on the Method of Rightly Conducting One's Reason and Seeking the Truth in the Sciences*, in *PWD*, vol. 1, pp. 132–9; AT, vol. 6, 41–55.

was nothing certain in natural philosophy but every man's experiments to himself.[16]

It seemed quite plausible to advocates of the mechanical philosophy to construct a physics of the heavens that regarded planets as material objects whose motions in space were amenable to mathematical description. A world consisting only of matter and motion appeared to be accessible to both observation and mathematical analysis, while the substantial forms and occult qualities of the Aristotelians had come to seem obscure. Moreover, a mechanical philosophy of nature described a homogeneous universe, all the parts of which were governed by the same laws of nature, a uniformity of nature throughout space that Koyré called "the destruction of the Cosmos."[17]

The first half of the seventeenth century witnessed the development of a community of thinkers who shared a fairly explicit concern to formulate a mechanical philosophy to provide metaphysical foundations for these developments in natural philosophy. Important members of this group included Isaac Beeckman (1588–1637), Marin Mersenne (1588–1648), Thomas Hobbes (1588–1679), Pierre Gassendi (1592–1655), René Descartes (1596–1650), Sir Kenelm Digby (1603–65), and Walter Charleton (1620–1707). These men knew each other personally and reacted to each other's work.[18]

16.	Thomas Hobbes, *Elements of Philosophy: The First Section, Concerning Body* (1655), translated into English, in *The English Works of Thomas Hobbes of Malmesbury,* edited by Sir William Molesworth, 11 vols. (London, 1839–45; reprinted, Aalen: Scientia, 1962), vol. 1, pp. viii–ix.

17.	Alexandre Koyré, "Galileo and Plato," in *Metaphysics and Measurement: Essays in the Scientific Revolution* (London: Chapman & Hall, 1968), p. 19; first published in *Journal of the History of Ideas,* 4 (1943): 400–28. See also Jean Jacquot, "Harriot, Hill, Warner, and the New Philosophy," in John W. Shirley, *Thomas Hariot: Renaissance Scientist* (Oxford University Press, 1974), p. 108.

18	For the relations among these people, see Robert Hugh Kargon, *Atomism in England from Hariot to Newton* (Oxford University Press, 1966), chaps. 6-8; Samuel I. Mintz, *The Hunting of Leviathan: Seventeenth Century Reactions to the Materialism and Moral Philosophy of Thomas Hobbes* (Cambridge University Press, 1969), chap. 1; Robert Lenoble, *Mersenne ou la naissance du mécanisme,* 2d edition (Paris: J. Vrin, 1971), chap. 1; Michael Foster, "Sir Kenelm Digby (1603-1665) as Man of Religion and Thinker– I. Intellectual Formation." *Downside Review,* 106 (1988): 101-25; and Lindsay Sharp, "Walter Charleton's Early Life, 1620-1659, and the Relationship to Natural Philosophy in mid-Seventeenth Century England." *Annals of Science, 30* (1973): 311-40.

Despite considerable differences in political context and religious affil-
iation, these mechanical philosophers formed a European community that
crossed both national boundaries and confessional lines. Beeckman, a
member of the Dutch Reformed church, was a school rector in the Netherlands.[19] Mersenne, Gassendi, and Descartes were all Catholics living in
the France of Louis XIII and the Fronde. Both Mersenne and Gassendi
were priests. Charleton, personal physician to Charles I, was a Royalist
and an Anglican and probably spent the early 1650s in France.[20] Digby
was an English Catholic who spent time in Paris.[21] Hobbes, born into a
Protestant family and educated at a Puritan college at Oxford, had a
reputation as a materialist and possibly an atheist. He served as tutor to
the son of William Cavendish, in whose household he mingled with many
of the central figures of the "new philosophy."[22] He spent many years in
Paris, where he was personally acquainted with Mersenne, Gassendi,
Descartes, and Digby.[23]

Whatever their differences in politics, nationality, and faith, these men
formed an international, self-consciously intellectual community. Although all of them were educated in Aristotelianism, they were united in
their opposition to it and in their support of a mechanical philosophy to
replace it. They shared an admiration for Galileo and a commitment to the
"new science" more generally. With the exception of Beeckman, each
published at least one major work, spelling out his own version of the new
philosophy.[24] Despite many important differences in detail, these books
resemble each other in important ways. They defended the mechanical
philosophy and argued against Aristotelian and occult alternatives. They
included sections describing the ultimate components of the world, matter

19. Hooykaas, "Beeckman," p. 566.
20. Sharp, "Walter Charleton's Early Life," pp. 311–27.
21. Foster, "Sir Kenelm Digby," pp. 42–3.
22. Mintz, *Hunting of Leviathan,* pp. 3–5. On the role of the Cavendish circle in
    English natural philosophy, see Kargon, *Atomism in England,* chap. 7.
23. Mintz, *Hunting of Leviathan,* chap. 1.
24. The works in question are as follows: Marin Mersenne, *Quaestiones celeber-
    rimae in Genesim* (1623), *L'impiété des déistes* (1624), *La vérité des sciences*
    (1625), and *Traité de l'harmonie universelle* (1627); René Descartes, *Prin-
    cipia philosophiae* (1644); Sir Kenelm Digby, *Two Treatises. In the One of
    Which, The Nature of Bodies; in the Other, the Nature of Mans Soule; is
    Looked Into: In Way of Discovery, of the Immortality of Reasonable Soules*
    (1644); Thomas Hobbes, *De corpore,* Part I of *The Elements of Philosophy*
    (1655); Walter Charleton, *Physiologia Epicuro-Gassendo-Charltoniana, or
    a Fabrick of Science Natural Upon the Hypothesis of Atoms* (1654); and
    Pierre Gassendi, *Syntagma philosophicum* (1658).

and motion. They explained various phenomena in mechanical terms, namely, the impact of material particles. They included lists of all the known qualities of bodies and showed how they could be explained in mechanical terms. They devoted considerable attention to explaining human perception. And central to their accounts was the doctrine of primary and secondary qualities, the view that material bodies actually possess only a few primary qualities and that the observed qualities of bodies result from the interaction of the primary qualities with our sense organs. They thus mechanized the natural world and human perception, declaring that qualities are subjective, being relative to the human perceiver. A cursory look at the tables of contents of their expositions of the mechanical philosophy reveals this commonality of concern.

As a community of thinkers, these natural philosophers were struggling with a related set of issues. The skeptical crisis manifested itself in the attention each of them devoted to questions about method. While not all of them were as philosophically imaginative as Descartes in attempting to preserve the traditional certainty of *scientia* or as pragmatically innovative as Gassendi in his "mitigated scepticism," they all considered the epistemological challenge of scepticism.[25] With the possible exception of Hobbes,[26] they all were concerned to avoid the materialism and atheism that was traditionally associated with Greek atomism. Consequently, their treatises on the mechanical philosophy contain sections establishing the existence of God, the nature of his providential relationship to the creation, human freedom, and the immortality of the human soul.

In this study I focus on the thought of two members of this group, Gassendi and Descartes, whose solutions to these problems had the greatest influence in the long run.[27] Gassendi and Descartes established two significantly different versions of the mechanical philosophy. Agreeing on the fundamental tenet of the mechanical philosophy – that all natural phenomena can be explained in terms of matter and motion – as well as rejecting the Aristotelian and occult philosophies of the day, they disagreed about virtually everything else: the nature of matter, the episte-

25. Popkin, *History of Scepticism,* chaps. 7, 9, 10.
26. The problem of interpreting Hobbes' theological position is complex. On the relationship between his thought and Reformation theology, see Leopold Damrosch, Jr., "Hobbes as Reformation Theologian: Implications of the Free-Will Controversy," *Journal of the History of Ideas,* 40 (1979): 339–52.
27. Kargon considers Gassendi, Descartes, and Hobbes to be "the three most important mechanical philosophers of the mid-seventeeth century." See *Atomism in England,* p. 54.

mological status of scientific knowledge, and particular mechanical explanations of individual phenomena. Gassendi, following the ancient models of Epicurus and Lucretius, maintained that indivisible atoms and the void are the ultimate components of nature. Atoms possess magnitude, figure, and heaviness, properties that cannot be fully known by reason alone. He advocated an empiricist theory of scientific knowledge, claiming, in addition, that only individuals exist and that it is impossible to have knowledge of the essences of things. Descartes maintained that the universe is a plenum and that the matter filling it is infinitely divisible. According to Descartes, matter is identical with geometrical space, and its only property is extension, an attribute that can be understood rationally, without any appeal to observation or experience.[28] Although his theory of scientific knowledge indeed required appeal to empirical methods, he claimed that the first principles of natural philosophy can be known a priori and can lead to knowledge of the essences of things.

I have chosen to focus on the views of Gassendi and Descartes because of their demonstrably strong influence on the further development of the mechanical philosophy.[29] Although both thinkers advocated the mechanical account of nature, their differences were evident to their contemporaries. The next generation of natural philosophers, who accepted the mechanical philosophy in general, felt that they had to choose between Gassendist and Cartesian versions or find some accommodation between the two.[30] For example, the young Isaac Newton (1642–1727) constructed thought experiments in an attempt to decide between Cartesian and Gassendist explanations of particular phenomena.[31] Similarly, when Robert Boyle (1627–91) claimed to "write for Corpuscularians in general

---

28. Daniel Garber, *Descartes' Metaphysical Physics* (Chicago: University of Chicago Press, 1992), pp. 117–20.
29. I have included somewhat more material on Gassendi than Descartes in this book because Gassendi's work has been relatively neglected in the scholarly literature.
30. On this debate in the philosophical community, see Thomas M. Lennon, *The Battle of the Gods and Giants: The Legacies of Descartes and Gassendi, 1655–1715* (Princeton, N.J.: Princeton University Press, 1993).
31. Richard S. Westfall, *Never at Rest: A Biography of Isaac Newton* (Cambridge University Press, 1980), pp. 96–7. See also Richard S. Westfall, "The Foundations of Newton's Philosophy of Nature," *British Journal for the History of Science, 1* (1962): 171–82. For the full text of Newton's early notebook, see J. E. McGuire and Martin Tamny, *Certain Philosophical Questions: Newton's Trinity Notebook* (Cambridge University Press, 1983).

[rather] than any party of them," he had the Cartesians and the Gassendists in mind as the two main parties of mechanical philosophers.[32]

The differences between these two versions of the mechanical philosophy produced, in the latter part of the seventeenth century, different styles of scientific thought, one emphasizing an empiricist approach to science, the other a more rationalistic, mathematical approach. In this study, I explore in detail and attempt to explain the ways in which Descartes' and Gassendi's versions of the mechanical philosophy differed from each other. I argue that these differences were related to differences in their underlying theological assumptions about God's relationship to the creation,[33] specifically, the issue of how binding God's act of creation is on his future interactions with the world.

Is God bound by his creation, or is he always free to change whatever he created in the world? The salient theological assumptions are expressions of the role of contingency and necessity in the universe. The language that Gassendi and Descartes used to articulate answers to these questions was originally developed in the thirteenth and fourteenth centuries as an outgrowth of the reintroduction of Aristotle's works into mainstream philosophical thought. Contingency and necessity had been interpreted at that time in terms of the dialectic between God's omnipotence and his omniscience. There was a delicate balance in medieval theology between the rationality of God's intellect and his absolute freedom in exercising his power and will. Theologians who emphasized God's rationality were more inclined to accept elements of necessity in the creation than those who emphasized his absolute freedom and concluded that the world is utterly contingent. The necessity at stake for these thinkers was both metaphysical and epistemological. Metaphysical necessity is expressed in the relations between the essences and qualities of substances, in the relations between one entity and another, and in the relations between causes and their effects. In this metaphysical sense, the state of one being entails the corresponding state in another. Metaphysical necessity of this kind often provided foundations for epistemological necessity, the capacity to know with demonstrative certainty one state of affairs in the world on the basis of knowledge of another. In the seventeenth century,

32. Robert Boyle, *The Origin of Forms and Qualities According to the Corpuscular Philosophy,* in *The Works of the Honourable Robert Boyle,* edited by Thomas Birch, 6 vols (London, 1772; reprinted, Hildesheim: Georg Olms, 1965), vol. 3, p. 7.

33. On the central role of theology in seventeenth-century natural philosophy, see Andrew Cunningham, "How the *Principia* Got Its Name: Or, Taking Natural Philosophy Seriously," *History of Science,* 29 (1991): 377–92.

these ideas about God's relationship to the creation were transformed into views about the metaphysical and epistemological status of human knowledge and the laws of nature. Necessity found expression in the view that the laws of nature describe the essences of things and can be known a priori, while the empiricist and probabilist interpretations of scientific knowledge provided a way of thinking about the contingency of a world that no longer contained essences in a Platonic or Aristotelian sense.

I begin by examining the sources of these two conceptualizations of divine power, intellectualism and voluntarism, in their thirteenth- and fourteenth-century settings. Briefly, voluntarism is the view that the creation is absolutely contingent on God's will. Intellectualism is the view that there are some elements of necessity in the creation. Whether God created these elements of necessity or whether they exist independent of him, he cannot override them. These are complicated and subtle theological positions, and I discuss them more fully in Chapter 1. I have chosen to use the views of Aquinas and Ockham as paradigm cases of these two theological positions. Following this background, I examine the role of theological presuppositions in the thought of both Gassendi and Descartes in Chapters 2 through 6. The relationship between God and the creation occupied an important place in their writings. Gassendi was a theological voluntarist, and Descartes was a kind of intellectualist. I then proceed in Chapters 7 through 9 to examine the role that these presuppositions played in their respective formulations of the mechanical philosophy. Concepts, originally developed in one area of discourse, theology, were translated or transplanted into another domain, natural philosophy, where they took on a life of their own and had far-reaching ramifications.[34] Gassendi's empiricism and antiessentialism were tied to his voluntarism. The rationalist components of Descartes' theory of knowledge found metaphysical foundations in his intellectualist understanding of the deity. Furthermore, the theory of matter each adopted was intimately related to his epistemological predilections. Descartes equated matter with geometrical extension, endowing it with intelligible properties that can be known in a purely a priori way. By contrast, Gassendi endowed matter with some properties – such as solidity and weight – which can

34. Amos Funkenstein develops the concept of the "dialectical anticipation of a new theory by an old one." He speaks, in this context, of the "transplantation" of existing categories to a new domain. His conceptualization here captures the same idea I am trying to express when talking about the translation of concepts from one domain to another. See Funkenstein, *Theology and the Scientific Imagination from the Middle Ages to the Seventeenth Century* (Princeton, N.J.: Princeton University Press, 1986), pp. 14–17.

only be known empirically. I conclude by suggesting that the two versions of the mechanical philosophy, established by Gassendi and Descartes, respectively, developed later in the seventeenth century into two styles of scientific pursuit, styles that differed in the emphasis they placed on empirical evidence and mathematics and in their interpretations of mechanical models of natural phenomena.

I

Theology and the philosophy of nature

1

Divine power and divine will in the Middle Ages: Historical and conceptual background

> What indeed has Athens to do with Jerusalem: What concord is there between the Academy and the Church?
>
> *Tertullian, On Prescription Against Heretics*[1]

If the development of science in Western Europe can be understood, at least in part, as resulting from the coupling of Greek ideas about rationality with biblical notions of God's power, one troublesome offspring of this union was the problem of reconciling two of God's primary attributes, his omnipotence and his omniscience. Although many classical thinkers had considered the cosmos to be governed by rational principles of some kind, their discussions of chance, fate, and fortune had implicitly acknowledged the fact that human life depends on forces beyond rational control.[2] Medieval Christian theologians faced the difficult task of reconciling the Old Testament God of Abraham, Isaac, and Jacob, who created the world and rules it freely according to his own will, with Greek ideas about the self-sufficiency and rationality of cosmic principles – an endeavor memorably characterized by Lovejoy as "perhaps the most extraordinary triumph of self-contradiction."[3] Tensions between these classical and

1. Tertullian, *On Prescription Against Heretics,* in *The Ante-Nicene Fathers,* translated and edited by Alexander Roberts and James Donaldson, revised by A. Cleveland Coxe, 10 vols. (Edinburgh: T. & T. Clark, 1885–7; reprinted Grand Rapids, Mich.: Eerdmans, 1986–7), vol. 3, p. 246.
2. See, e.g., Marcus Tullius Cicero, *De fato,* translated by H. Rackham (Cambridge, Mass: Loeb Classical Library, 1942). For a discussion of these issues in an ethical context, see Martha C. Nussbaum, *The Fragility of Goodness: Luck and Ethics in Greek Tragedy and Philosophy* (Cambridge University Press, 1986).
3. Arthur O. Lovejoy, *The Great Chain of Being: A Study of the History of an Idea* (Cambridge, Mass.: Harvard University Press, 1936), p. 157.

biblical ideas formed the intellectual context of medieval and early modern discussions about God's relationship to the creation.

In the present chapter, I explore the way God's relationship to the creation was considered in the thirteenth and fourteenth centuries, when it came to be formulated in terms of the dialectic between the absolute and ordained power of God. The problem became particularly acute at this time as a consequence of the direct encounter between Augustinian theology and the philosophy of Aristotle, following the translation and reception of the Aristotelian texts in the eleventh and twelfth centuries.[4] The language used to describe divine power – the medieval term was the *potentia absoluta et ordinata Dei* (the absolute and ordained power of God) – remained in use until the seventeenth century when it was transformed into discourse about such seemingly far-flung topics as the epistemological and metaphysical status of the laws of nature, the proper method for science, and the nature of matter. I pick up the threads of these developments in the thirteenth and fourteenth centuries in part to delineate their historical origins and in part to create the conceptual background necessary for understanding their impact on seventeenth-century natural philosophy.

Embryonic discussions of these ideas can be found in early Christian writings. Expressing his horror at the rapacity of the invading Goths who had sacked Rome in 410, Saint Jerome (d. 420) questioned whether God can raise up a virgin after she has fallen. He was more concerned with the enormity of the invading barbarians' crimes than with the extent of divine power, but his exclamation implicitly posed the question of whether God can alter a course of events, once it has occurred. In a seminal ninth-century treatise on the Eucharist, *De corpore et sanguine Domini*, Paschasius Radbertus (d. 865) argued for real presence in the elements of the Eucharist on the grounds that God can violate the natural order, as he did when blood and water flowed from a wound inflicted on the dead Christ's body.[5] In so doing, Paschasius was beginning to consider the relationship between God's omnipotence and the natural order of the created world.

4. On the translations of Aristotle's works in the Middle Ages and their impact, see Bernard G. Dod, "Aristotles Latinus," and C. H. Lohr, "The Medieval Interpretation of Aristotle," both in *The Cambridge History of Later Medieval Philosophy*, edited by Norman Kretzmann, Anthony Kenny, and Jan Pinborg (Cambridge University Press, 1982), pp. 45–79, pp. 80–98.
5. For the early history of these ideas, see Francis Oakley, *Omnipotence, Covenant, and Order: An Excursion in the History of Ideas from Abelard to Leibniz* (Ithaca, N.Y.: Cornell University Press, 1984), chap. 2. Paschasius Radbertus discusses the power and will of God in *De corpore et sanguine Domini*, in *Sancti Paschasii Radberti Abbatis Corbeiensis Opera Omnia*, in *Patrologiae Latinae*, edited by J.-P. Migne (Turnhoti: Typographi Brepols

Increasing contact between Greek philosophy and Christian theology in the twelfth and thirteenth centuries led to the formulation of two different understandings of the relationship between divine power and the natural order God created. These theological positions are most commonly called "voluntarism" and "intellectualism" or "rationalism," and their meaning is explained in terms of the relationships between two of God's attributes, his intellect and his will.[6] Voluntarists stress the primacy of God's will over his intellect, and intellectualists emphasize God's intellect over his will. Because theologians from both camps agree that God has both intellect and will, the differences between the two positions are rather subtle, and discussions in these terms can easily become confused.[7]

Rather than defining these theological positions in terms of the relationship between the divine attributes, I have found it useful to think of the

Editores Pontificii, 1844–1902), vol. 120, columns 1267–72. I am grateful to Celia Chazelle for providing me this reference. For a summary of Paschasius Radbertus' views on the Eucharist, see Jaroslav Pelikan, *The Christian Tradition: A History of the Development of Doctrine*, 5 vols. (Chicago: University of Chicago Press, 1978), vol. 3, pp. 74–80. For Paschasius' views on the miraculous properties of the Eucharist, see Adolph Harnack, *History of Dogma*, translated by James Millar, 7 vols. (London: Williams & Norgate, 1898), vol. 5, pp. 312–18.

6. I follow Foster's usage here. "What I shall term 'rationalism' in theology is the doctrine that the activity of God is an activity of reason." M. B. Foster, "Christian Theology and Modern Science of Nature (II.)," *Mind*, 45 (1936), 1. See Richard Taylor in "Voluntarism," in *The Encyclopedia of Philosophy*, edited by Paul Edwards, 8 vols. (New York: Macmillan and Free Press, 1967), vol. 8, pp. 270–2.

7. There are terminological problems in naming these positions. While "voluntarism," at least in the theological realm, is fairly unambiguous, the terms "intellectualism" and "rationalism" both have connotations that could burden the discussion with unintended meanings. "Intellectualism," which I take to refer to the primacy of God's intellect, also has the meaning of the opposite of "antiintellectualism," a meaning that has nothing to do with the theological issues at stake here. Similarly, "rationalism" in theology can refer to the Enlightenment argument that religion does not require revelation but can be based on rational grounds; it also can refer to a school of German biblical criticism in the period 1740 to 1840, closely allied to Kant's antisupernaturalism. I intend neither of these extraneous meanings. Likewise, "rationalism" has a precise philosophical meaning, describing an epistemology that allows for the possibility of synthetic a priori knowledge and that is usually contrasted to "empiricism." I argue that epistemological rationalism is a concomitant of rationalism in theology, but these meanings are quite different and separate. See Bernard Williams, "Rationalism," in *The Encyclopedia of Philosophy*, edited by Edwards, vol. 7, p. 69.

differences between the two positions in terms of God's power considered in relation to the creation. Medieval theologians used the terms *potentia Dei absoluta* and *potentia Dei ordinata* – God's absolute and ordained power – to denote that relationship.[8] These terms do not refer to two powers that God possesses, but rather to two ways of understanding divine power. God's absolute power refers to what is theoretically possible for him to do providing it does not involve a logical contradiction. It refers to divine power apart from any particular acts he has chosen to perform. In willing the creation of the world, God exercised his absolute power. In so doing, he did not exhaust the creative possibilities open to him. He could have created different worlds containing different creatures and operating according to different laws from those found in this world. Such alternative worlds might have contained twenty planets that revolve around the moon. In this sense God's power is unbounded.[9] *Potentia ordinata* refers to divine power with respect to what he has actually chosen to do in establishing the present order.[10] God has created a world containing only five planets that circle the earth, and he has created the laws governing the relationships among them. His *potentia ordinata* is his power vis-à-vis this creation, his governance of the world in accordance with the way he created it and his ability to intervene in the creation.

The difference between rationalist and voluntarist theology can be understood in terms of the distinction between *potentia absoluta* and *potentia ordinata*. In these terms, the difference can be expressed by asking how binding the created order is on God's present and future acts. Rationalists or intellectualists were prepared to accept the existence of some necessity in the world, while voluntarists regarded the present order as utterly contingent. Viewed in this way, there were many different kinds of intellectualists. Some, like Peter Abelard (1079–1142), took an almost Platonic position, claiming that a rational order independent of God exists.[11] Platonic forms and Scholastic uncreated essences, such as those discussed by the sixteenth-century Jesuit Aristotelian Francisco Suárez are both

8. See William J. Courtenay, *Capacity and Volition: A History of the Distinction of Absolute and Ordained Power* (Bergamo: Pierluigi Lubrina, 1990). See also Richard P. Desharnais, "The History of the Distinction Between God's Absolute and Ordained Power and Its Influence on Martin Luther," Ph.D. dissertation, Catholic University of America, 1966.

9. William J. Courtenay, "The Dialectic of Omnipotence," in *Divine Omniscience and Omnipotence in Medieval Philosophy*, edited by Tamar Rudavsky (Dordrecht: Reidel, 1985), p. 243.

10. Ibid., p. 247.

11. Oakley, *Omnipotence, Covenant, and Order*, p. 45.

examples of such entities existing independently of God. As such, they exist outside God's creative act and limit the scope of his power to act freely. Other intellectualists, like Thomas Aquinas, took a more moderately rationalist position, allowing that while nothing exists that God did not create freely, nevertheless the laws of nature or essences that he did create embody some necessary relations.[12] These necessary relations render a priori demonstrative knowledge of the creation possible.[13]

In contrast to rationalists or intellectualists, voluntarists have insisted on God's omnipotence and his absolute freedom of will: Nothing exists independently of him, and nothing that he created can bind or impede him. Certainly a rational order independent of God would reduce the scope of his absolute power and would limit his freedom of action in the world. Even the necessity of laws he created freely would restrict the exercise of his power over the creation. Arguing in this vein, Peter Damian (1007–72) contended that God in his omnipotence had created everything, even the laws of logic, which he can change if he so wills.[14] Apart from Damian's extreme position, most voluntarists regarded the law of noncontradiction as the only exception to God's absolute freedom lest such potentially devastating absurdities as God's willing himself to cease existing could occur.[15]

The voluntarists' emphasis on the contingency of the creation had important epistemological implications. Without necessary relations, it is not possible to attain a priori, demonstrative knowledge of the creation. Synthetic a priori knowledge – to use a modern term – is not possible because any guarantee that the contents of the human mind must correspond to the world would involve the existence of some kind of necessary relations, unacceptable to the voluntarists' emphasis on God's absolute power. Any regularities that are observed in the world are simply that,

12. On these varieties of intellectualism, see Norman J. Wells, "Descartes and the Scholastics Briefly Revisited," *New Scholasticism*, 35 (1961): 172–90.

13. "A rationalist theology involves both a rationalist philosophy of nature and a rationalist theory of knowledge of nature. If God made the world according to reason, the world must embody the ideas of his reason; and our reason, in disclosing to us God's ideas will at the same time reveal to us the essential nature of the created world." Foster, "Christian Theology and Modern Science of Nature (II.)," p. 10.

14. Courtenay, "The Dialectic of Omnipotence," pp. 244–5; Taylor, "Voluntarism," p. 271. See also Oakley, *Omnipotence, Covenant, and Order*, pp. 42–4.

15. And there are even exceptions to that stipulation, as, e.g. Peter Damian. See Oakley, *Omnipotence, Covenant, and Order*, p. 44, and Courtenay, "The Dialectic of Omnipotence," pp. 244–5.

observed regularities. The natural order may be regular, but it is also completely contingent. Since God's freedom suffers no restraint, he can alter the observed regularities at will, a possibility to which miracles attest. Consequently, our knowledge of the creation is fallible and has only observation as it source, for there is no guarantee that the course of nature will be constant or must correspond to the limited capacities of human understanding. It follows for voluntarists that all human knowledge of nature must be ultimately empirical. Moreover, nothing about the creation can ever be known with certainty: God's absolute power renders our knowledge fallible and, in this sense, only probable. Either nominalism or conceptualism – in any case, a metaphysics that denies independent ontological status to universals – is also the regular concomitant of voluntarist theology. Only particular individuals exist in the world, and there are no necessary relations connecting them. Consequently, any relations we attribute to groups of them – such as similarity – are the product of our minds, they do not exist independently in the world.

In discussing these interpretations of divine power and its relationship to the world God created, I focus on Thomas Aquinas (c. 1225–74) and William of Ockham (c. 1285–1349) as paradigm cases of the two positions whose seventeenth-century destiny I will pursue in subsequent chapters. Not only are their views particularly clear and well-articulated examples of intellectualist and voluntarist theologies, but they are historically important as well. Both influenced later generations, and ideas descended from theirs were still to be found in the seventeenth-century texts that are the focus of this book.

Thomas Aquinas devoted himself to the project of combining Aristotelian philosophy with Christian theology. His synthesis employed Aristotelian metaphysics as the basis for explaining various points of religion and theology. His agenda was theological rather than purely philosophical, and in his masterpiece, the *Summa theologiae,* he "changed the water of philosophy to the wine of theology."[16] Despite his attraction to the philosophy of Aristotle, various Aristotelian doctrines were problematic from a Christian perspective: the eternity of the world, the deterministic nature of the physical world, and the necessitarianism that characterized some of Aristotle's arguments, particularly as interpreted by the Spanish Muslim Ibn Rushd, or Averroës (c. 1126–98) who had an active following in the thirteenth century. It was Aquinas' attempt to avoid the neces-

---

16. Étienne Gilson, *History of Christian Philosophy in the Middle Ages* (London: Sheed & Ward, 1955), p. 365.

sitarianism that had become associated with Averroism that led him to formolate his version of the two powers of God.[17]

Aquinas was absolutely unambiguous in maintaining God's omnipotence. God's will is totally free, and nothing exists that God did not will into existence. His freedom implies that he could have created things so that the world and its history would have had an entirely different course from the present.[18] Aquinas used the language of *potentia absoluta* and *potentia ordinata* to describe God's power:[19]

> What is attributed to His power considered in itself, God is said to be able to do in accordance with His absolute power (*secundum potentiam absolutam*). . . . What is, however, attributed to the divine power, according as it carries into execution the command of a just will, God is said to be able to do by his ordinary power (*de potentia ordinata*).[20]

17. F. C. Copleston, *Aquinas* (Harmondsworth: Penguin Books, 1955), pp. 141–8; Étienne Gilson, *The Philosophy of St. Thomas Aquinas*, translated by Edward Bullough (New York: Dorset, 1929), chaps. 1 and 7.

18. St. Thomas Aquinas, *Summa theologica*, translated by the Fathers of the English Dominican Province, 3 vols. (New York: Benziger, 1947), bk, 1, qu. 25, art. 5, p. 140.

19. For an analytical description of Aquinas' understanding of *potentia absoluta et ordinata dei*, see Marilyn McCord Adams, *William Ockham*, 2 vols. (Notre Dame, Ind.: University of Notre Dame Press, 1987), vol. 2, pp. 1187–90.

20. Aquinas, *Summa theologica*, bk., I, qu. 25, art. 5, p. 140. I am following this translation for consistency. However, I find the translation of "*potentia ordinata*" as "ordinary power" misleading. The Latin text of the sentence reads, "Quod autem attribuitur potentiae divinae secundum quod exequitur imperium voluntatis justae, hoc dicitur Deus posse facere de potentia ordinata." The Blackfriars edition of the *Summa theolgiae* translates the phrase as "ordinate power," but notes, "Ordinate power; nowadays called *potentia ordinaria*," which they note is commonly translated "ordinary power." St. Thomas Aquinas, *Summa Theologiae*, Latin text and English translation, 61 vols. (Blackfriars, in conjunction with London: Eyre & Spottiswoode, and New York: McGraw-Hill, 1964), vol. 5, p. 173. Deferrari and Barry cite this passage as reading "*de potentia ordinaria*." Roy J. Defarrari and Sister M. Inviolata Barry, with Ignatius McGuiness, *A Lexicon of St. Thomas Aquinas, Based on the* Summa Theologica *and Selected Passages of His Other Works* (Washington, D.C.: Catholic University of America Press, 1948), p. 779. The phrase "*potentia ordinaria*" appears in the sixteenth-century writings of Suárez and the Coimbrian commentators. See Étienne Gilson, *Index scolastico-cartésien*, 2d edition (Paris: J. Vrin, 1979), pp. 88–9, 250–1.

Despite Aquinas' conviction that the intellect is logically primary, he believed that the divine will and intellect are united in the divine essence.[21] Considered absolutely, the divine will is not subject to any necessity. "For God does things because He wills to do so . . . God is bound to nobody but Himself."[22] The only necessity or restrictions on him are relative to the order he created by his absolute power.[23] From the standpoint of his absolute power, God can create any other order that he wants. However, if God once wills something to be, he cannot afterward will it not to have been:

> God's will, however, is immutable. . . . [T]he will would be changed, if one should begin to will what before he had not willed; or cease to will what he had willed before. This cannot happen, unless we presuppose change either in the knowledge or in the disposition of the substance of the willer. . . . Now it has already been shown that both the substance of God and His knowledge are entirely unchangeable. Therefore His will must be entirely unchangeable.[24]

Divine immutability thus introduces a kind of necessity into the created world that Aquinas called "necessity of supposition."[25] Before exercising his will, God is absolutely free to will or not to will. But once he has willed something to be, his immutability precludes him from willing it not to be.[26] Necessity of supposition follows from God's creation of essences or forms. As an Aristotelian, Aquinas considered all essences to be universals. Individuals, as particulars, have natures or essences that necessarily determine certain of their characteristics: "If God . . . wills something other than Himself, it is necessary that He will for this object whatever is necessarily required by it. Thus, it is necessary that God will the rational soul to exist, supposing that He wills man to exist."[27] God is free to create

21.   St. Thomas Aquinas, *Summa contra Gentiles,* translated by Anton C. Pegis, 5 vols. (Notre Dame, Ind.: University of Notre Dame Press, 1956), bk. 1, chaps. 72–4, vol. 1, pp. 239–45.
22.   Aquinas, *Summa theologica,* bk. 1, qu. 25, art. 5, p. 140.
23.   Ibid., pp. 140–1.
24.   Ibid., bk. 1, qu. 19, art. 7, p. 109.
25.   Aquinas, *Summa contra Gentiles,* bk. 1, chap. 83, ¶1, vol. I, p. 263. Aquinas' distinction between absolute necessity and necessity by supposition seems to parallel Aristotle's distinction between absolute and conditional necessity, concepts he developed to express the contingency of certain events in the world. See William Lane Craig, *The Problem of Divine Foreknowledge and Future Contingents from Aristotle to Suarez* (Leiden: E. J. Brill, 1988), p. 48.
26.   Aquinas, *Summa contra Gentiles,* bk. I, chap. 83, ¶4, vol. I, p. 264.
27.   Ibid., ¶5, vol. I, p. 264.

a non-rational animal, but that creature would not be a man. If God chooses to create a man, then it is necessary that the man be rational. Aquinas maintained that, despite his omnipotence, God could not will what is impossible, and this impossibility applies to the attributes predicated by natures.[28] The law of noncontradiction combined with the existence of Aristotelian essences prevents God from creating substances with contradictory characteristics.

"The will of God, therefore, cannot be of that which is of itself impossible."[29] The list of things of themselves impossible is fairly extensive. God "cannot make one and the same thing to be and not to be; He cannot make contradictories to exist simultaneously." Contradiction arises from qualities as well as essences, so God cannot create opposite qualities in the same subject simultaneously. He cannot make a thing lacking any of its essential attributes, as a man without a rational soul. Again,

> since the principles of certain sciences – of logic, geometry, and arithmetic, for instance – are derived exclusively from the formal principles of things, upon which their essence depends, it follows that God cannot make the contraries of those principles; He cannot make the genus not to be predicable of the species, nor lines drawn from a circle's center to its circumference not to be equal, nor the three angles of a rectilinear triangle not to be equal to two right angles.[30]

This inability is an impossibility based on the supposition that God has created the present order. It is not an impossibility "with respect to His power or will considered absolutely."[31]

In addition to things that are necessary by supposition, there are other components of the universe that are "simply and absolutely necessary":[32]

> It is because created things come into being through the divine will that they are necessarily such as God willed them to be. Now, the fact that God is said to have produced things voluntarily, and not of necessity, does not preclude His having willed certain things to be which are of necessity and others which are contingently, so that there may be an ordered diversity in things. Therefore, nothing prevents certain things from being necessary.[33]

According to Aquinas, such absolute necessity in the created world does not infringe on God's absolute power. God can will certain things to be

---

28. Ibid., chap. 84, ¶2, vol. I, pp. 264–5.
29. Ibid., ¶3, vol. I, p. 265.
30. Ibid., bk. II, chap. 25, ¶14, vol. II, pp. 74–5.
31. Ibid., ¶¶15–25, vol. II, pp. 75–6.
32. Ibid., chap. 30, ¶1, vol. II, p. 85.
33. Ibid., ¶4, vol. II, p. 86.

necessary, even though he wills them freely. Once he has willed them to be necessary, however, they do introduce necessity into the created order.

Similarly, certain things are absolutely impossible, because they entail a contradiction. For example, it would entail a contradiction for God to make the past not to have been:

> For as it implies a contradiction to say that Socrates is sitting and is not sitting, so does it to say that he sat and did not sit. But to say that he did sit is to say that it happened in the past. To say that he did not sit, is to say that it did not happen. Whence that the past should not have been does not come under the scope of divine power. . . . if the past thing is considered as past, that it should not have been is impossible, not only in itself, but absolutely since it implies a contradiction.[34]

Unlike necessity and impossibility of supposition, absolute impossibility infringes on God's power:

> [For the past not to have been] is more impossible than the raising of the dead; in which there is nothing contradictory, because this is reckoned impossible in reference to some power, that is to say, some natural power; for such impossible things do come beneath the scope of divine power.[35]

God created the natures of things, which are thus subject to his will. Natural impossibility is the negative form of necessity by supposition. Absolute impossibility, however, limits God's power:

> As God, in accordance with the perfection of the divine power, can do all things, and yet some things are not subject to His power, because they fall short of being possible; so, also, if we regard the immutability of the divine power, whatever God could do, He can do now. Some things, however, at one time were in the nature of possibility, whilst they were yet to be done, which now fall short of the nature of possibility, when they have been done. So is God said not to be able to do them, because they themselves cannot be done."[36]

At this point, the nature of Aquinas' intellectualism becomes clear. God's act of creation or act of willing is absolutely free. It is when things are considered "in relation to their proximate principles" that "they are found to have absolute necessity."[37] God is capable of freely creating necessary connections in the universe. The death of a particular animal follows necessarily from the fact that it is composed of contraries, "although it was not absolutely necessary for it to be composed of contr-

---

34. Aquinas, *Summa theologica*, qu. 25, art. 4, vol. I, p. 139.
35. Ibid.
36. Ibid.
37. Aquinas, *Summa contra Gentiles*, bk. II, chap. 30, ¶7, vol. II, p. 87.

aries." That is to say, God was free to have created it otherwise. But, given that he created it from contraries, its mortality necessarily follows. Likewise, God created certain natures by his free will; but once these natures have been created, individuals possessing them must have the properties that flow from these natures.[38] "A saw, because it is made of iron, must be hard; and a man [because he is rational] is necessarily capable of learning."[39]

Necessity in created things arises from their essences, the natures or forms that make them what they are.[40] Implicit in Aquinas' intellectualism is an Aristotelian essentialist ontology.[41] The world God created consists of individuals that are composed of matter and form. For Aquinas, as for Aristotle, forms do not exist apart from individual substances but always exist in conjunction with matter, making things what they are rather than something else. It is from these forms that certain properties necessarily flow and thereby introduce necessary causal relations into the created world.

Forms also play a central role in Aquinas' theory of knowledge. Since all knowledge is knowledge of forms, the ontology of matter and form provides the metaphysical foundation for his claim to necessary knowledge of the material world. "For the intellect knows bodies by understanding them, not indeed through bodies, nor through material and corporeal species; but through immaterial and intelligible species, which can be in the soul by their own essence."[42]

How does the intellect, which is immaterial, receive the forms of material bodies? "In the present state of life in which the soul is united to a passible [susceptible to sensation, liable to change] body, it is impossible for our intellect to understand anything actually, except by turning to the phantasms."[43] However phantasms, or sense representations, are not intelligible themselves. The active intellect must transform them into intelligible forms. "Our intellect understands material things by abstracting

38. Ibid.
39. Ibid., ¶9, vol. II, pp. 87–8.
40. Aquinas, *Summa theologica*, bk. I, qu. 86, art. 3, vol. I, p. 442.
41. Frederick Copleston, *A History of Philosophy*, 9 vols. (Garden City, N.Y.: Doubleday, 1962; first published 1950), vol. 2, pt. 2, p. 46.
42. Aquinas, *Summa theologica*, bk. I, qu. 86, art. 3, vol. I, p. 422. See Julius R. Weinberg, *A Short History of Medieval Philosophy* (Princeton, N.J.: Princeton University Press, 1964), pp. 186, 207. Copleston, *A History of Philosophy*, vol. 2, Pt. 2, p. 110.
43. Aquinas, *Summa theologica*, bk I, qu. 84, art. 7, p. 429.

[intelligible species] from the phantasms."[44] How does the intellect accomplish this abstraction?

> Abstraction may occur in two ways: First by way of composition and division; thus we may understand that one thing does not exist in some other, or that it is separate therefrom. Secondly, by way of simple and absolute consideration: thus we understand one thing without considering the other. . . . [T]he things which belong to the species of a material thing, such as a stone, or a man, or a horse, can be thought of apart from the individualizing principles which do not belong to the notion of the species. This is what we mean by abstracting the universal from the particular, or the intelligible species from the phantasm; that is, by considering the nature of the species apart from its individual qualities represented by the phantasm.[45]

The form of the material body, then, acts on the senses to produce a sensible representation or phantasm, which the intellect transforms, by an act of abstraction, into an intelligible form in the mind.[46]

Knowledge is really knowledge of universals or forms. Because necessary connections exist between forms and the properties flowing from them, we are able to know that in nature certain facts necessarily follow from the real essences of things. Consequently for Aquinas, as for Aristotle, science consists of demonstrative knowledge of the real essences of things.[47] Properly developed, such knowledge attains certainty.

Scientific knowledge is not necessary in every respect, however. Because matter is the principle of individuation of material substances,[48] knowledge of singular things is contingent. That is, "prime matter, the ultimate property-bearer in composite substances, combines with quantitative dimensions to individuate":[49]

> Contingency arises from matter, for contingency is a potentiality to be or not to be, and potentiality belongs to matter; whereas necessity results from form, because whatever is consequent on form is of

44. Ibid., qu. 85, art. 1, p. 432.
45. Ibid.
46. Ibid., art. 3, p. 435.
47. Aristotle, *Posterior Analytics*, translated by Jonathan Barnes, in *The Complete Works of Aristotle*, edited by Jonathan Barnes, 2 vols. (Princeton, N.J.: Princeton University Press, 1984), vol. 1, pp. 114–15.
48. Copleston, *Aquinas*, pp. 95–6. Aquinas' views on the question of the individuating principle of substances that are not purely physical, such as men, is far more complex. See Gilson, *The Philosophy of St. Thomas Aquinas*, p. 219 n5.
49. Marilyn McCord Adams, "Universals in the Early Fourteenth Century," in *The Cambridge History of Later Medieval Philosophy*, edited by Kretzmann, Kenny, and Pinborg, p. 411.

necessity in the subject. But matter is the individualizing principle: whereas the universal comes from the abstraction of the form from the particular matter.[50]

Rational knowledge is possible only of the universal, while knowledge of particular things must come from the senses. Consequently, all demonstrative knowledge of necessary things can be known a priori, but all knowledge of contingencies is a posteriori:

> The contingent considered as such, is known directly by sense and indirectly by the intellect; while the universal and necessary principles of contingent things are known only by the intellect. Hence if we consider the objects of science in their universal principles, then all science is of necessary things. But if we consider the things themselves, thus some sciences are of necessary things, some of contingent things.[51]

Aquinas concluded that scientific knowledge is demonstrative knowledge of necessary relations embodied in universals. The rationalist components of his epistemology are directly tied to his essentialist ontology, which, in turn, is closely connected to his understanding of the relationship between the absolute and ordained powers of God.

While the language of God's absolute and ordained power had been formulated as early as 1245, it was only after the condemnations of 1277 that the dialectic between the two aspects of divine power came to play a central role in philosophy and theology.[52] In that year, the bishop of Paris, Étienne Tempier, condemned 219 propositions that various members of the philosophical community at the University of Paris had espoused.[53] The object of many of the condemnations was the Averroistic interpretation of Aristotle for which there were a number of vocal proponents in Paris during the thirteenth century. In fact, Averroës' influence and authority were so great that he was often simply referred to as "The Com-

50. Aquinas, *Summa theologica,* bk. I, qu. 86, art. 3, p. 442.
51. Ibid.
52. Courtenay, "The Dialectic of Omnipotence," p. 247.
53. For historical accounts of the condemnations, see the following: Gordon Leff, *Paris and Oxford Universities in the Thirteenth and Fourteenth Centuries* (New York: Wiley, 1968); John F. Wippel, "The Condemnations of 1270 and 1277 at Paris," *Journal of Medieval and Renaissance Studies,* 7 (1977): 169–201; Edward Grant, "The Condemnation of 1277, God's Absolute Power, and Physical Thought in the Late Middle Ages," *Viator: Medieval and Renaissance Studies,* 10 (1979): 211–44. See also R. Hooykaas, "Science and Theology in the Middle Ages," *Free University Quarterly, 3* (1954): 77–163. I am grateful to H. Floris Cohen for bringing this article to my attention.

mentator," whose commentary, of course, was on the work of "The Philosopher," Aristotle.[54] Averroës and his Parisian followers taught, among
other things, that the four elements and the terrestrial substances composed of them fall under the influences of the movements of the celestial
spheres. For this reason, the Averroistic cosmos was strictly determined by
necessary causes,[55] a conclusion that seemed to deny, or at least restrict,
the activity of divine providence in the creation.[56] The Latin Averroists
had also interpreted Aristotle's physics and metaphysics in a thoroughly
deterministic and necessitarian manner.[57] For example, where Aristotle
had argued in *De caelo* that the earth *is* stationary at the center of the
universe, the Averroists had interpreted this argument to prove that the
earth *must* be stationary at the center of the universe. Accordingly, many
of the condemned propositions were ones that seemed to restrict God in
the free exercise of his will by positing some kind of necessity in the
creation.

The condemnations of 1277 marked a watershed in discussions of
divine omnipotence and will: After 1277 it became more difficult to assert
necessity in the creation or in human knowledge of the creation than it
had been before. One consequence of the condemnations was that the
assertion of the absolute power of God became the primary focus for
many philosophers and theologians.[58] William of Ockham's thought
illustrates this aspect of the reaction to the condemnations of 1277 and
supplies a paradigmatic example of voluntarism and its philosophical
consequences. One of the most influential philosophers of the fourteenth
century,[59] Ockham understood the distinction between *potentia Dei ab-*

54. See, e.g., St. Thomas Aquinas, *Commentary on Aristotle's* Physics, translated
    by Richard J. Blackwell, Richard J. Spath, and W. Edmund Thirlkel (Lon-
    don: Routledge & Kegan Paul, 1963), p. 5. On Averroes' life and thought,
    see Dominique Urvoy, *Ibn Rushd (Averroes),* translated by Olivia Stewart
    (London: Routledge, 1991).
55. Gilson, *History of Christian Philosophy in the Middle Ages,* p. 224.
56. See, e.g., Aquinas, *Summa contra Gentiles,* bk. 3, chap. 76, ¶¶ 7–11, vol. 3,
    pp. 256–8.
57. On the Latin Averroists and their most famous spokesman, Siger of Brabant,
    see Copleston, *A History of Philosophy,* vol. 2, pt. 2, chap. 42; Gilson,
    History of Christian Philosophy in the Middle Ages, pt. 9, chap. 1.
58. Edward Grant, "Hypotheses in Late Medieval and Early Modern Science,"
    Daedalus, 91 (1962): 599–616; Edward Grant, *Physical Science in the Mid-
    dle Ages* (New York: Wiley, 1971), chap. 3; Leff, *Paris and Oxford
    Universities.*
59. Weinberg, *A Short History of Medieval Philosophy,* p. 235.

soluta and *potentia Dei ordinata* in much the same terms as had Aquinas, but he spelled out the two concepts with far more clarity:[60]

> God is able to do certain things by his ordained power and certain things by his absolute power. This distinction should not be understood to mean that in God there are really two powers, one of which is ordained and the other of which is absolute. For with respeut to things outside himself there is in God a single power, which in every way is God himself. Nor should the distinction be understood to mean that God is able to do certain things ordinately and certain things absolutely and not ordinately. For God can do anything inordinately.
>
> Instead, the distinction should be understood to mean that 'power to do something' is sometimes taken as 'power to do something in accordance with the laws that have been ordained and instituted by God', and God is said to be able to do these things by his ordained power. In an alternative sense, 'power' is taken as 'power to do anything such that its being does not involve a contradiction', regardless of whether or not God has ordained that he will do it.[61]

Despite the similarity of Aquinas' and Ockham's language, they drew very different implications from the distinction between God's absolute and ordained powers. Although Ockham accepted the limits that the law of noncontradiction imposed on divine freedom,[62] he radically reduced the domain of absolutely necessary relations in the creation and of necessity in human knowledge.[63]

According to Ockham, the only restriction on God's power is the principle of noncontradiction. "Anything is to be attributed to the divine

60. Oakley, *Omnipotence, Covenant, and Order,* pp. 51–5; Courtenay, "Nominalism," pp. 40–1; David J. Clark, "Ockham on Human and Divine Freedom," *Franciscan Studies, 38* (1978): 150. On Ockham and divine power, see Adams, *William Ockham,* vol. 2, pp. 1198–1207, 1218–31.

61. Ockham, *Quodlibet,* 6, q. 1, in William of Ockham, *Quodlibetal Questions,* translated by Alfred J. Freddoso and Francis E. Kelley, 2 vols. (New Haven, Conn.: Yale University Press, 1991), pp. 491–2. See Courtenay, "The Dialectic of Omnipotence," pp. 254–5.

62. For example, he agreed that "God cannot make what is past not to be past without its afterwards being true to say that it was past." See William Ockham, *Predestination, God's Foreknowledge, and Future Contingents,* translated by Marilyn McCord Adams and Norman Kretzmann (New York: Appleton–Century-Crofts, 1969), p. 85. See also Adams, *William Ockham,* pp. 1218–28.

63. Stated another way, "a larger number of aspects of the created order are absolutely necessary for Thomas than for Ockham." Courtenay, "Nominalism," p. 40.

power when it does not contain a contradiction."[64] An immediate implication of God's absolute power, considered in this way, is that the
created order is absolutely contingent.[65] Not even the causal structure of
the world can interfere with divine freedom. God is not bound to follow
the laws he created. He can always bypass those laws and produce the
same effects directly. This is what it means to say that "anything can be
attributed to divine power. . . . Whatever God can produce by means of
secondary causes, He can directly produce and preserve without them."[66]

Since, contrary to Aquinas, there are no necessary relations in nature,
every reality – "be it a substance or an accident" – exists apart from every
other reality. God can "cause, produce, and conserve" each individual
separately from any other.[67] There are no necessary relations linking
them. The contingency of the creation is the true meaning of Ockham's
account of *potentia absoluta* and *potentia ordinata*. Everything other
than God exists contingently. Since the world is not logically necessary,
God equally could have chosen not to create it.[68]

It should be noted that there were other types of necessity that Ockham
was willing to countenance. For example, although all knowledge about
the world is contingent, such as my seeing Socrates, the contingency of the
event of my seeing Socrates does not eliminate the necessity that when I do
see Socrates, I must be seeing him.[69] Also, analytic propositions are necessary.[70] Another kind of necessity in Ockham's world resulted from the
fact that if God wills a course of events, then the occurence of that course
of events is necessary. Every event must occur as God wills it, even though
God can change his mind and will a different course of events, which will
then become necessary.[71] But these kinds of necessity in no way limit

64.   Ockham, *Ordinatio*, qu. i, n *sqq.*, in Ockham, *Philosophical Writings*, edited
      and translated by Philotheus Boehner (Edinburgh: Nelson, 1957), p. 25.
65.   Paul Vignaux contends that the dialectic of the two powers of God is central
      to Ockham's views in metaphysics and epistemology. See his *Nominalisme
      au XIVe siècle* (Montréal: Institut d'Études Mediévales, and Paris: J. Vrin,
      1948), pp. 12–38.
66.   Ibid. Weinberg argues that Duns Scotus was one of the first philosophers to
      state this principle, which appeared with increasing frequency in the four-
      teenth century. See *A Short History of Medieval Philosophy*, pp. 225–6.
67.   Ockham, *Philosophical Writings*, p. xx.
68.   Heiko Augustinus Oberman, *The Harvest of Medieval Theology: Gabriel
      Biel and Late Medieval Nominalism* (Cambridge, Mass.: Harvard University
      Press, 1963), pp. 38–9.
69.   Ockham, *Ordinatio*, q. i, n *sqq.*, in *Philosophical Writings*, p. 27.
70.   See Adams, "Intuitive Cognition," p. 394.
71.   Ockham, *Quodlibet*, 1, q. 13, in *Quodlibetal Questions*, vol. 1, p. 66.

God's absolute power. They do not presuppose an ontology of necessary connections in the world as Aquinas had. There is, in essence, no difference between the domain of God's absolute power and that of his ordained power. No part of the creation is exempt from the continued exercise of God's absolute power.

Ockham appealed to miracles as historical examples of the contingency of the world order. Miracles, according to Ockham, demonstrate that the principles underlying nature – the second causes – are contingent. They can be suspended, and God can act directly to produce effects quite different from the usual course of nature. That fire burns is a contingent fact that is not necessary – this is the point for Ockham of the miracle of the fiery furnace in the Book of Daniel.[72]

For Ockham the implications of divine omnipotence swept away the underpinnings for a priori, necessary knowledge that Aquinas had found in a metaphysics of essences and internal relations. All knowledge, according to Ockham, begins with perception of singular entities, sensory

72. Courtenay, "Nominalism," pp. 42–3. This interpretation of the miracle contrasts with the interpretations of both Aquinas and of seventeenth-century natural philosophers like Robert Boyle. Although Aquinas interpreted some miracles as examples of God doing directly what he ordinarily does by second causes, he also understood miracles as examples of God's being able to do things "which nature could never do," such as "that two bodies should be coincident; that the sun reverse its course, or stand still; that the sea open up and offer a way through which people may pass." Another kind of miracle, for Aquinas, occurs when "God does something which nature can do, but not in this order [i.e., not in the normal temporal sequence]," such as causing an animal "to live after death or see after becoming blind." (Aquinas, *Summa contra Gentiles*, bk. 3, chap. 101, ¶¶ 2–4, vol. 4, pp. 82–3.) In these passages, Aquinas was primarily concerned to define the boundary between the natural and the supernatural. Boyle and other seventeenth-century physico-theologians invoked miracles for yet another reason, to prove the activity of divine providence in order to overcome the perceived threat of atheism and materialism. See Robert Boyle, *Some Motives and Incentives to the Love of God, Pathetically Discoursed of in a Letter to a Friend*, in *The Works of the Honourable Robert Boyle*, edited by Thomas Birch, 6 vols. (London, 1772; reprinted, Hildesheim: Georg Olms, 1965), vol. 1, p. 268; Robert Boyle, *Some Considerations about the Reconcileableness of Reason and Religion* (1675), in *Works*, vol. 4, p. 158. On Boyle's views on miracles, see Richard S. Westfall, *Science and Religion in Seventeenth-Century England* (New Haven, Conn.: Yale University Press, 1958), pp. 87–93.

representations, which he called intuitive cognitions.[73] Intuitive cognitions are the ultimate source of our knowledge of contingent truths about the world:

> Intuitive cognition of a thing is cognition that enables us to know whether the thing exists or does not exist, in such a way that, if the thing exists, then the intellect immediately judges that it exists and evidently knows that it exists, unless the judgment happens to be impeded through the imperfection of this cognition.[74]

Such knowledge is singular in the sense that it has as its subject only the particular, contingent facts observed.[75] It is necessary: A person having an intuitive cognition is necessarily having one. But it does not imply that any necessary relations exist in nature or that there is a necessary connection between the intuitive cognition and its mind-independent referent. Although intuitive cognitions are evident and necessarily true, they depend on epistemologically prior nonintrospective experience of the material world.[76]

Universals exist only in the mind as cognitions abstracted from our experience of particulars:[77] "No universal is a substance regardless of how it is considered. On the contrary, every universal is an intention of the mind which, on the most probable account, is identical with the act of understanding."[78] We cannot acquire knowledge about contingent facts from abstractive cognitions because abstractive cognitions are mental concepts, abstracted from perception: "Abstractive cognition . . . is that knowledge by which it cannot be evidently known whether a contingent fact exists or does not exist. In this way abstractive cognition abstracts from existence and non-existence."[79] Thus, abstractive cognitions do not enable us to know whether the concepts in question refer to actually

---

73. Ockham, *Summa totius logicae*, pt. I, chap. 15, in *Ockham's Theory of Terms. Part I of the Summa Logicae*, edited and translated by Michael J. Loux (Notre Dame, Ind.: University of Notre Dame Press, 1974), pp. 79–82. For the details of Ockham's philosophy, see Gordon Leff, *William of Ockham: The Metamorphosis of Scholastic Discourse* (Manchester: Manchester University Press, 1975).

74. Ockham, *Ordinatio*, q. i, n *sqq.*, in *Philosophical Writings*, p. 26.

75. Ockham, *Quodlibet*, 1, q. 13, in *Quodlibetal Questions*, translated by Freddoso and Kelley, vol. 1, p. 65.

76. Marilyn McCord Adams, "Intuitive Cognition, Certainty, and Scepticism in William Ockham," *Traditio*, 26 (1970): 393.

77. Ockham, *Summa totius logicae*, pt. I, chap. 15, pp. 79–82.

78. Ibid., p. 81. See also Vignaux, *Nominalisme*, pp. 26–8.

79. Ockham, *Ordinatio*, q. i, n *sqq.*, in Ockham, *Philosophical Writings*, p. 26.

existing things.[80] Since universals – which are abstractive cognitions – do not exist outside the mind, it is not possible to reason from universal concepts to knowledge about the world. That is to say, because Ockham did not accept the extramental reality of universals, he thought that there are no necessary truths about natural kinds. In this way, Ockham's epistemology differed greatly from that of Aquinas, who had argued that by the process of abstraction, we come to know the real essences of things, knowledge of which is precisely what provides grounds for necessary knowledge about the world.

Ockham denied that it is possible to reason from concepts to material reality. To confuse mental contents and the external world is the source of many errors and confusions. This is the central point of his epistemology:[81]

> All knowledge has to do with a proposition or propositions. . . . Now the fact is that the propositions known by natural science are composed not of sensible things or substances, but of mental contents or concepts that are common to such things. . . . Hence, properly speaking, the science of nature is not about corruptible and generable things nor about natural substances nor about movable things, for none of these things is subject or predicate of any conclusion known by natural science. Properly speaking, the science of nature is about mental contents which are common to such things, and which stand precisely for such things in many propositions, though in some propositions these concepts stand for themselves.[82]

The subject matter of science is therefore abstractive cognitions.[83] But Ockham had already stated that we cannot make inferences about what actually exists in the world from abstractive cognitions. Thus, even if it is possible to discover necessary relations among our abstractive cognitions, it is not possible to infer that these necessary relations also exist in the world.

Ockham explicitly stated that God can create in our minds intuitive cognitions of nonexistents, on the sole condition that they are not self-contradictory. He based this claim on one of his fundamental assumptions:

80. Ibid., pp. 26–7.
81. Adams, "Intuitive Cognition," p. 393. See also Leff, *The Dissolution of the Medieval Outlook* (New York: Harper & Row, 1976), pp. 57, 62.
82. Ockham, *Expositio super viii libros Physicorum*, in Ockham, *Philosophical Writings*, p. 12.
83. On Ockham's physics, both method and content, see André Goddu, *The Physics of William of Ockham* (Leiden: E. J. Brill, 1984).

> Whatever God produces by the mediation of secondary causes, he can
> immediately produce and conserve in the absence of such causes. . . .
> But he is able to produce an intuitive cognition of a corporeal thing by
> the mediation of a [corporeal] object. Therefore he is able to produce
> this cognition immediately by himself.[84]

Consequently, there is no necessary connection between the ideas in our
minds and things in the natural world. Moreover, "God can cause an act
of believing through which I believe a thing to be present that is [in fact]
absent."[85] It follows that we cannot reason from the ideas in our minds to
knowledge of the external world. Actual observation is required, and so
knowledge of the world, for Ockham, can only be empirical.

The absence of necessary connections in the world does not, however,
result in chaos. Although God cannot be bound by any of his creations, he
can freely choose

> to follow a stable pattern in dealing with his creation in general and
> with man in particular. If God has freely chosen the established order,
> he *has* so chosen, and while he can dispense with or act apart from the
> laws he has decreed, he has nonetheless bound himself by his promise
> and will remain faithful to the covenant that of his kindness and
> mercy, he has instituted with man.[86]

Faith in God's covenant provides grounds for believing that we can have
fairly reliable, if not certain, knowledge of the creation.

Exploration of medieval discussions of the dialectic of God's absolute
and ordained power has exposed two different ways of expressing the
relationship between contingency and necessity in the world. One approach emphasizes the apparent intelligibility and stability of the world
and is optimistic about the human capacity to acquire reliable knowledge
by rational methods. The other underscores the contingency of the natural
order and the limits of human knowledge. Similar patterns of thinking
appear in seventeenth-century natural philosophy in the context of discussions about the status of the laws of nature and approaches to formulating
methodological prescriptions for the new science. These patterns can be
understood as transformations of the discourse about the absolute and
ordained power of God into assumptions about the power of natural
philosophy to penetrate the secrets of nature.

Despite important similarities between medieval and seventeenth-
century discussions of theology and natural philosophy, these discussions

84. Ockham, *Quodlibet*, 6, q. 6, in *Quodlibetal Questions*, vol. 2, p. 506.
85. Ockham, *Quodlibet* 5, q. 5, in *Quodlibetal Questions*, vol. 2, p. 416.
86. Oakley, *Omnipotence, Covenant, and Order*, p. 62. See also Courtenay,
    "Nominalism," p. 51; and Oberman, *The Harvest of Medieval Theology*,
    p. 39.

occurred within significantly different intellectual contexts. For the medieval thinkers the context was straightforwardly theological. Controversies about God's absolute and ordained power, the status of universals, and the foundations of knowledge took place within the framework of the debate about the relative authority of reason and revelation in the decades following the translation and introduction of Aristotle's writings into the mainstream of European intellectual life. The medieval thinkers were concerned with finding theologically acceptable ways to use Greek philosophy, but the aim of their enterprise remained primarily theological. By the early seventeenth century, the focus of discussion had changed. Even though the mechanical philosophers were Christians for whom theology still played a major role, their goal was to provide epistemological and metaphysical foundations for the new science. They wanted to create a philosophically sound account of the natural world, one that conformed to certain theological presuppositions. The theological concerns remained genuine, but more as boundary conditions than as the primary focus of interest. Attention had shifted from God to nature and from theology to natural philosophy. Nevertheless, old theological preoccupations continued to be reflected, not only in their explicit theological presuppositions, but also in the styles of thinking that formed the different versions of the mechanical philosophy.

2

Baptizing Epicurean philosohy: Gassendi on divine will and the philosophy of nature

As for the Epicurean Commentaries, you seem to hesitate lest I go wrong in religion. But I am opposed to anything which could conflict with it. You insist on Providence: truly I defend it against Epicurus.

Gassendi to Tommaso Campanella, November 1632[1]

Gassendi's Epicurean project

Pierre Gassendi is most frequently remembered for introducing the philosophy of the ancient atomist Epicurus into the mainstream of European thought.[2] His revival of Epicureanism can be understood in the context of the early-seventeenth-century search for a new philosophy of nature. The young Gassendi was an active member of the intellectual community of

1. In Pierre Gassendi, *Opera omnia*, 6 vols. (Lyons, 1658; facsimile reprint, Stuttgart-Bad Cannstatt: Friedrich Frommann Verlag, 1964), vol. 4, p. 54. All translations are mine unless otherwise noted.
2. See, e.g., Gaston Sortais, *La philosophie moderne depuis Bacon jusqu'à Leibniz* 2 vols. (Paris: Paul Lethielleux, 1920-2), vol. 2, pp. 1–179; Bernard Rochot, *Les travaux de Gassendi sur Épicure et sur l'atomisme, 1619–1658* (Paris: J. Vrin, 1944); Olivier René Bloch, *La philosophie de Gassendi: Nominalisme, matérialisme, et métaphysique* (The Hague: Martinus Nijhoff, 1971); Marie Cariou, *L'atomisme: Gassendi, Leibniz, Bergson, et Lucrèce* (Paris: Aubier Montaigne, 1978); Richard H. Popkin, "Gassendi, Pierre," in *The Encyclopedia of Philosophy,* edited by Paul Edwards, 8 vols. (New York: Macmillan and Free Press, 1967), vol. 3, pp. 269–73; and Howard Jones, *Pierre Gassendi, 1592–1655: An Intellectual Biography* (Nieukoop: B. de Graaf, 1981); Antonina Alberti, *Gassendi e l'atomismo Epicureo* (Firenze: Instituto Universitario Europeo, 1981); and Barry Brundell, *Pierre Gassendi: From Aristotelianism to a New Natural Philosophy* (Dordrecht: Reidel, 1987).

Provence and, eventually, all of France.³ He conducted experiments and discussed many scientific topics, particularly in astronomy, with his patron and correspondent, Nicolas-Claude Fabri de Peiresc (1580–1637).⁴ He was a friend of Hobbes, whom he encountered in the circle around Mersenne in Paris.⁵ He corresponded with Beeckman, Galileo, Hevelius, and other astronomers and natural philosophers. He performed the famous experiment of dropping a heavy object from the top of a moving ship's mast, an experiment previously imagined by Galileo as proving the possibility of the earth's motion.⁶ And he wrote a treatise defending

3. For Gassendi's place in the scientific community of his day, see Lynn Sumida Joy, *Gassendi the Atomist: Advocate of History in an Age of Science* (Cambridge University Press, 1987), chaps. 1–4; Lisa T. Sarasohn, "French Reaction to the Condemnation of Galileo, 1632–1642," *Catholic Historical Review*, 74 (1988): 34–54; and Lisa Tunick Sarasohn, "Epicureanism and the Creation of a Privatist Ethic in Early Seventeenth-Century France," in *Atoms, Pneuma, and Tranquillity: Epicurean and Stoic Themes in European Thought*, edited by Margaret J. Osler (Cambridge University Press, 1991), pp. 175–95.

4. On Peiresc as patron, see Lisa T. Sarasohn, "Nicolas-Claude Fabri de Peiresc and the Patronage of the New Science in the Seventeenth Century," *Isis, 84* (1993): 70–90. See also Anne Reinbold, ed., *Peiresc ou la passion de connaître: Colloque de Carpentras, novembre 1987* (Paris: J. Vrin, 1990). Gassendi wrote a biography of Peiresc after his death in 1637, entitled *Viri illustris Nicolai Claudii Fabricii de Peiresc, senatoris aqvisextiensis vita*, reprinted in *Opera omnia*, vol. 5, pp. 237–62. This book was translated into English in the seventeenth century. See Petrus Gassendus, *The Mirrour of True Nobility and Gentility. Being the Life of the Renowned Nicolaus Lord of Peiresck, Senator of the Parliament of Aix*, translated by W. Rand (London: 1657). See also Pierre Gassendi, *Peiresc, 1580–1637: Vie de l'illustre Nicolas-Claude Fabri de Peiresc, conseiller au Parlement d'Aix*, translated by Roger Lasalle with the collaboration of Agnès Bresson (Paris: Belin, 1992).

5. Frithiof Brandt, *Thomas Hobbes' Mechanical Conception of Nature*, translated by Vaughan Maxwell and Annie I. Fausbøll (Copenhagen: Levin & Munksgaard, and London: Librairie Hachette, 1928), p. 179; Samuel I. Mintz, *The Hunting of Leviathan: Seventeenth-Century Reactions to the Materialism and Moral Philosophy of Thomas Hobbes* (Cambridge University Press, 1962), p. 20; Brundell, *Pierre Gassendi*, pp. 108, 151; Steven Shapin and Simon Schaffer, *Leviathan and the Air-Pump: Hobbes, Boyle, and the Experimental Life* (Princeton, N.J.: Princeton University Press, 1985), pp. 82–3.

6. Stillman Drake, *Galileo at Work: His Scientific Biography* (Chicago: University of Chicago Press, 1978), p. 404. Gassendi described this experiment in a letter to Louis-Emmanuel de Valois in June 1641, *Opera omnia*, vol. 6, pp. 108–9.

Galileo's new science of motion, *De motu impresso a motore translato* (1642). As an active member of this community of natural philosophers, Gassendi predictably rejected the philosophy of Aristotle. There is evidence that he was interested in Epicureanism as a replacement for Aristotelianism from at least the mid-1620s.

Before European intellectuals could embrace the philosophy of Epicurus, however, the views of the ancient atomist had to be purged of the accusations of atheism and materialism that had dogged them since antiquity.[7] Epicurus was not, in fact, an atheist. He believed that the gods exist, but he thought that as blessed and immortal beings they took part in neither natural nor human affairs. Thus, he denied any providential interpretation of the world. He was a materialist, believing that everything in the world is composed of atoms moving in infinite void space.[8] Atoms, for Epicurus, were infinite in number and had existed forever. He denied the immortality of the soul and the existence of any immaterial being except for the void.[9]

One might well ask why Gassendi, a Catholic priest, undertook the task of reviving and Christianizing such a flagrantly pagan philosophy. As a humanist, Gassendi naturally sought an ancient model on which to build a new philosophy of nature.[10] Gassendi's humanism is evident in two

7. For Epicurus's views on the gods and religion, see J. M. Rist, *Epicurus: An Introduction* (Cambridge University Press, 1972), pp. 140–63, 172–5.
8. On atoms and void in Epicurus, see David Furley, "Knowledge of Atoms and Void in Epicureanism," in *Cosmic Problems: Essays on Greek and Roman Philosophy of Nature,* edited by David Furley (Cambridge University Press, 1989), pp. 161–71. See also David Furley, *The Greek Cosmologists, vol. 1. The Formation of the Atomic Theory and Its Earliest Critics* (Cambridge University Press, 1987), chap. 9. For relevant primary sources, see A. A. Long and D. N. Sedley, *The Hellenistic Philosophers,* 2 vols. (Cambridge University Press, 1987).
9. On Epicurean theology, see Rist, *Epicurus,* chap. 8; and A. A. Long, *Hellenistic Philosophy: Stoics, Epicureans, Sceptics,* 2d edition (Berkeley: University of California Press, 1986), pp. 41–9. For translations of the relevant texts, see A. A. Long and D. N. Sedley, *The Hellenistic Philosophers,* vol. 1, pp. 139–49.
10. I follow Kristeller, here, in defining Renaissance humanism, not in terms of particular philosophical view, but in terms of the regard for the authors of classical Greece and Rome. "It was the novel contribution of the humanists to add the firm belief that in order to write and to speak well it was necessary to study and to imitate the ancients." Paul Oscar Kristeller, *Renaissance Thought and Its Sources,* edited by Michael Mooney (New York: Columbia University Press, 1979), pp. 24–5. See also Paul Oskar Kristeller, "Human-

principal ways: his insistence on finding an ancient model for his philosophical views; and his style of writing, which is marked by frequent allusions to and quotations from classical authors.[11] The classical options available to Gassendi included Aristotle, Plato, the Stoics, and the Epicureans. In tune with other proponents of the new science, he was passionately anti-Aristotelian, a view he voiced in his earliest surviving letter: "I do not ignore the errors, contradictions, and tautologies . . . that abound in Aristotle."[12] He spelled out these objections in vituperative detail in his first published work, the *Exercitationes paradoxicae adversus Aristoteleos* (1624), a skeptical attack on the Stagirite's philosophy.[13]

What attracted Gassendi to Epicureanism among the remaining classical schools of philosophy? Gassendi's own writings are not very helpful on this score. The first indications of his interest in restoring Epicurus come from the early 1620s. In his earliest preserved letter, he mentioned that he possessed a copy of Lucretius, Epicurus' Roman expositor, although he said nothing about having read the book.[14] The *Exercitationes* not only revealed his acquaintance with Epicureanism, but also expressed his desire to expound it – at least in ethics – in proposed future writings.[15] The reasons for his interest remain somewhat obscure, although scholars

ism," in *The Cambridge History of Renaissance Philosophy,* edited by Charles B. Schmitt, Quentin Skinner, and Eckhard Kessler (Cambridge University Press, 1988), pp. 113–37. On the interpretation of Renaissance humanism, see Wallace K. Ferguson, *The Renaissance in Historical Thought* (Boston: Houghton Mifflin, 1948). For the various historical interpretations of humanism in modern scholarship, see Donald Weinstein, "In Whose Image and Likeness? Interpretations of Renaissance Humanism," *Journal of the History of Ideas,* 33 (1972): 165–76. On Gassendi's humanism and its difference from the earlier humanism of Lorenzo Valla, see Lynn S. Joy, "Epicureanism in Renaissance Philosophy," *Journal of the History of Ideas,* 53 (1992): 573–83. See also Joy, *Gassendi the Atomist,* chaps. 1–4; and Brundell, *Pierre Gassendi,* pp. 48–51.
11. See Margaret J. Osler, "Ancients, Moderns, and the History of Philosophy: Gassendi's Epicurean Project," in *The Rise of Modern Philosophy,* edited by Tom Sorell (Oxford University Press, 1993), pp. 129–43.
12. Gassendi to Pybrac, April 1621, in Gassendi, *Opera omnia,* vol. 6, p. 2.
13. Pierre Gassendi, *Dissertations en forme de paradoxes contre les Aristotéliciens (Exercitationes paradoxicae adversus Aristoteleos* [1624]), translated into French by Bernard Rochot (Paris: J. Vrin, 1959). On the anti-Aristotelian strain in Gassendi's thought, see Brundell, *Pierre Gassendi.*
14. Gassendi to Pybrac, April 1621, in Gassendi, *Opera omnia,* vol. 6, p. 1.
15. Rochot, *Les travaux de Gassendi sur Épicure,* pp. vii, 22–4, 34–40. Gassendi, *Exercitationes,* pp. 14–15; in *Opera omnia,* vol. 3, pp. 103–4.

have generated a number of hypotheses to explain it. Emphasizing Gassendi's scientific interests, Bernard Rochot suggested that Gassendi's attraction to Epicurus was stimulated by his contact with Isaac Beeckman in Holland in 1628. Beeckman had been an atomist for a long time and had come to see Epicureanism as an attractive replacement for Aristotelianism, which he had already rejected.[16] Rochot's view is reinforced by Barry Brundell, who argues that Gassendi explicitly developed Epicureanism to replace the Aristotelian philosophy that had dominated the schools.[17] Lynn Joy thinks that Gassendi abandoned the skeptical project he had embarked on in the *Exercitationes* when Mersenne's rebuttal to the skeptics in *La verité des sciences contre les sceptiques ou Pyrrhoniens* (1625) convinced him that the skeptical approach would lead only to an endless series of attacks and counterattacks without any grounds for resolution.[18] Lisa Sarasohn has suggested that Epicureanism provided Gassendi with a way of responding to skepticism without actually rejecting the possibility of knowledge of the physical world.[19]

Plausible as these explanations appear, they fail to come to grips with the question of what it was about Epicureanism per se that attracted Gassendi. The fact that there has been no sustained discussion of Gassendi's choice of Epicurus is a reflection of the fact that we live in an intellectual world that has been molded by views deriving directly from Gassendi's Epicureanism[20] Consequently, to us, atomism seems the natu-

16. Rochot, *Les travaux de Gassendi sur Épicure*, p. 34. Rochot continued: "Gassendi wanted to show that the thought of Epicurus was on the one hand without danger to religion and most of Ethics and on the other in accord with the most recent progress in science" (p. 58); my translation. Howard Jones echoes Rochot without further comment. See Jones, *Pierre Gassendi*, p. 29.
17. Brundell, *Pierre Gassendi*.
18. Joy, *Gassendi the Atomist*, pp. 24–38.
19. Louise [Lisa] Tunick Sarasohn, "The Influence of Epicurean Philosophy on Seventeenth Century Ethical and Political Thought: The Moral Philosophy of Pierre Gassendi," Ph.D. dissertation, University of California, Los Angeles, 1979, p. 118. See also Henri Berr, *Du scepticisme de Gassendi*, translated by Bernard Rochot (Paris: Éditions Albin Michel, 1960). The most important discussion of skepticism in this period is Richard H. Popkin, *The History of Scepticism from Erasmus to Spinoza*, revised and expanded edition (Berkeley: University of California Press, 1979; first published 1960).
20. David Fate Norton, "The Myth of 'British Empiricism,'" *American Philosophical Quarterly*, 1 (1981): 331–44. For example, Dijksterhuis displayed presentist assumptions when he wrote, "It was this conception [the conception of the world called mechanical or mechanistic] that first led to the

ral choice as a foundation for natural philosophy. Yet as Gassendi's lengthy and sometimes torturous discussion demonstrates, many aspects of atomism were problematic, particularly in the seventeenth-century theological context. Stoicism was another possible philosophy of nature, which because of its emphasis on providence could well have been seen as prima facie more consistent than Epicureanism with Christian theology. Indeed, a generation before Gassendi, Justus Lipsius (1547–1606) had attempted to produce a Christianized version of Stoicism to provide foundations for the new science.[21] Gassendi owned a copy of Lipsius, along

methods of research and treatment that have caused the great flourishing of physical science – a term used here to include the whole science of inanimate nature, chemistry and astronomy as well as physics – of which we are reaping the fruit in our own day: experiment as the source of knowledge, mathematical formulation as the descriptive medium, mathematical deduction as the guiding principle in the search for new phenomena to be verified by experimentation. It was the subsequent success of this conception which made the development of technology possible, thus leading to the sweeping industrialization without which the life of modern society has long since become inconceivable." E. J. Dijksterhuis, *The Mechanization of the World Picture,* translated by C. Dikshoorn (Oxford University Press, 1961), p. 3. Similar positivistic assumptions, without quite as many Comtian overtones, run through Marie Boas' classic discussion "The Establishment of the Mechanical Philosophy," *Osiris, 10* (1952): 412–541. Jan V. Golinski makes a similar point about the historiography of chemistry and the mechanical philosophy. He discusses how Hélène Metzger and others wrote about the adoption of the mechanical philosophy as requiring no other explanation than that it was epistemologically and psychologically privileged. See Jan V. Golinski, "Chemistry in the Scientific Revolution: Problems of Language and Communication," in *Reappraisals of the Scientific Revolution,* edited by David C. Lindberg and Robert S. Westman (Cambridge University Press, 1990), pp. 368–71; and Jan V. Golinksi, "Hélène Metzger and the Interpretation of Seventeenth-Century Chemistry," *History of Science, 25* (1987): 85–97.

21. See Jason Lewis Saunders, *Justus Lipsius: The Philosophy of Renaissance Stoicism* (New York: Liberal Arts Press, 1955). Lipsius gave a systematic presentation of Stoic natural philosophy *Physiologiae Stoicorum Libri Tres* (Antwerp, 1604). Its Stoicism notwithstanding, this book is comparable in scope to Gassendi's *Syntagma philosophicum* and other early presentations of the mechanical philosophy such as Kenelm Digby's *Two Treatises* (1644), ,0Walter Charleton's *Physiologia Epicuro-Gassendo-Charltoniana* (1654), and Thomas Hobbes' *De corpore* (1655). See also J. H. M. Salmon, "Stoicism and Roman Example: Seneca and Tacitus in Jacobean England," *Journal of the History of Ideas, 50* (1989): 199–227.

with his copy of Lucretius.[22] Recent scholarship has demonstrated that there were in fact many Stoic elements in seventeenth-century science.[23]

Another possible explanation for Gassendi's choice of Epicurus lies in the seemingly paradoxical fact that he found it easier to reconcile Epicureanism than any other ancient philosophy to his theology and his ethics.[24] "I observe from his [Epicurus'] physics of atoms and the void and from his morality of pleasure that many difficulties are removed and are more easily explained than by the positions of the other philosophers."[25] Gassendi did not think that all of Epicurus' propositions "are agreeable to religion," but rather that they can be modified to conform to it.[26] Standing early in the *Syntagma philosophicum*, this statement provides a guidepost as we make our way through the complexities of his text.

Although there is evidence that Gassendi was attracted to the philosophy of Epicurus as early as 1628,[27] he did not defend Epicureanism in

22. Gassendi to Pybrac, April 1621, in *Opera omnia*, vol. 6, p. 1.
23. For the much neglected story of the Stoic side of early modern science, see Peter Barker, "Jean Pena and Stoic Physics in the Sixteenth Century," and B. J. T. Dobbs, "Newton and Stoicism," both in *Recovering the Stoics*, edited by Ronald H. Epp, *Southern Journal of Philosophy*, 23 Supplement (1985): 93–108, 109–24; Peter Barker and Bernard R. Goldstein, "Is Seventeenth-Century Physics Indebted to the Stoics?" *Centaurus*, 27 (1984): 148–64; Peter Barker, "Stoic Contributions to Early Modern Science," and B. J. T. Dobbs, "Stoic and Epicurean Doctrines in Newton's System of the World," both in *Atoms, Pneuma, and Tranquillity*, edited by Osler, pp. 135–54, 221–38.
24. Lisa Sarasohn stresses the importance of ethics in Gassendi's thought in the following works: "The Ethical and Political Philosophy of Pierre Gassendi," *Journal of the History of Philosophy*, 20 (1982): 239–60; "Motion and Morality: Pierre Gassendi, Thomas Hobbes, and the Mechanical World-View," *Journal of the History of Ideas*, 46 (1985): 363–80; "Epicureanism and the Creation of a Privatist Ethic in Early Seventeenth-Century France," in *Atoms, Pneuma, and Tranquillity*, edited by Osler, pp. 175–95; *Freedom in a Deterministic Universe: Gassendi's Ethical Philosophy* (Ithaca, N.Y.: Cornell University Press, forthcoming). See also Gianni Paganini, "Épicurisme et philosophie au XVIIème siècle. Convention, utilité et droit selon Gassendi," *Studi filosofici*, 12–13 (1989–90): 5–45.
25. Gassendi, *Syntagma philosophicum*, in *Opera omnia*, vol. 1, p. 30.
26. Ibid.
27. Rochot, *Les travaux de Gassendi sur Épicure*, p. vii; Jones, *Pierre Gassendi*, p. 205. The standard evidence for this statement is his letter to Erycius Puteanus, April 1628, in which he stated his intention to present the philosophy of Epicurus as a continuation to his *Exercitationes*. See Gassendi, *Opera omnia*, vol. 6, p. 11.

print until the 1640s. During the intervening years, his Epicurean project
grew from the straightforward humanist task of translating Book X of
Diogenes Laertius' *Lives of Eminent Philosophers*,[28] a major source for
knowledge of Epicurus' writings and ideas, to a full-fledged rehabilitation
of Epicureanism. By 1631, Gassendi had sketched an outline of his
project. In a long letter to Peiresc, Gassendi described his Epicurean
project in terms that anticipate all of his later work:

> I am revising my philosophy of Epicurus and am about half way into
> it, but what is already done considerably exceeds what you have seen.
> The cause of this excess is that I am not only adjusting all the Greek
> passages, for which I did not have the books at our place, with others
> found since and am interposing them again with the responses and
> mollifiers suitable to the points which touch on our faith, but beyond
> that I am trying to enrich the whole work by comparing all that we
> remember of ancient philosophy with the doctrine of Epicurus.[29]

This letter contains an outline of most of the material that would later
appear in the "Logic" and "Physics" of the posthumous *Syntagma philosophicum*, including the sections that contain his discussions of theological matters.[30] These sections are the following: (1) On the existence and
attributes of God – "De existentia naturæ divinæ," "De forma naturæ
divinæ, an humane similis est," "De immortalitate et foelicitate ejusdem
naturæ diviniæ"; (2) On God as creator – "De origine mundi"; (3) On
divine providence and free will – "De providentia, seu gubernatione mundi," "De fortuna et fato in mundo"; and (4) On the immortality of the
soul – "De animorum immortalitate." Although the order of the chapters
and their placement in the text would be altered in later versions – for
example, the chapter on fortune and fate would later find its place in Book
3 of the "Ethics" – all the issues central to Gassendi's theology were
present in this early sketch. He reiterated his concern to "defend providence against Epicurus" in a letter to Thomas Campanella written in
November 1632.[31] The sketch of 1631, the letter to Campanella, and
other documents from this period provide strong evidence against Bloch's
contention that Gassendi interpolated the theological material into the
"Physics" in the early 1640s, and then only in order to dissimulate his

28. Gassendi to Peiresc, 11 September 1629, in *Lettres de Peiresc*, edited by
    Philippe Tamizey de Larroque, in *Documents inédits sur l'histoire de France*,
    7 vols. (Paris: Imprimerie Nationale, 1888–98), vol. 4, p. 213.
29. Gassendi to Peiresc, 28 April 1631, in ibid., p. 249.
30. Gassendi's outline of 1631 is so significant in establishing his early concep-
    tion of his project and the presence of its theological components that I have
    reproduced it in the Appendix at the end of this chapter.
31. Gassendi, *Opera omnia*, vol. 6, p. 54.

materialism of which he was becoming increasingly aware.[32] From the
earliest stages of his Epicurean project, Gassendi was concerned to baptize
his ancient model, Epicurus.

Although the question of whether Gassendi's letter to Peiresc is an
outline of an existing draft of his Epicurean commentary or whether it is
simply a statement of his intentions is controversial, during the next few
years, Gassendi did produce a manuscript of his Epicurean commentary,
which he shared "*cahier*" by "*cahier*" with Peiresc and some other
friends.[33] By 1634, he completed a draft of what would later be published
as *De vita et moribus Epicuri* (1647).[34] At each stage of this process, he
was explicitly concerned to Christianize Epicureanism.

Gassendi's mature thinking about Epicureanism appears in several
texts. During 1641 and 1642, he wrote a series of letters containing a
sketch of his Epicurean project to his patron, the new governor of
Provence, Louis-Emmanuel de Valois. In 1647, he wrote *De vitae et moribus Epicuri* defending Epicurus against allegations of decadence and im-
morality. He published one other Epicurean work during his lifetime,
Animadversiones in decimum librum Diogenes Laertii (1649). Two other
works appeared only in his posthumously published *Opera omnia* (1658),
the brief *Philosophiae Epicuri syntagma*, originally published as an appendix to the *Animadversiones,* and the massive *Syntagma philosophicum.*[35] Gassendi worked and reworked his Epicurean project for most of
his intellectual life.[36] The *Syntagma philosophicum* is the culmination of
this process. Although consistent in spirit with his earlier Epicurean
works, it contains more contemporary scientific material than the earlier
works, material more suited to an exposition of Epicureanism than to a
commentary on an ancient text, as the earlier *Animadversiones* had
been.[37] Consisting of the traditional three parts of philosophy entitled

32. Bloch, *La philosophie de Gassendi,* pp. 430, 476–81.
33. See René Pintard, *La Mothe le Vayer, Gassendi, Guy Patin: Études de bibli-
    ographie et de critique suivies de textes inédits de Guy Patin* (Paris: Boivin,
    1943), pp. 35–44; Rochot, *Les travaux de Gassendi sur Épicure,* pp. 43–51.
    For a discussion of this scholarly controversy as it bears on the "Ethics," see
    Sarasohn, "The Influence of Epicurean Philosophy on Seventeenth Century
    Ethical and Political Thought," chap. 5.
34. Rochot, *Les travaux de Gassendi sur Épicure,* pp. 59–83.
35. For the publication history of Gassendi's works, see Bernard Rochot, "Gas-
    sendi (Gassend), Pierre," in *Dictionary of Scientific Biography,* edited by
    Charles Coulton Gillispie, 16 vols. (New York: Scribner, 1972), vol. 5, pp.
    284–90.
36. Rochot, *Les travaux de Gassendi sur Épicure,* recounts this process in detail.
37. Ibid., pp. 191–2.

"Logic," "Physics," and "Ethics," the *Syntagma philosophicum* contains a complete exposition of philosophy and the history of philosophy.[38] In this respect, the *Syntagma philosophicum* is comparable to other early-seventeenth-century expositions of the mechanical philosophy.

To accomplish his goal, Gassendi had to confront the theologically objectionable components of Epicureanism: polytheism, a corporeal conception of the divine nature, the negation of all providence, the denial of creation ex nihilo, the infinitude and eternity of atoms and the universe, the plurality of worlds, the attribution of the cause of the world to chance, a materialistic cosmogony, the denial of all finality in biology, and the corporeality and mortality of the human soul.[39] It is interesting, as Gassendi himself noted, that many of the theological repairs that he felt obliged to make to Epicureanism were similar to those that Thomas Aquinas had found necessary in order to adopt Aristotelianism.[40]

What, if anything, was left of Epicureanism, after Gassendi completed all these repairs? Rather than reviving the philosophy of Epicurus, did he actually drown his ancient model in the baptismal font? The answer to these questions is not unequivocal. On the one hand, Gassendi retained the fundamental feature of Epicurean natural philosophy, namely, an atomistic explanation of the natural world, and a hedonistic ethics redefined in terms of Christian salvation.[41] On the other hand, his insistence on a creationist, providential account of nature and human life was a complete and explicit repudiation of Epicurus' intentions, which had been to create a natural and ethical philosophy that would free people from worries about the whims of the gods and the fear of punishment and reward in life after death.[42]

The question of Gassendi's religious convictions has been the subject of

38. For one interpretation of the historical aspect of Gassendi's work, see Joy, *Gassendi the Atomist*. For Gassendi's ethical and political views, see especially Sarasohn, "The Ethical and Political Philosophy of Pierre Gassendi," and "Motion and Morality." See also Marco Messeri, *Causa e spiegazione: La fisica di Pierre Gassendi* (Milan: Franco Angeli, 1985); and Jones, *Pierre Gassendi*. For a general historiographical overview of scholarship on Gassendi, see Brundell, *Pierre Gassendi*, pp. 5–15.
39. Bloch, *La philosophie de Gassendi*, p. 300.
40. Gassendi, *Syntagma philosophicum*, in *Opera omnia*, vol. 1, pp. 280–1. See Brundell, *Pierre Gassendi*, pp. 55–6. Dijksterhuis makes this point. See *The Mechanization of the World Picture*, pp. 231–2.
41. Sarasohn, "The Ethical and Political Philosophy of Pierre Gassendi," pp. 239–60.
42. See Long, *Hellenistic Philosophy*, pp. 41–2, 49–50.

considerable scholarly dispute.[43] Some scholars, notably René Pintard and J. S. Spink place him among the skeptics and freethinkers, the *libertins érudits,* who played a prominent role in French intellectual life in the early seventeenth century.[44] Both Pintard and Spink disagree with Henri Busson who had argued that the libertine movement, rather than being a harbinger of deism and the Enlightenment critique of religion, was a holdover from sixteenth-century naturalism that was opposed by a number of thinkers inspired by the forces of the Counter-Reformation, among whom Busson numbered Gassendi.[45] Other scholars have cited Gassendi's virtuous life and comportment as a priest as evidence that he was really a holy, perhaps even a saintly man.[46] More recent studies tend to acknowledge the presence of theological considerations within Gassendi's thought and writings, but not all writers give them much significance. Richard Popkin interprets Gassendi's thought as primarily, philosophical, an attempt to respond to the skeptical crisis of the day. By enunciating a "mitigated skepticism," Popkin argues, Gassendi was able to react to the skeptical critique of dogmatic knowledge without falling into yet another dogmatism.[47] Bernard Rochot believed that Gassendi took orthodoxy and theology seriously and that they played an important role in his philosophy. Of particular importance, Rochot argued, was Gassendi's attempt to eliminate the theologically objectionable components of Epicureanism so that atomism might become acceptable to the seventeenth-century community of natural philosophers.[48] Olivier René Bloch regards

43. Summaries of most of these views can be found in Raymond Collier, "Gassendi et le spiritualisme ou Gassendi était-il un libertin?" in *Actes du Congrès du Tricentenaire de Pierre Gassendi (1655–1955)* (Digne: CNRS, 1957), pp. 97–133; and Brundell, *Pierre Gassendi,* pp. 5–15.
44. See René Pintard, *Le libertinage érudit dans la première moitié du XVIIe siècle,* 2 vols. (Paris: Boivin, 1943); and J. S. Spink, *French Free-Thought from Gassendi to Voltaire* (London: Athlone, 1960).
45. Henri Busson, *La pensée religieuse française de Charron à Pascal* (Paris: J. Vrin, 1933).
46. Abbé A. Martin, *Histoire de la vie et des écrits de Pierre Gassendi* (Paris, 1854). Gassendi's orthodoxy was warmly defended against detractors by Charles Jeannel, *Gassendi spiritualiste* (Montpellier: Félix Seguin, 1859). I am grateful to Francine Michaud for giving me a copy of this little, polemical work.
47. Popkin, *History of Scepticism,* chap. 7.
48. Rochot, *Les travaux de Gassendi sur Épicure,* p. 58. Similar interpretations are found in L. Mandon, *Étude sur le Syntagma philosophicum de Gassendi* (1858; reprinted, New York: Burt Franklin, 1969), chap. 5, and Robert Hugh Kargon, *Atomism in England from Hariot to Newton* (Oxford University Press, 1966), pp. 67–8.

Gassendi's theological objections to his opponents, both ancient and modern, as "the seasoning of a polemic whose point was purely secular and scientific."[49] Bloch sees Gassendi as a closet materialist who used methods of dissimulation similar to those of Bayle and the Encyclopedists and argues that although Gassendi used the language of orthodoxy, he used it in a superficial way to mask the profane materialism that he really espoused. To support this contention, Bloch cites what he interprets as various anomalies and juxtapositions of contradictory theses within Gassendi's writings. Since Gassendi the priest could not openly endorse materialism, the tensions in his thought must, according to Bloch, be interpreted as evidence of deliberate dissimulation.[50]

I do not agree with Bloch's assessment. The grounds for my disagreement will become evident in the following discussion. Nevertheless, the state of Gassendi's soul and the sincerity of his religious convictions are simply irrelevant to my argument. Lengthy theological passages in the *Syntagma philosophicum* contain a recognizable voluntarist position that is coherent with Gassendi's metaphysical and epistemological views. Whether this theology was something to which he was passionately committed or whether it is simply a statement of views he had uncritically absorbed, the fact is that it provided the intellectual background for the other elements of his philosophy, which were infused with concepts drawn from medieval theology. Understanding his thought in the context of the long history of the dialectic between the absolute and ordained power of God elucidates the relationships among the various parts of his philosophy.

Gassendi's main expositions of theology occur in both the "Physics" and the "Ethics."[51] In the "Physics," Gassendi used his theology as one of the general principles upon which his new philosophy of nature was to be founded. In an introductory section entitled "On the Nature of Things Universally," he considered the nature of the universe: Is it unique? What is its size and shape? Is it endowed with a soul? Did it have a beginning, and will it come to an end? Following these general questions, Gassendi spelled out the ontological foundations of his physics, the ultimate terms of explanation for his new philosophy of nature. He discussed

49. Bloch, *La philosophie de Gassendi*, p. 312; my translation.
50. Ibid., chaps. 9–11.
51. Gassendi, *Syntagma philosophicum*, in *Opera omnia*, vol. 1, pp. 283–337, vol. 2, pp. 821–60. The section of Bernier's *Abrégé* on the "Ethics" was translated into English in the seventeenth century. See François Bernier, *Three Discourses of Happiness, Virtue, and, Liberty, collected from the works of the Learn'd Gassendus,* translated out of French (London: Awnsham & John Churchil, 1699).

the nature of space and time and argued for the existence of the void. He considered the nature of the material principle and concluded that the ultimate material components of the universe are physical atoms, the attributes of which he described in detail. Finally, he discussed the efficient principle of things, claiming that God is the first cause of the world and that all second causes can be reduced to the motions of atoms. Gassendi's long discussion of the efficient principle contains an extended account of God's existence, his attributes, and his providential relationship to the creation. After establishing these conceptual foundations for his physics, Gassendi devoted the remainder of the "Physics" to a detailed account of the natural world – celestial and terrestrial, animate and inanimate – in terms of the motions and interactions of atoms in the void. The "Physics" concludes with his proof of the immortality of the human soul.

The organization of the "Physics" parallels the structure of courses in physics in seventeenth-century French universities:

> Initially the professor introduced the student to the fundamental prin-
> ciples that governed the existence and mutations of natural bodies.
> Next he would demonstrate in what way these principles could ex-
> plain the structure of the universe and the behaviour of various ter-
> restrial phenomena. Finally he examined the nature of life and
> compared the different types of organism, plants, animals, and man,
> one with another.[52]

That Gassendi intended the massive and scholarly *Syntagma philosoph-icum* to serve as a university textbook is doubtful. But clearly he did intend its contents to replace the contents of the established Aristotelian curriculum point by point.

Gassendi's voluntarist theology

If Gassendi's primary goal was to baptize Epicureanism, the theological framework within which he did so was voluntarist. Voluntarism was the unifying thread that bound his natural philosophy, theory of knowledge, and ethics into a coherent whole. It also provided him with a rationale for choosing among the various ancient philosophies available as foundations for the new science.

Gassendi's discussion of God, his existence and his attributes, com-prises most of Book IV of the First Part of the "Physics," a section entitled "On the Efficient Principle or the Causes of Things." His account of God,

---

52. L. W. B. Brockliss, "Philosophy Teaching in France, 1600–1740," *History of Universities*, 1 (1981): 146.

his nature, and his role in the world occurs within the context of arguments against Epicurus' materialistic view of a universe run by chance. In this fifty-page section of the *Syntagma philosophicum*, Gassendi undertook to explicate his concept of causality, particularly in relation to God's role in the creation. For Gassendi, the subject matter of physics is the efficient principle. He identified cause with the efficient principle, explicitly reducing all causality to Aristotle's efficient cause and demonstrating how the three other Aristotelian causes can be eliminated in favor of the efficient cause.[53] Although elsewhere he explicitly argued for the existence of final cuses in nature and for the fact that they are an appropriate and important part of the subject matter of natural philosophy, he rejected any kind of immanent finality in the Aristotelian sense.[54] What Gassendi called final causes are actually divine intentions reflected in the design of the creation. Thus, the purposiveness found in nature is, for Gassendi, externally imposed by God. The natural order itself is ruled only by efficient causes, including God.

Our knowledge of causes must be based on the appearances, our only source of knowledge about the world.[55] All events have causes, although we cannot know the causes of all events in the same way that we cannot know whether the number of the stars is odd or even. Gassendi interpreted the causal order of nature and the evident fact that various parts of things – particularly plants and animals – are designed for certain ends as evidence of the intelligence of the creator. Just as the clockmaker applies his intelligence to efficient causes to produce an elegant timepiece, so God utilizes efficient causes in designing the world.[56] God, who creates the second causes he uses, differs from the clockmaker, who makes use of materials found in the world. Gassendi drew the common seventeenth-century distinction between two kinds of efficient causes. The first cause is God, the most general cause of all.[57] Second causes are the physical causes

---

53. Gassendi, *Syntagma philosophicum*, in *Opera omnia*, vol. 1, pp. 283–7. Brundell finds Gassendi's reduction of the four Aristotelian causes to the efficient cause significant evidence of his overriding anti-Aristotelian concerns. Brundell, *Pierre Gassendi*, pp. 69–76.

54. See Pierre Gassendi, *Disquisitio metaphysica seu dubitationes et instantiae adversus Renati Cartesii metaphysicam et responsa*, edited and translated into French by Bernard Rochot (Paris: J. Vrin, 1962), pp. 396–9, in *Opera omnia*, vol. 3, p. 359.

55. Gassendi, *Syntagma philosophicum*, in *Opera omnia*, vol. 1, p. 286. See Chapter 4, this volume.

56. Ibid., p. 285.

57. Ibid., pp. 286–7.

of specific phenomena in the natural world, the motions and collisions of the atoms from which everything in the world is composed.[58]

Having introduced the deity into his discussion of causality, Gassendi proceeded to establish the existence of God and to describe his attributes. Arguing for the existence of God, he explicitly opposed Epicurus. "Let it be said that Epicurus erred in his description of the nature of the divine; but he seems to have committed the lapse not from malice, but from ignorance."[59] According to Gassendi, belief in God is universal, a fact that proves that he exists. In an attempt to circumvent the obvious fact that there are some people who are atheists, Gassendi argued that these exceptions do not invalidate the general principle that belief in God is universal. His use of the term "universal" here seems to be in the sense of "natural" rather than "true of each and every individual," for he went on to assert that the fact that some are born or become atheists does not detract from the general principle that people by nature have the idea of God impressed upon them.[60] Just as "it can be inferred that since all men, if you except the few who are blind, must have the facility to see light and say that it exists, so it is that God exists, since all men, if you except the few who are atheists, have the notion (*anticipatione*). They know that God exists and they acknowledge it."[61] Likewise, the fact that there is a

58. Ibid., pp. 338–71.
59. Ibid., p. 290.
60. Ibid.
61. Ibid., pp. 290–1. Lewis and Short translate the word "*anticipatio*" as "innate idea." Charlton T. Lewis and Charles Short, *A Latin Dictionary* (Oxford University Press, 1879). To do so here would seriously contradict Gassendi's epistemology, in which he unambiguously states that all our ideas originate in the senses. In the "Logic," he made the following two statements: (1) "Omnis, quae in Mente est Anticipatio, seu Praenotio dependet a Sensibus; idque vel incursione, vel proportione, vel similitudine, vel compositione;" and (2) "Anticipatio est ipsa rei notio, sive definitio, sine qua quidquam quaerere, dubitare, opinari, imo & nominare non licet." Gassendi *Syntagma philosophicum*, in *Opera omnia*, vol. 1, p. 54. These statements occur in a section entitled "The Logic of Epicurus." Brundell argues that Gassendi explicitly adopted the Epicurean canonic as a replacement for Aristotelian dialectic, which he had criticized in the *Exercitationes*. See Brundell, *Pierre Gassendi*, pp. 84–6. Significantly, in the *Disquisitio metaphysica*, Gassendi explicitly rejected Descartes' claim that the idea of God is innate, arguing that even the idea of God comes to us from the outside. See Gassendi, *Disquisitio metaphysica*, pp. 250–3 (in *Opera omnia*, vol. 3, p. 326). In the *Syntagma philosophicum*, he stated that *anticipationes* originate in the senses: " . . . vero Anticipationem ortum habere a sensibus . . . " *Opera omnia*, vol. 1, p. 292.

great diversity of opinion about the nature of God does not damage the claim that all people have some idea of God.[62]

Knowledge of God, according to Gassendi, like all knowledge, comes from the senses. The idea of God originates in the fear of natural forces such as thunder and lightning.[63] God revealed himself directly to Adam and Eve, who thus experienced him immediately and received his gift of faith. Their knowledge has been conveyed to succeeding generations through teachers, prophets, and further revelations.[64] Although our knowledge of God is empirical, we do not have a sensory image of God as Epicurus had thought.[65] According to that theory, such an image would be formed by the passage of material simulacra from the surfaces of the gods to our organs of vision, a theory presuming that the gods are material and thus mortal. Gassendi rejected this Epicurean theory, claiming instead that things comprehended through the senses are occasions that lead us to form an anticipation or mental image of God. Such occasions can occur from any of the senses – for example, the sense of hearing, as in the experience of Adam and Eve to whom God spoke directly, or the sense of sight, which reveals God's intelligent design in the world.[66]

The argument from design played a central role in Gassendi's thought, providing evidence both for the existence of God and for his providential relationship to the creation.[67] Whereas the "Sacred Faith" informs us of God's existence, observation of his wisdom in the creation teaches us by our own experience that the world was created by an intelligent designer: "The paths of the stars, the vicissitudes of storms, the succession of generations, the order and use of parts – everything, in a word, that is in the world announces order and declares that the world is a most orderly system."[68]

Even the smallest things, such as snowflakes, are signs of intelligent design.[69] Where there is order, there must be an orderer. For Epicurus,

62. Ibid., p. 291.
63. Ibid., p. 292.
64. Ibid., p. 293.
65. On this difficult point, see Long and Sedley, *The Hellenistic Philosophers,* vol. 1, pp. 144–9.
66. Gassendi, *Syntagma philosophicum,* in *Opera omnia,* vol. 1, p. 293.
67. I disagree with Brundell who speaks of Gassendi's "perfunctory use of the argument from design." Brundell, *Pierre Gassendi,* p. 71. Gassendi appealed to design repeatedly, usually to support a providential creationism.
68. Gassendi, *Syntagma philosophicum,* in *Opera omnia,* vol. 1, p. 294.
69. Ibid., p. 315.

the effectrix of all things consists of atoms, moved perpetually by their own innate weight. Forming one configuration after another, they create the world as we see it by chance.[70] Chance or fortune, however, cannot be the source of the order observed in the world, for "chance and fortune are indeed nothing." They are blind, "not sharing in the plans, not understanding the order."[71] The harmony and elegance of the world – the parts of animals, their generation, and the general order of the world – indicate that it was created by hands other than fortune.[72] An orderly world could no more be the product of chance than could a work of architecture.[73] Similarly, the world cannot have an innate or immanent order. This observed order is the work of "reason and planning," but the world cannot act by plan and reason on its own. The source of this reason and planning must be distinct from the order itself. The first cause is "what we call God and what can be called the first cause, prime mover, fount of all being and origin of all perfection, the highest being and prince of the world."[74] Observation of the world leads inexorably to the conclusion that God exists and is the first cause and designer of the world.

What are the attributes of this God? He cannot be represented to the mind as he is in himself because his perfection surpasses any possible representation we might attempt. Consequently, our knowledge of God is extremely limited.[75] We do know, however, that any perfections observed in other things may also be attributed to God, since all perfections flow from him.[76] Thus, we know that he is immense, eternal, omniscient, and omnipotent. It is by virtue of his intelligence that he can create intellect, which exists in the world. We also know that he is unique, since multiplicity would render him liable to corruption and would therefore be an imperfection.[77] In short, echoing the language of the ontological argument, which however he did not accept as valid,[78] Gassendi proclaimed that God is that than which nothing greater can be conceived.[79]

70. Ibid., p. 312.
71. Ibid.
72. Ibid., pp. 312–16.
73. Ibid., p. 316.
74. Ibid., p. 295.
75. Ibid., p. 302.
76. Ibid., p. 303.
77. Ibid.
78. See Gassendi, *Disquisitio metaphysica*, pp. 250–1, 302–3, 327–9, 390–3, (in *Opera omnia*, vol. 3, pp. 326, 337, 343, 357–8).
79. Gassendi, *Syntagma philosophicum*, in *Opera omnia*, vol. 1, p. 305.

God knows all things: the present, the past, the future, and the concourse of causes. He knows not only all that he has created, but the infinite other possibilities that he did not create.[80] Here the concepts of God's absolute and ordained power inform Gassendi's discussion. Gassendi interpreted God's omnipotence as freedom from any necessity or limits. "There is nothing in the universe that God cannot destroy, nothing that he cannot produce; nothing that he cannot change, even into its opposite qualities."[81] In other words, like Ockham, Gassendi believed that God's absolute power is in no way constrained by the creation, which contains no necessary relations that might limit God's power or will. Even the laws of nature lack necessity. Like everything else God created, he can negate them. "He is free from the laws of nature, which he constituted by his own free will."[82] Indeed, God can do anything short of violating the law of noncontradiction.[83] God was totally free in choosing to create the world: He could have abstained from creating it just as freely as he chose to create it.[84] Moreover, God could have created an entirely different natural order, if it had pleased him to do so.[85] That is to say, God could have created black snow, and he could have created a cold fiery substance just as he can raise the dead and heal the crippled. The order that God created is thus utterly contingent on his absolute power. Even though he has chosen to create this universe – the one containing white snow and hot fire – he has created nothing in it that he cannot change at will. An implicit assumption here (one that Gassendi makes explicit elsewhere)[86] is that there are no essences in the world God created. There are no necessary connections linking fire and heat or whiteness and snow. God could not have made white snow black or hot fire cold, for such combinations of attributes are contradictory. God is only limited by the law of noncontradiction. He could have created substances with properties very different from the ones that presently exist. And by implication, he is free to do so at any time. This is what Gassendi meant when he said, "There is nothing in the universe that God cannot destroy, nothing that he cannot produce; nothing that he cannot change, even into its opposite qualities."[87] The inability to create contradictories lies not in a restriction

80. Ibid., p. 307.
81. Ibid., p. 308.
82. Ibid., p. 381; also p. 234.
83. Ibid., p. 309.
84. Ibid., p. 318.
85. Ibid., vol. 2, p. 851.
86. See Chapter 4.
87. Gassendi, *Syntagma philosophicum*, in *Opera omnia*, vol. 1, p. 308.

to God's power but in a repugnance in the things themselves because they are contradictory.[88]

Even the fact that God responds to our prayers in no way limits his freedom:

> When God is beseeched, he acts because he determines to; and when it is said he is bent by prayers, the mode of speaking is vulgar . . . for we only give him prayers that he constituted himself. Therefore, God, as he is the highest good, is the most free; and he is not bound, as he can do whatever he knows, whatever he wishes.[89]

Within the stipulation that nothing God creates can impede his absolute power, God makes use of second causes to carry out the ordinary course of nature:

> It is his general providence that establishes the course of nature and permits it to be served continuously. From which it follows, as when either lightning or other wonderful effects are observed, God is not on that account suddenly summoned, as if he alone were its cause and nothing natural had intervened. . . . [A]side from him, particular [causes] are required that are . . . not thought to be uncreated; but are believed to be hidden from our skill and understanding.[90]

Second causes, as part of the created order, do not restrict God's freedom, because he can dispense with second causes altogether if he chooses. "Indeed, to the extent that an artisan has ingenuity, to that extent his work requires less matter and services; thus the divine artisan will be the most skilled since he requires neither matter nor ministers nor organs nor equipment."[91] Here Gassendi sounds much like Ockham: "Whatever God can produce by means of secondary causes, He can directly produce and preserve without them."[92] "And truly he is free, since he neither is confined by anything nor imposes any laws on himself that he cannot violate if he pleases. . . . Therefore, God . . . is the most free; and he is not bound as he can do whatever . . . he wishes."[93] The natural order, which God created by his absolute power, is utterly contingent on his will.

Like Ockham and unlike Aquinas, Gassendi essentially erased the line separating God's absolute and ordained powers. According to this volun-

---

88. Ibid.
89. Ibid., p. 309.
90. Ibid., p. 326.
91. Ibid., p. 317.
92. William of Ockham, *Ordinatio,* qu. i, n *sqq.,* in Ockham, *Philosophical Writings,* translated and edited by Philotheus Boehner (Edinburgh: Nelson, 1957), p. 25.
93. Gassendi, *Syntagma philosophicum,* in *Opera omnia,* vol. 1, p. 309.

tarist theology, nothing that God has created binds him in the way that the essences of things limit Aquinas' God. According to the intellectualist Aquinas, the relations between essences and properties of created substances are necessary in both a metaphysical and an epistemological sense: An object made of iron must be heavy, and we can know that it has the property of heaviness on the basis of an a priori inference from our knowledge that it possesses the essence of iron. Gassendi denied both kinds of necessity. That is the thrust of his statements that God can create black snow and that, more generally, God can always do directly what he usually does by second causes. Snow happens to be white as a contingent fact about this world. But snow does not possess an essence that necessitates that it be white, and consequently, we cannot know anything about the color of snow a priori. The created order contains nothing to limit the continued exercise of God's absolute power. For Gassendi, as for other voluntarists, the domain of God's absolute power completely overlaps the domain of his ordained power.

There are some passages in Gassendi's text that so emphasize the regularity and order of nature that they seem to be contradict his voluntarism. A notable example occurs in his discussion of astrology, where he wrote that some astrologers

> *exempt the Stars from the Jurisdiction of Fate, and . . . suppose them under the government only of God* who made them: and yet when they pretend this for an excuse and support of a most vain and indeed prophane Art; I dare not say, they have done either discreetly or piously. For they made as if they were certain, that such an effect would inevitably ensue, as they have foretold; unless God by some special resolve of his providence is pleased to avert it: as if it were better, that God should be supposed to pervert the General Order of Causes and Effects, that his infinite wisedome ordained and instituted from all Eternity; than that their foolish Aphorismes should be suspected of uncertainty and deceipt: and as if it were more reasonable to recurr to a miracle, than to confess the fallibility of their rules.[94]

Whether or not God chooses to interrupt the "General Order of Causes and Effects" by producing a miracle, the fact is that he has the freedom and power to do so if he wishes. God instituted the natural tenor of things

94. Petrus Gassendus, *The Vanity of Judiciary Astrology. Or Divination by the Stars. Lately written in Latin by that Great Schollar and Mathematician, the Illustrious PETRUS GASSENDUS; Mathematical Professor to the King of FRANCE*, translated into English by a Person of Quality (London: Humphrey Moseley, 1659), pp. 75–6. This work is a contemporary English translation of the "Physics," sec. II, bk. VI, "De effectibus siderum," in *Opera omnia*, vol. 1, pp. 713–52.

by his absolute power. It is the contingency of the creation that must be emphasized, not the actual occurrence of miracles. This contingency by no means entails chaos or disorder, simply the absence of necessity.

Although Gassendi did not explicitly speak in terms of the dialectic between the absolute and ordained powers of God, these ideas provided the conceptual background for his voluntarist theology. That Gassendi did not cite Ockham or any other Scholastic authors in the *Syntagma philosophicum* was characteristic of his preference to cite ancient and patristic writers but never medieval ones and rarely his own contemporaries. The lack of citations of medieval writers must not be taken as evidence either that Gassendi was not acquainted with these authors or that they did not influence his thinking.[95] He was in fact aware of their writings and in the *Exercitationes* explicitly numbered himself among those who admit "this mad opinion of the Nominalists" about the status of universals.[96]

Gassendi's providential worldview

Divine providence and human freedom were fundamental components of Gassendi's conception of the world. Agreeing with Epicurus that the ultimate goal of natural philosophy is to produce tranquillity by explaining the natural causes of things, Gassendi disagreed with the ancient atomist's materialistic and antiprovidential outlook. Rather, Gassendi argued, natural philosophy leads to true religion:

> Natural philosophy (*Physiologia*) is the contemplation of the natural universe of things from the magnitude, variety, disposition, and beauty of its wonders. . . . Our natural reason deduces from it that there exists a most wise, powerful, and good, divine will (*Numen*) by which it is governed . . . , so that we acknowledge this divine will for the greatness of his excellence and beneficence. And reverence, which is the true religion, must be cultivated.[97]

This religion teaches that "God is the cause that created the world, and he rules it with general providence and also special providence for humanity."[98] Gassendi believed that Epicurus committed his gravest error in asserting that chance, not God, is the cause of the world.[99] To refute this

95. Gassendi, *Exercitationes*, pp. 278–9, in *Opera omnia*, vol. 3, p. 159. On the humanists' reluctance to cite medieval sources, see Edward Grant, *Much Ado About Nothing: Theories of Space and Vacuum from the Middle Ages to the Scientific Revolution* (Cambridge University Press, 1981), p. 182.
96. Gassendi, *Exercitationes*, pp. 280–3 (in *Opera omnia*, vol. 3, p. 159).
97. Gassendi, *Syntagma philosophicum*, in *Opera omnia*, vol. 1, p. 128.
98. Ibid., p. 311.
99. Ibid., pp. 320–1.

error, Gassendi countered with a battery of arguments for a providential understanding of nature:

God created the world, and he does not neglect it now:

> It depends no less certainly on its author than a light depends on its source; wherefore, as light cannot be observed without the sun from which it was created, the world cannot be preserved . . . without God by whom it was produced. . . . The world that would be nothing without God, has nothing from itself whereby it could subsist on its own and stand without God. . . . [T]he world would be reduced to nothing if God were to cease supporting it.[100]

God rules the creation like the captain of a ship, the driver of a chariot, the conductor of a chorus, the head of a household, the prince in a realm, the general in an army, and the soul in the body.[101] In this frequently reiterated string of metaphors, Gassendi emphasized God's direct and absolute rule of the world. Gassendi noted that these metaphors provide only a limited understanding of God's relationship to his creation, since unlike his human counterparts, God is intimately present to the whole world.[102]

God not only rules, he rules wisely. Nthing demonstrates God's providence more directly than the fact that he provides for the things he created.[103] Against those who object that care of all his creatures would pollute God with the sadness in the world, Gassendi replied, "They ought to say that the sun is polluted by his ministers by whom he illumines the sewers, gives birth to toads, [and] produces worms."[104] What appears polluting to us, may, from another standpoint, signify God's wisdom and providence. Against those who say that the multiplication of second causes would diminish God's power, Gassendi gave the standard voluntarist response: "All things – even the least, which are in the world – can be easily done by God by his providence alone."[105] Even the use of second causes and the regular functioning of the world point to God's providential care. Things in the world constantly follow their natural tenor, which was instituted by a wise plan and not by chance. If nature acted blindly, then its beginning could be referred to chance. But since everything is planned, the parts of nature point to wisdom.[106] God created the natural

---

100. Ibid., p. 323.
101. Ibid.
102. Ibid., p. 324.
103. Ibid.
104. Ibid., p. 325.
105. Ibid.
106. Ibid.

causes. Knowledge of this divine wisdom and power mitigates our fears and produces tranquillity.[107]

Epicurus had thought of the gods as perfectly tranquil beings, remote from any human concerns, and had argued that they cannot be angered or propitiated by anything we do and that they leave us to our own devices.[108] He had cited the presence of evil in the world as putative evidence against the existence of a providential God. At times, he had noted, God seems to care for the animals more than he does for humans.[109] Contrary to Epicurus, Gassendi affirmed that God "rules with special providence for humanity."[110] Of all creatures, only man knows his creator and the benefits he derives from him. Observation of the heavens and everything else in nature reveals that all these things have been provided for the benefit of man. "And why indeed is there so much beauty in the universe unless it is for man who considers it for himself? By observing, he sings a hymn to the Author."[111] As for the problem of evil, God has provided man with an immortal soul in order to compensate by rewards and punishments in the afterlife for evil and lack of justice in this life. No one's happiness can be judged before death.[112]

Gassendi's preoccupation with providence and design was characteristic of seventeenth-century natural philosophy, which was shot through with similar discussions. The specter of Epicurean materialism and atheism haunted the mechanical philosophers.[113] In *L'impiété des déistes* (1624), Mersenne wrote against the atheists, among whom he numbered the ancient Epicurus and his contemporary Lucilio Vanini (1586–1619), who was burned at the stake in Toulouse for impiety.[114] Even after natural philosophers generally accepted the mechanical view of nature, they continued to devote hundreds of pages to demonstrating God's providential activity in a world consisting of matter and motion.[115] Like Gassendi,

107. Ibid., p. 326.
108. Ibid.
109. Ibid., p. 327.
110. Ibid.
111. Ibid., p. 329.
112. Ibid., pp. 329–30. For Gassendi's arguments for the immortality of the soul, see the next section.
113. On the complex question of what "atheism" meant in the seventeenth century, see Michael J. Buckley, S.J., *At the Origins of Modern Atheism* (New Haven, Conn.: Yale University Press, 1987), chaps. 1 and 2.
114. Robert Lenoble, *Mersenne ou la naissance du mécanisme*, 2d edition (Paris: J. Vrin, 1971; first published 1943), p. 173 n8.
115. Richard S. Westfall, *Science and Religion in Seventeenth-Century England* (New Haven, Conn.: Yale University Press, 1958), chap. 4.

Boyle saw design everywhere in nature and appealed to it in order to defend natural philosophy against the charge that it ascribes the world to chance and therefore leads to atheism.[116] Similarly, Newton was deeply concerned to ensure a role for providence in a mechanical universe, a concern that stimulated his alchemical and theological interests.[117] Even after the publication of the *Principia* (1687), which Newton considered to be a powerful argument for design,[118] the existence and nature of providence continued to be a serious issue for natural philosophers. The proper interpretation of providence and design, after all, was one of the central themes of the Leibniz–Clarke correspondence.[119]

The limits of mechanization: Gassendi on the immortality of the soul

> Here was the error of Epicurus, not that he called void an incorporeal nature, but that he admitted no other incorporeal things, such as those we endorse, like the divine, the angelic, and the human soul.
> *Gassendi to Valois*[120]

One of the aspects of Epicureanism most troubling to Gassendi was its denial of the immortality of the soul. The Epicureans had considered fear of death and anxiety about punishment or reward in the afterlife as important sources of distress. By giving a materialist account of a *mortal* soul and, thus eliminating the possibility of any life after death, they believed that they could allay these fears and promote tranquillity. They had con-

116. Robert Boyle, *Some Considerations Touching the Usefulness of Experimental Natural Philosophy* (1663), in *The Works of the Honourable Robert Boyle,* edited by Thomas Birch, new edition, 6 vols. (London, 1772; reprinted, Hildesheim: Georg Olms, 1965), vol. 2. See also Timothy Shanahan, "God and Nature in the Thought of Robert Boyle," *Journal of the History of Philosophy,* 26 (1988): 547–69.
117. B. J. T. Dobbs, *The Janus Faces of Genius: The Role of Alchemy in Newton's Thought* (Cambridge University Press, 1991), pp. 33–7.
118. *Four Letters from Sir Isaac Newton to Doctor Bentley, containing Some Arguments in Proof of a Deity* (London, 1754), facsimile reprint in I. Bernard Cohen (ed.), *Isaac Newton's Papers and Letters on Natural Philosophy* (Cambridge, Mass.: Harvard University Press, 1958).
119. H. G. Alexander (ed.), *The Leibniz-Clarke Correspondence* (Manchester: University of Manchester Press, 1956), pp. xv, 11–12, 13–14.
120. November 1642, in Gassendi, *Opera omnia,* vol. 6, p. 157.

sidered the soul, like everything else in the cosmos, to be composed of atoms and the void.[121] In order to legitimate his Christian adaptation of Epicurean atomism, Gassendi insisted on the existence of an *immortal, incorporeal* human soul. In the process of arguing for the immortality of the soul, he spelled out the limits of his mechanization of nature. He was not a materialist.[122] Even as he reconstructed Epicurean atomism, he insisted that limits be imposed on mechanization. God, angels, and the immortal human soul were deliberately excluded from his mechanical philosophy.

The Epicurean theory of the soul is not compatible with orthodox Christian belief, which includes as articles of faith the survival of the soul after death, divine punishment and reward in the afterlife, the Resurrection of Christ, and the possibility of human resurrection at the time of the Second Coming.[123] Epicurean philosophy was traditionally regarded as heterodox. Dante had placed Epicurus in the Sixth Circle of Hell: "In this part Epicurus with all his followers, who make the soul die with the body, have their burial place" (*Inferno*, X, 14). Interestingly, Dante had placed

121. Epicurus, "Letter to Herodotus," 63–7, in Long and Sedley, *The Hellenistic Philosophers,* vol. 1, pp. 65–6. See also Rist, *Epicurus,* chap. 5.
122. In making this statement, I disagree with Bloch, the thrust of whose argument is to establish that Gassendi was a materialist. See Bloch, *La philosophie de Gassendi,* esp. p. 285.
123. Strictly speaking, an atomic account of the soul might not be incompatible with the doctrine of resurrection: At the appropriate time, God simply reassembles the indestructible atoms into their previous configuration, and the soul can enjoy eternal bliss or damnation. This view is not one that Gassendi would have countenanced, but positions like it had some currency in seventeenth-century England. For example, Digby wrote: "Methinkes it is but a grosse conception, to think that every *Atome* of a burned *Cadaver,* scattered by the wind throughout the world, and after numerous variations changed peradventure into the body of another man, should at the sounding of the last *Trumpet* bee raked together againe from all the corners of the earth, and bee made up anew into the same *Body* it was before of the first man. Yet if we will be *Christians,* and rely upon Gods promises, wee must beleeve that we shall rise againe with the same body, that walked about, did eate, drinke, and live here on earth; and that wee shall see our Savior & *Redeemer,* with the very same eyes, wherewith we now look upon the fading *Glories* of this contemptible world." Kenelm Digby, *Observations upon Religio medici* (London, 1644; facsimile reprint, Menston: Scolar Press, 1973), pp. 78–9. See Norman T. Burns, *Christian Mortalism from Tyndale to Milton* (Cambridge, Mass.: Harvard University Press, 1972).

"Democritus, who puts the world on chance," only in limbo along with the virtuous pagans (*Inferno*, IV, 136).[124]

Although the immortality of the soul had been adopted as a virtually unquestioned article of Christian doctrine during the Middle Ages, particularly by the followers of Augustine, discussions of this issue increased following the Aristotelian revival of the thirteenth century. The writings of Averroës and those of some of his followers questioned Aristotle's commitment to the personal immortality of the human soul.[125] Because of the close ties between Christian theology and Aristotelian philosophy in the thought of Thomas Aquinas, Averroës' interpretation of Aristotle was considered to be subversive to true belief.[126]

Pietro Pomponazzi (1462–1525), the Aristotelian naturalist, challenged Augustinian assumptions when he taught that natural reason and experience were the only criteria of truth for the natural philosopher. He rejected the Averroist view that all people share one immortal soul, a theory Averroës had falsely ascribed to Aristotle and one that Pomponazzi claimed reduced humans to the level of animals. He also criticized the Thomist position on individual immortality as indefensible on rational grounds alone, arguing that it too was based on a misinterpretation of Aristotle.[127] Pomponazzi ultimately concluded that the human soul is "the highest material form, attaining in its most elevated operations something beyond materiality."[128] Because it is capable of intellectual knowledge, the human soul is different from the souls of animals and the forms

124. Dante Alighieri, *The Divine Comedy*, translated by Charles S. Singleton, 3 vols. (Princeton, N.J.: Princeton University Press, 1970), vol. 1, pp. 43, 99.

125. Emily Michael and Fred S. Michael, "Two Early Modern Concepts of Mind: Reflecting Substance vs. Thinking Substance," *Journal of the History of Philosophy*, 27 (1989): 29–30.

126. Étienne Gilson, "Autour de Pomponazzi: Problématique de l'immortalité de l'âme en Italie au début du XVIe siècle," *Archives d'histoire doctrinale et littéraire du Moyen Age*, 28 (1961): 164–9.

127. Ibid., pp. 185–6.

128. Eckhard Kessler, "The Intellective Soul," in *The Cambridge History of Renaissance Philosophy*, edited by Schmitt, Skinner, and Kessler, pp. 500–3. Although Pomponazzi did not publish these ideas until *De immortalitate animae* (1516), manuscript notes establish that he had been lecturing on this question at Padua at least since 1500. For an English translation of this text, see Pietro Pomponazzi, "On the Immortality of the Soul," translated by William Henry Hay II and John Herman Randall, Jr., and annotated by Paul Oskar Kristeller, in *The Renaissance Philosophy of Man*, edited by Ernst Cassirer, Paul Oskar Kristeller, and John Herman Randall, Jr. (Chicago: University of Chicago Press, 1948).

of plants and stones. Strongly influenced by Alexander of Aphrodisias (fl. c. 220 A.D.), Pomponazzi thought that "the human soul, as material, is thus a perishable substance which its perfection renders capable of participating in intelligible knowledge."[129] As such, the soul is mortal, although it is the subject of knowledge that, by its nature, transcends mortality.[130] Pomponazzi's reasoning thus undermined not only the immortality of the human soul, but also the generally accepted idea that Aristotelianism is perfectly compatible with Christian theology.[131]

Pomponazzi's statements on the mortality of the soul were condemned by the Fifth Lateran Council in 1513, which went on to establish the immortality of the soul as official dogma and asked Christian philosophers to "'use all their powers' to demonstrate that the immortality of the soul can be known by natural reason, not by faith alone."[132] Pomponazzi's controversial views and his subsequent condemnation by the church were the intellectual background from which many seventeenth-century discussions of the immortality of the soul developed.[133] In addition, mechanical philosophers felt a pressing need to argue for the immortality of soul in order to defend themselves against charges of materialism and atheism.

Gassendi understood that the nature of the soul was a serious problem for a Christian rendition of atomism, and he devoted many pages of the *Syntagma philosophicum* to this issue.[134] In rejecting Aristotelianism, he had to find a new philosophical proof for the immortality of the soul, since the traditional Thomist proof presupposed the Aristotelian metaphysics of matter and form. His most comprehensive treatment of the problem appears, significantly, in the final book of the "Physics," which is entitled "On the Immortality of Souls." Beginning with the nature of the soul in general, he then considered the souls of animals, both their nature and their function, and finally the human soul. He concluded by demonstrating the immortality of the human soul. His desire to refute the Epicurean theory of a mortal, material soul was central, for he devoted an entire

129. Gilson, "Autour de Pomponazzi," p. 189.
130. Ibid.
131. Charles H. Lohr, "Metaphysics," in *The Cambridge History of Renaissance Philosophy*, edited by Schmitt, Skinner, and Kessler, p. 603.
132. Michael and Michael, "Two Early Modern Concepts of Mind," p. 31.
133. Ibid., pp. 29–34.
134. Gassendi, *Syntagma philosophicum*, pt. II, sec. III, posterior part, bks. III, IX, XIV, XV, in *Opera omnia*, vol. 2.

chapter to a point-by-point refutation of Epicurus' objections to the immortality of the soul.[135]

Most of Gassendi's discussion of the soul occurs in the section of the "Physics" entitled "On Living Earthly Things, or on Animals." This section of 440 double-column pages of dense prose deals mostly with topics in physiology and the theory of sensation, but after a general discussion "On the Variety of Animals" and "On the Parts of Animals," Gassendi addressed the nature of the soul. The soul, he said, is what distinguishes living things from inanimate things.[136] Adopting the distinction between *anima* and *animus* directly from Lucretius,[137] he said, "The *anima* is that by which we are nourished and by which we feel; and the *animus* is that by which we reason."[138] Gassendi agreed with Lucretius that the *anima*, or sentient soul, is present throughout the body; he disagreed with Lucretius on the locus of the *animus*, or rational soul, placing it in the head rather than the chest.[139] Gassendi also disagreed with the Epicurean argument that since there is nothing incorporeal except the void, both the *anima* and the *animus* are corporeal. Drawing on his voluntarist theology, Gassendi rejected Epicurus's theory. "The actions of God are not necessary"; even if nothing incorporeal can be imagined except the void, it does not follow that God's creative act is restricted by the limited scope of human imagination.[140]

---

135. Ibid., vol. 2, pp. 633–50. The major commentators on Gassendi tend to underplay this issue. Bloch, for example, devotes only three pages to Gassendi's extensive discussion. See Bloch, *La philosophie de Gassendi*, pp. 397–400. Joy, *Gassendi the Atomist*, does not deal with it at all. Brundell, *Pierre Gassendi* touches on it in passing, but does not give it sustained attention.

136. Gassendi, *Syntagma philosophicum*, in *Opera omnia*, vol. 2, p. 237.

137. Lucretius claimed that the soul consists of two parts, which he called the *anima*, or irrational part responsible for vitality and sensation, and the *animus*, or rational part. Titus Carus Lucretius, *De rerum natura*, translated by Cyril Bailey, 3 vols (Oxford University Press, 1947), vol. 1, p. 309.

138. Gassendi, *Syntagma philosophicum*, in *Opera omnia*, vol. 2, p. 237.

139. Ibid., pp. 445–6.

140. Ibid., p. 246. This is an example of the sort of argument that Bloch regards as evidence for the fact that Gassendi allowed faith to dictate the content of philosophy, even when the conclusions of that philosophy taken on their own stood opposed to the conclusions known by faith: "It is clear that the immateriality of the rational soul is simply postulated in philosophy in the name of theological exigency." Bloch, *La philosophie de Gassendi*, pp. 369, 374. I do not agree. Just because arguments based on the divine attributes

Gassendi did not, however, object to the corporeality of the *anima,*
which he equated with the animal soul.[141] Humans differ from animals in
possessing rational souls, and Gassendi approached the ultimate problem
of the *animus* by contrasting it to the souls of animals, which correspond
to the irrational part of the human soul.[142] Gassendi incorporated the
Aristotelian theory of a tripartite soul as consisting of vegetative, sensitive, and rational parts. Man alone possesses all three; animals possess
sensitive and vegetative souls; plants possess only the vegetative soul.[143]
The *anima* is "something which can be said to live in the body of the
animal which is alive, leaving it at death. Clearly, life is its presence in the
body of the animal, and death is its absence."[144] Because of the fact that it
is in various parts of the body while the animal is alive, "the *anima* seems
to be something very fine."[145] Denying that it is a form or merely a
symmetrical disposition of matter, as various Peripatetics had maintained,
Gassendi claimed that "the soul [of animals] seems to be a very tenuous
substance, just like the flower of matter (*florem materiae*) with a special
disposition, condition, and symmetry holding among the crasser mass of
the parts of the body."[146] The *anima* is the principle of organization and

do not seem adequate in light of modern philosophical analysis, it does not
follow that Gassendi did not take such arguments seriously or that certain
theological claims did not provide the underlying assumptions that formed
his philosophical views.

141.   For an extensive discussion of Gassendi's views on the animal soul, see
Sylvia Murr, "L'âme des bêtes chez Gassendi," *Corpus, 16–17* (1991): 37–
63.

142.   Gassendi, *Syntagma philosophicum,* in *Opera omnia,* vol. 2, p. 250.

143.   Aristotle, *De anima,* translated by J. A. Smith, in *The Complete Works of
Aristotle,* edited by Jonathan Barnes, 2 vols. (Princeton, N.J.: Princeton
University Press, 1984), 414a2–4, 415a3–6, 435a12. See also W. D. Ross,
Aristotle: A Complete Exposition of His Works and Thought, 5th edition
(New York: Meridian, 1959), pp. 128–9.

144.   Gassendi, *Syntagma philosophicum,* in *Opera omnia,* vol. 2, p. 250.

145.   Ibid.

146.   Ibid. Gassendi had used very similar language in talking about the principle
of motion in individual objects such as boys or atoms: "For when a boy
runs to an apple offered to him, what is needed to account for the apple's
attraction to the boy is not just a metaphorical motion, but also most of all
there must be a physical, or natural, power inside the boy by which he is
directed and impelled toward the apple. Hence it may apparently be said
most plainly that since the principle of action and motion in each object is
the most mobile and active of its parts, a sort of bloom of every material
thing (*quasi flos totius materiae*) and which rarefied tissue of the most
subtile and mobile atoms – it may therefore be said that the prime cause of

activity for the organism. It is the source of the animal's vital heat, a phenomenon that can be explained by the subtlety and activity of its constituent atoms:

> Such a substance seems to be made of a most subtle texture, extremely mobile or active corpuscles, not unlike those of fire or heat. . . . [T]hey seem from their own motion and penetration through bodies to create the heat that is in the animal. Since for this reason they are said to be excited or to feel hot and the heat of animals manifestly depends . . . on the motion or actions of the soul, it follows that cold and death are the cessation of such action. Finally, the soul seems to be like a little flame or a most attenuated kind of fire, which . . . thrives or remains kindled while the animal lives, since, if it no longer thrives or is extinguished, the animal dies.[147]

The fact that vital heat is a sign of life is evidence for the claim that the animal soul is like a little flame; so is the fact that "just as a snuffed out candle is repeatedly rekindled . . . so a suffocated and strangled animal, not yet dead, having been led from the water or heavy smoke or released from the halter" can be revived by being repeatedly brought to inhale the air."[148] Like a flame, the *anima* is in constant motion, not only when the animal is awake, but also in sleep, as dreaming demonstrates. It is the "principle of vegetation, sensation, and every other vital action."[149]

According to Gassendi, the animal soul is transmitted from one generation to the next at the moment of conception. Extending his analogy between the animal soul and a flame, he claimed that the animated portion of the semen kindles the soul in the embryo in the same way that a burning torch can kindle the flame of a new one.[150] "Thus from the creation of animals at the beginning of the world . . . the soul (*anima*) was propagated . . . , and when the first [souls] were extinguished, there were always others being made, and so it remains even now in those that

motion in natural things is the atoms, for they provide motion for all things when they move themselves through their own agency and in accord with the power they received from their author in the beginning; and they are consequently the origin, and principle, and cause of all the motion that exists in nature." Ibid., vol. 1, p. 337, translated by Craig B. Brush, in *The Selected Works of Pierre Gassendi* (New York: Johnson Reprint, 1972), pp. 421–2. Bloch interprets Gassendi's talk of the *flos materiae* as an unacknowledged influence of the animism of Telesio and Campanella. See Bloch, *La philosophie de Gassendi*, pp. 228–30.

147. Gassendi, *Syntagma philosophicum*, in *Opera omnia*, vol. 2, pp. 250–1.
148. Ibid., p. 252.
149. Ibid.
150. Ibid.

exist."[151] Thus, for the souls of animals, only one act of creation was necessary. Ever since the initial creation, the souls of animals have been transmitted from one generation to the next by the biological process of reproduction.

The souls of animals and the animal part of the human soul presented no problem for Gassendi, the mechanical philosopher. Maintaining that all natural phenomena must be explicable in terms of matter and motion, Gassendi perceived no special problem in the facts of biology and perception, as they could be readily explained – or so he thought – in terms of the motions of atoms.[152] His insistence on the special tenuousness, subtlety, and mobility of the atoms comprising the animal soul signals the difficulty that these phenomena present to any reductionist materialism. Gassendi was, however, content to assert that they were so reducible.

The animal soul is only one part of the human soul. Another part, unique to humankind, is the subject of theological as well as philosophical concern:[153]

> The human soul is composed of two parts: . . . the irrational, embracing the vegetative and sensitive is corporeal, originates from the parents, and is like a medium or fastening (*nexus*) joining reason to the body; and . . . reason, or the mind, which is incorporeal, was created by God, and is infused and unified as the true form of the body.[154]

In saying that the soul is the form of the body, Gassendi drew on the same Scholastic ideas as Descartes, who, despite his opposition to real qualities and substantial forms in general, continued to think of the human soul as the substantial form of the body.[155]

It is with regard to the rational soul that man can be said to have been made in the image of God; the material, sensitive soul is, by contrast,

151. Ibid., p. 253.
152. See Emily Michael and Fred S. Michael, "Gassendi on Sensation and Reflection: A Non-Cartesian Dualism," *History of European Ideas,* 9 (1988): 583–95; Emily Michael and Fred S. Michael, "Corporeal Ideas in Seventeenth-century Psychology," *Journal of the History of Ideas,* 50 (1989): 31–48. It hardly needs to be said that Gassendi was overly optimistic on this count.
153. Gassendi, *Syntagma philosophicum,* in *Opera omnia,* vol. 2, p. 255.
154. Ibid., p. 256.
155. On the Scholastic doctrine, see Gilson, "Autour de Pomponazzi," p. 167. On Descartes' view of the soul as the substantial form of the body, see Daniel Garber, *Descartes' Metaphysical Physics* (Chicago: University of Chicago Press, 1992), pp. 89, 99, 104, 276.

simply communicated to the fetus by means of the father's semen.[156] In an interesting aside, Gassendi noted that if one were to hold the position that the soul develops along with the body, as the Epicureans maintained, then one would not "commit homicide according to either civil or canon law [by procuring] an abortion in the first days after conception."[157] Arguments of this kind are still used in many countries to determine at what fetal age abortion is acceptable. To the seventeenth-century Catholic priest, this possibility was patently absurd.

One of Gassendi's chief arguments for the claim that the soul consists of two parts, "one rational, the other sensitive," is that it is a doctrine that is consonant with Scripture:

> The theologians distinguish . . . two parts of the soul, one higher, the other lower. They prove this distinction especially from that place in the Apostle – "I see another law in my limbs repugnant to the law of my mind." Evidently, since one simple thing cannot be opposed to itself, it is argued from the battle that exists between sense and mind that sense and mind, or the rational and sensitive souls, are different things.[158]

The theory of the two souls explains inner conflict between morality and bodily drives. If humans did not have rational souls, Gassendi argued, there would be no accounting for the fact that they are "a little less than the angels and remain the same after death"; otherwise, they would not differ from the brutes in either life or death.

Although, as Gassendi noted, in common parlance we frequently speak of the human soul as unitary, we are, at such times, really designating only the rational soul:

> This is certainly what must be understood when Christ the Lord said, "Refuse to fear them who strike the body and cannot strike the soul"; indeed, he understood only the rational soul, for otherwise the sensitive [soul], as not separate from the body, would be dead . . . from the striking.[159]

The rational soul is very different from the sensitive soul:

> In agreement with the Holy Faith, we say that the mind, or that superior part of the soul (which is appropriately rational and unique to man) is an incorporeal substance, which is created by God, and infused into the body; . . . it is like an informing form.[160]

---

156. Gassendi, *Syntagma philosophicum*, in *Opera omnia*, vol. 2, p. 256.
157. Ibid.
158. Ibid., p. 257.
159. Ibid., p. 258.
160. Ibid., p. 440.

Gassendi's discussion shifted, at this point, from assertion of faith to philosophical argument. First, he established that "the intellect is distinct from the imagination."[161] The imagination, or phantasy, as understood by seventeenth-century philosophers, is the faculty by means of which we have images of the objects of our thought.[162] Rejecting Aristotle,[163] Gassendi claimed that not all human thought involves the use of imagination: "I use an example . . . concerning the magnitude of the sun. Having been led by reasoning, we understand that the sun is 160 times larger than the earth; yet the imagination is frustrated, and however much we try . . . , our imagination cannot follow such vastness."[164] Similar examples illustrate the proposition that there are things that "we understand what cannot be imagined, and thus the intellect is distinct from the phantasy."[165] Since the phantasy imagines with material species, Gassendi concluded that the intellect must be incorporeal.[166]

Gassendi began his second argument for the incorporeality of the intellect with the observation that the intellect can know itself. This ability goes "beyond all corporeal faculties, since something corporeal is in a certain place . . . so that it cannot proceed toward itself but only toward something different." For this reason, it makes no sense to speak of sight seeing itself or knowing its own vision. The ability to reflect on itself is unique to the human intellect and a sure sign of its immateriality.[167] The function of this argument is similar to Descartes' use of the *cogito* to draw an absolute distinction between mind and matter.

Gassendi's third argument for the incorporeality of the soul is that not only do we form concepts of universals, but we also perceive the reason for their universality.[168] This ability distinguishes us from animals. Even if animals on occasion seem to form universal notions – for example, a dog

161.   Ibid.
162.   Ibid., pp. 398–424. According to the *Oxford English Dictionary:* "Fantasy, Phantasy. 1. In scholastic philosophy: a. mental apprehension of an object of perception; the faculty by which this is performed . . . b. The image impressed on the mind by an object of sense. . . . 4. Imagination: the process and faculty of forming mental representations of things not actually present."
163.   Aristotle, *De anima*, 403a5–25, in *The Complete Works of Aristotle*, edited by Barnes, vol. 1, pp. 642–3.
164.   Gassendi, *Syntagma philosophicum*, in *Opera omnia*, vol. 2, p. 440.
165.   Ibid. For a discussion of this point, see Michael and Michael, "Gassendi on Sensation and Reflection," pp. 584–90.
166.   Gassendi, *Syntagma philosophicum*, in *Opera omnia*, vol. 2, p. 441.
167.   Ibid.
168.   Ibid.

can recognize a stranger as a human being – nevertheless, animals never apprehend the universal purely abstractly, but always with some degree of individuality and concreteness. Unlike humans, they do not understand the nature of universality per se, because they possess only a corporeal faculty, the phantasy.[169]

Having established to his own satisfaction that the "rational soul or human mind" is incorporeal, Gassendi proceeded to argue that "it has God as its author by whom it was brought into being or who created it from nothing."[170] The passage from nothing into something is infinitely great; execution of this transformation, like the original Creation of the universe, requires more than the finite force that belongs to natural things, "but an infinite [force], which is God alone, so that the rational soul, which is an incorporeal substance, can only know God as author."[171] But to assert that there is a cause surpassing all of nature that can create something ex nihilo is to violate physical principles. Appealing to his voluntarism, he continued:

> Is that not what Thales, Pythagoras, Plato, the Stoics, and many great philosophers have asserted? Therefore, it is not abhorrent that there is a power in the author of the world that surpasses all of nature; and which, forming and ordering particular things, produces the individual forces by which they operate from the beginning. Why therefore is it not a physical dogma that there is a power in the maker of nature by which something can be made from nothing?[172]

Indeed, to say that it is impossible for something to be created from nothing because we have never observed such an event would be as absurd as asserting that "Daedalus had no power within himself except such as he observed in his automata."[173] God's power surpasses the observable domain. Rejecting the principle "*ex nihilo nihil fit*,"[174] the starting point of Epicurean – and almost all Greek – physics, Gassendi maintained that there is nothing that does not proclaim that there "was an acting infinite force in the great maker of all, who alone depends on nothing else, who is not limited by the action of any force."[175] It is no reproach to physics to resort to the author of nature, particularly in the case of the rational soul. "Since it is immaterial and unless [it is created from nothing], cannot be

169. Ibid., p. 442.
170. Ibid.
171. Ibid., p. 443.
172. Ibid.
173. Ibid.
174. Lucretius, *De rerum natura*, I, 156–73.
175. Gassendi, *Syntagma philosophicum*, in *Opera omnia*, vol. 2, p. 443.

created by any other cause than God."[176] Since the soul is created by God, like all other things in the world, "it is within the order of things that God constituted in nature and protects by his providence."[177] In this sense, the soul is nothing extraordinary or beyond the natural order, "since wherever and whenever a man is born [God] creates a rational soul, which he infuses into his body."[178] Unlike material things, however, which are all composed of atoms and the void and are produced by second causes, each person's rational soul is individually and directly created by God.

"How can something incorporeal be joined with the body . . . and [how can it] be the informing form?"[179] Even if it is argued that the sensitive soul – like Descartes' pineal gland – serves as an intermediary, linking the body and the rational soul, the problem remains; for the sensitive soul "is nevertheless corporeal and infinitely distant from the incorporeal."[180] The difficulty of explaining the connection between the incorporeal rational soul and the body remains, however fine and attenuated the body may be. Unlike angels or pure intelligences, which subsist independently as pure acts of intelligence, the sensitive soul is such that "its nature has a destination and inclination to the body and the rational soul."[181] If the sensitive soul can be regarded as the form of the body, then the rational soul, even though it is a substance, can be regarded as the form of the individual person.[182] This point has a long history in Scholastic discussions of the nature of the soul. Christian philosophers were concerned to preserve simultaneously the substantiality of the soul, its immortality, and the unity of the person, doctrines that are not prima facie compatible.[183]

Like Descartes, Gassendi drew a line of demarcation between body and mind, and then, like him, faced the difficult problem of explaining how they are related. Gassendi's distinction between the material world, which

176. Ibid.

177. Ibid.

178. Ibid. Gassendi's views on the different origins of the sensitive and rational souls are similar to those of Thomas Aquinas: "The souls of brutes are produced by some power of the body, whereas the human soul is produced by God." Thomas Aquinas, *Summa Theologica*, translated by the Fathers of the English Dominican Province, 3 vols. (New York: Benziger, 1947), pt. I, qud 75, art. 6, vol. 1, p. 368.

179. Gassendi, *Syntagma philosophicum*, in *Opera omnia*, vol. 2, pp. 443–4.

180. Ibid., p. 444.

181. Ibid.

182. Ibid.

183. On these issues, see Gilson, "Autour de Pomponazzi," and Anton Charles Pegis, *St. Thomas and the Problem of the Soul in the Thirteenth Century* (Toronto: Pontifical Institute of Medieval Studies, 1934).

includes the corporeal *anima,* which is composed of extended atoms and is diffused throughout the body, and the incorporeal *animus,* the seat of ratiocination, parallels Descartes' distinction between *res extensa* and *res cogitans.* Although Gassendi considered animals to possess considerable mental life in contrast to Cartesian automata,[184] he drew a sharp line between the intellectual capabilities of animals and humans. Both philosophers spoke of the rational soul as the substantial form of the person. Nonetheless, the problem of linking mind and body existed for both philosophers, and its solution eluded them both. Descartes' use of the pineal gland as a link between mind and body was no less arbitrary or question-begging than Gassendi's use of the sensitive soul as intermediary between the body and the rational soul.[185]

Gassendi's claim that the rational soul, in contrast to the animal soul, is incorporeal established one of the boundaries of his mechanization of the world. In this respect, Gassendi's philosophy resembles that of Descartes and differs from that of Hobbes. Descartes' radical distinction between *res extensa* and *res cogitans*[186] creates a boundary that falls along the same lines as Gassendi's distinction between the animal and rational souls. Hobbes, however, sought to explain even the human soul and social processes in terms of matter and motion alone.[187] In this respect, the difference between Hobbes, on the one hand, and Gassendi and Descartes, on the other, is a difference in defining the domains to which mechanization applies.[188]

Gassendi finally addressed the question of the soul's immortality. He considered this issue to be the "crown of the treatise" and the "last touch of universal physics."[189] His strategy was to prove the immortality of the soul from faith, physics, and morality. As a statement of faith, Gassendi declared:

> [The rational soul] survives after death or remains immortal; and as it bore itself in the body, either it will be admitted to future happiness in Heaven, or it will be thrust down unhappy into Hell, and it will regain

184. Murr, "L'âme des bêtes chez Gassendi," p. 37.
185. René Descartes, *Treatise of Man,* translated by Thomas Steele Hall (Cambridge, Mass.: Harvard University Press, 1972), pp. 79–85. See also René Descartes, *Traité de l'homme,* in *Oeuvres de Descartes,* edited by Charles Adam and Paul Tannery, 11 vols. (Paris: J. Vrin, 1897–1983) (hereafter AT), vol. 11, pp. 171–7.
186. See Chapter 9.
187. Brandt, *Hobbes' Mechanical Conception of Nature,* p. 355.
188. See Sarasohn, "Motion and Morality," pp. 363–80.
189. Gassendi, *Syntagma philosophicum,* in *Opera omnia,* vol. 2, p. 620.

its own body in the general resurrection, just as it was in itself and will receive its good or evil.[190]

Although "the divine light shines for us from this Sacred Faith, theologians have been accustomed to discuss arguments for and against the immortality of the soul."[191] Gassendi joined the theologians in their arguments for the immortality of the soul. His support of an article of faith with philosophical and physical arguments was Gassendi's response to the Fifth Lateran Council's call on philosophers to "use all their powers," including natural reason, to defend the immortality of the soul.[192]

On the basis of physics, Gassendi endorsed the argument that "the rational soul is immaterial; therefore it is immortal." An immaterial thing is also immortal or incorruptible because, "lacking matter, it also lacks mass and parts into which it can be divided and analyzed. Indeed, what is of this kind neither has dissolution in itself nor fears [it] from another."[193] Gassendi's line of argument – that the immortality of the soul follows from its immateriality – was a common seventeenth-century ploy. Digby adopted the same strategy in his *Two Treatises* (1644), in which he devoted the first treatise, *The Nature of Bodies,* to an account of the properties of matter so that he could then demonstrate in the second, *The Nature of Mans Soule,* that the soul shares no properties with bodies. He then argued from the immateriality of the soul to its immortality:

> Looking into the causes of mortality, we saw that all bodies round about us were mortall: whence perceiving that mortality extended it selfe as farre as corporeity, we found our selves obliged, if we would free the soule from that law, to shew that she is not corporeall.[194]

Henry More (1614–1687), the Cambridge Platonist, appealed to a similar argument to prove the immortality of the soul. The soul is one of many

---

190. Ibid., p. 627.
191. Ibid. On Gassendi's attitude toward the relationship between truths of reason and truths of faith, see Sylvia Murr, "Foi religieuse et *libertas philosophandi* chez Gassendi," *Revue des sciences philosophiques et théologiques,* 76 (1992): 85–100.
192. Michael and Michael, "Two Early Modern Concepts of Mind," p. 31. The Michaels note that Descartes explicitly refered to the Fifth Lateran Council in his epistle to the faculty of theology at the University of Paris, which introduced *The Meditations.* See, René Descartes, *Meditationes de prima philosophia,* in AT, vol. 7, p. 3.
193. Gassendi, *Syntagma philosophicum,* in *Opera omnia,* vol. 2, p. 628.
194. Kenelm Digby, *Two Treatises in One of which The Nature of Bodies; in the other The Nature of Mans Soule; is looked into: in Way of Discovering, of the Immortality of Reasonable Soules* (Paris: Gilles Blaizot, 1644; facsimile reprint, New York: Garland, 1978), p. 350.

spirits in the world. From the fact that spirit has different properties from matter – spirit is penetrable and "indiscerpible," while matter is impenetrable and "discerpible" – he concluded that spirit is incorporeal and thus immortal.[195]

Gassendi anticipated several objections to this apparently simple argument. One possible objection could be drawn from the Greeks, who believed that there are demons and other incorporeal spirits that both come into being and pass away. In support of this position, one might go on to maintain that "since only the author of nature is creative, for that reason he alone is said to have immortality. . . . [I]t must be disallowed that the whole world, even the incorporeal things that he produces from nothing . . . can absolutely be returned to nothing if he wishes it."[196] But, Gassendi objected, in spite of God's absolute power to annihilate all things that he created, "from the supposition that nothing works against the order of nature and this state of things that he instituted most wisely, likewise he preserves it constantly; and it is evident that incorporeal things are preserved eternally."[197] Divine providence explains the coherence of the natural order. Appealing to the language of the absolute and ordained power of God, Gassendi maintained that although everything in nature is absolutely subject to divine will, God freely chooses to preserve the constancy of the natural order.[198] The world may be regular, but its regularity is contingent.

195. Henry More, *The Immortality of the Soul, So Farre Forth As It Is Demonstrable from the Knowledge of Nature and the Light of Reason* (London: 1662); facsimile reprint in Henry More, *A Collection of Several Philosophical Writings* (1662), 2 vols. (New York: Garland, 1978), p. 21.
196. Gassendi, *Syntagma philosophicum*, in *Opera omnia*, vol. 2, p. 628.
197. Ibid.
198. Oakley emphasizes this aspect of voluntarist theology. "A considerable significance attaches to this fact. The only force, after all, capable of binding omnipotence without thereby denying it is the omnipotent will itself. Whereas God, therefore, cannot be said to be constrained by the natural order of things or bound by the canons of any merely human reason or justice, he is certainly capable by his own decision of binding himself to follow a stable pattern in dealing with his creation in general and with man in particular. If God has freely chosen the established order, he *has so* chosen, and while he can dispense with or act apart from the laws he has decreed, he has nonetheless bound himself by his promise and will remain faithful to the covenant that, of his kindness and mercy, he has instituted with man." Francis Oakley, *Omnipotence, Covenant, and Order: An Excursion in the History of Ideas from Abelard to Leibniz* (Ithaca, N.Y.: Cornell University Press, 1984), p. 62.

Another possible objection to the incorporeality of the soul is "that there are [church] fathers who linked immortality with corporeality," and there are "philosophers who claim that the heavenly bodies are incorruptible."[199] Some Christian theologians, most notably Tertullian, had indeed argued that the soul has a corporeal nature.[200] Gassendi replied that "whatever the meaning of those philosophers is, our consequent is not contradicted, . . . because there can be some incorruptible bodies no less than there are incorruptible, incorporeal things."[201] In other words, the possible existence of immortal, corporeal things does not weaken the argument that incorporeal things must be immortal.

To bolster his physical argument for the immortality of the soul, Gassendi asserted that the frequency of superstitions about death shows that there is universal agreement that the soul is immortal, despite differences about its state after death. Gassendi quoted Cicero with approval: " 'Everything about which all people agree must be considered a law of nature': it is indeed fitting that the sense of immortality is endowed by nature."[202] Thus, anyone who denies the soul's immortality is violating the principles of nature. Purported counterexamples of "certain . . . wild people . . . in the New World, in whom no opinion about immortality is inbred," do not stand close examination; for they all turn out to fear demons or nocturnal spirits, an indication that they really do believe in immortal souls.[203] Even if one could find a few genuine counterexamples, they would not negate the universality of the belief in immortality, any more than the fact that some people "are born with one leg or desire their own death" allows us to argue that "no men are bipeds or have a natural appetite for propagating life."[204] Not only is belief in the immortality of the soul universal, so too is the desire to survive death. Since nature does nothing in vain, people would not be endowed with such a desire if it were entirely unfounded. It follows that the soul must be immortal.[205]

Gassendi called his third line of argument for the immortality of the soul moral. He based it on an assumption that might be called the principle of the conservation of justice: "To the extent that it is certain that God exists, so it is certain that he is just. It is appropriate to the justice of God

---

199. Gassendi, *Syntagma philosophicum*, in *Opera omnia*, vol. 2, p. 179.
200. Frederick Copleston, *A History of Philosophy*, 9 vols. (Garden City, N.Y.: Doubleday, 1962), vol. 2, pt. 1, pp. 37–9.
201. Gassendi, *Syntagma philosophicum*, in *Opera omnia*, vol. 2, p. 629.
202. Ibid.
203. Ibid.
204. Ibid.
205. Ibid., p. 631.

that good happens to the good and evil to the wicked." But in this life, anyway, rewards and punishments are not so justly distributed. Consequently, "there must be another life in which rewards for the good and punishment for the evil are distributed."[206] One might object, he continued, that this reasoning would equally entail immortality for the souls of animals, since there are many examples of injustice in the animal kingdom, as when "a peaceful herd of sheep is mangled by a wolf or a simple dove, without evil, is attacked by a hawk."[207] But we need not conclude from these arguments that there is an afterlife for animals; the argument holds only for the human realm because God's special providence is unique to humankind. Moreover, "men alone, among animate beings, implore, know, . . . venerate, praise, and love their creator."[208] Humans also differ from animals in having knowledge of their future state, desiring pleasure, and fleeing pain, "for which reason it is not remarkable if providential justice distributes rewards and punishments to men that are not the same for the other animals."[209]

Gassendi objected to the Stoic criticism that the afterlife is not necessary to ensure a proper distribution of justice in the world, because virtue is its own reward and vice its own punishment, claiming that it is the incentive for future rewards and fear of future punishments that cause people to seek virtue and shun vice. If this were not the cause, it would not be necessary to pay soldiers before a difficult battle or a day laborer before the completion of the job. "No one, while he acts depraved, fears only that vice is its own punishment; rather he fears infamy, prison, torture, and the gallows."[210]

Gassendi concluded his painstaking discussion of the immortality of the soul by stating that his arguments lack the certainty of mathematics. Although they cannot replace "the Sacred Faith, as if it needs the light of reason," nevertheless they can support the truths of faith by overcoming some of the obstructions in its path.[211] Lack of mathematical certainty did not deprive Gassendi's arguments for the immortality of the soul of all foundation. On the contrary, they had the same epistemological status as he granted the rest of empirical knowledge, high probability.[212]

206. Ibid., p. 632.
207. Ibid.
208. Ibid.
209. Ibid.
210. Ibid.
211. Ibid., p. 650.
212. On the probability–certainty controversy in the seventeenth century, see Popkin, *The History of Scepticism*; Henry Van Leeuwen, *The Problem of*

How successful was Gassendi's project to baptize Epicureanism? The answer is not simple. Although his version of the mechanical philosophy influenced many natural philosophers, they were not entirely comfortable with adopting even a Christianized version of Epicurus, whose baptism they could not take for granted. For example, Robert Boyle, a pious natural philosopher who was a powerful advocate and admirer of Gassendi,[213] continued to find atomism theologically problematic. Advocating "corpuscularianism," a philosophy that sought to explain natural phenomena in terms of the properties and motions of small particles of matter, he claimed to "write rather for the Corpuscularians in general rather than any party of them" and remained agnostic on questions about the existence of indivisible atoms or the vacuum.[214] One reason for Boyle's reluctance to endorse atomism in print, although evidence exists that he leaned toward an atomist rather than a Cartesian interpretation of the mechanical philosophy, was the continuing association between atomism and atheism.[215] Newton, too, was not content with a purely mechanical account of nature and devoted immense energy to establish evidence of God's continued active role in the world.[216] For these and other natural philosophers who worked within a generally mechanical framework, Gassendi's efforts to baptize Epicureanism had not been entirely successful in

Certainty in English Thought, 1630–1690 (The Hague: Martinus Nijhoff, 1963); and Barbara J. Shapiro, *Probability and Certainty in Seventeenth-century England* (Princeton, N.J.: Princeton University Press, 1983).

213. Boyle wrote of Gassendi as the "best interpreter" of Epicurus. Robert Boyle, *The Excellency of Theology, Compared with Natural Philosophy . . . To Which are Annexed Some Occasional Thoughts about the Excellency and Grounds of the Mechanical Hypothesis* (1674), in *Works*, edited by Birch, vol. 4, p. 30.

214. Robert Boyle, *The Origins of Forms and Qualities According to the Corpuscular Philosophy*, in *Works*, edited by Birch, vol. 3, p. 7. On Boyle's theory of knowledge and his recognition of the limits of human reason, see Jan W. Wojcik, "Robert Boyle and the Limits of Reason: A Study in the Relationship Between Science and Religion in Seventeenth-Century England," Ph.D. dissertation, University of Kentucky, 1992.

215. J. J. MacIntosh, "Robert Boyle on Epicurean Atheism and Atomism," in *Atoms, Pneuma, and Tranquillity*, edited by Osler, pp. 197–219. On the relationship between Boyle and Gassendi, see Margaret J. Osler, "The Intellectual Sources of Robert Boyle's Philosophy of Nature: Gassendi's Voluntarism and Boyle's Physico-Theological Project," in *Philosophy, Science, and Religion, 1640–1700*, edited by Richard Ashcraft, Richard Kroll, and Perez Zagorin (Cambridge University Press, 1991), pp. 178–98.

216. Dobbs, *The Janus Faces of Genius*.

allaying their fears.[217] The twin specters of atheism and materialism remained.

Despite the ongoing concerns of Boyle and Newton, Gassendi's project was hardly a total failure. One measure of its success is the prominent role given to Epicurus in Thomas Stanley's immensely popular *History of Philosophy* (1655). Stanley devoted twice as many pages to Epicureanism to any other ancient school.[218] His lengthy account of the philosophy of Epicurus is virtually a translation of Gassendi's *Philosophiae Epicuri syntagma*.[219] Epicureanism had acquired canonical status in the history of philosophy, and – if page counting is a meaningful measure – it had supplanted the traditionally authoritative schools of Plato and Aristotle.

Gassendi's long excursus into theology in the "Physics" can be understood as part of his account of the ultimate terms of explanation for his new philosophy of nature. The philosophy of Epicurus, which attracted him because its atomism seemed to provide useful foundations for natural philosophy, required serious alterations before it would be acceptable to his voluntarist, Christian providentialism. Gassendi's discussion of these matters in the "Physics" articulated the theological boundary conditions within which the new philosophy of nature must lie.

217. Keith Hutchison, "Supernaturalism and the Mechanical Philosophy," *History of Science*, 21 (1983): 297–333.

218. Of the 1,091 pages in this work, Stanley devoted 80 to the Stoics, 194 to the Epicureans, 96 to the Pythagoreans, 54 to Plato, and 52 to Aristotle. See Thomas Stanley, *The History of Philosophy: Containing the Lives, Opinions, Actions and Discourses of the Philosophers of Every Sect,* 2d edition, 3 vols. (London: Thomas Bassett, 1687).

219. Kroll correctly demonstrates that Gassendi was the source for Stanley's account of Epicureanism. Kroll was mistaken, however, in confusing Gassendi's massive *Syntagma philosophicum,* which was only published posthumously in 1658, with his shorter *Syntagma Epicuri Philosophiae,* which was originally published as part of the *Animadversiones in decimum librum Diogenes Laertii* in 1649. See Richard W. F. Kroll, "The Question of Locke's Relation to Gassendi," *Journal of the History of Ideas,* 45 (1984): 350.

APPENDIX

Gassendi's 1631 outline of his Epicurean project[220]

L'*apologie* pour la vie d'Epicure contient ces chapitres icy:
  1. De serie vitæ Epicuri.
  2. De præcipuis authoribus a quibus quæsita Epicuro infamia.
  3. De objecta Epicuro impietate ac malitia.
  4. De objecta ingratitudine vanitate maledicentia.
  5. De objecta gula.
  6. De objecta venere.
  7. De objecto odio liberalium disciplinarum.
La *doctrine* ou *philosophie* est divisée en trois parties dont:
La canonique expliquée en un seul livre contient ces chapitres:
  1. De variis dijudicandæ veritatis criteriis.
  2. De canonibus vocum.
  3. De canonibus sensuum.
  4. De canonibus anticipationum.
  5. De canonibus passionum.
La *physique* est expliquée en quatre livres dont le premier, qui est De natura, contient ces chapitres:
  1. De natura rerum, seu de universo.
  2. De natura corporea, seu de atomis, an sint.
  3. De tribus atomorum proprietatibus magnitudine, figura, gravitate, ubi et de motu.
  4. De natura incorporea, hoc est inani, seu loco.
  5. De natura concreta, quænam principia, seu quæ elementa et quas caussas habeat.
  6. De ortu et interitu, seu generatione et corruptione naturæ concretæ.
  7. De mutationibus qualitatibusque naturæ concretæ.
  8. De imagine naturæ concretæ, quam etiam speciem visibilem, et intentionalem vocant.
  9. De existentia naturæ divinæ.

220. Gassendi to Peiresc, 28 April 1631, in Fabri de Peiresc, *Lettres de Peiresc*, edited by Philippe Tamizey de Larroque, in *Documents inédits sur l'histoire de France*, 7 vols. (Paris: Imprimerie Nationale, 1888), vol. IV, pp. 250–2. Bibliothèque nationale, fonds français, 6936, fol. 217. Autograph.

10. De forma naturæ divinæ, an humanæ similis sit.
11. De immortalitate et fœlicitate ejusdem naturæ divinæ.
12. De existentia naturæ dæmoniæ.
Le deuxiesme De mundo ceux cy:
 1. De structura et forma mundi.
 2. De origine mundi.
 3. De caussa productrice mundi.
 4. De serie ac modo quo productus mundus atque adeo primi homines.
 5. De providentia, seu gubernatione mundi.
 6. De fortuna et fato in mundo.
 7. De interitu mundi.
 8. De innumerabilibus mundis.
Le troisiesme dans lequel je suis, intitulé De sublimibus, ceux-cy:
 1. De cœli siderumque substantia.
 2. De variatate, positione et intervallis siderum.
 3. De magnitudine et figura siderum.
 4. De motu corporum cœlestium.
 5. De tempore, quod nonnulli volunt cœlestis motus esse consequens.
 6. De luce deque variis adspectibus siderum.
 7. De eclipsibus, deque varietate ortuum et occasuum.
 8. De proprietatibus siderum, quod ad effectus naturales spectat.
 9. De effectibus arbitrariis et fortuitis quos presciri posse astrologi profitentur.
10. De impressionibus igneis.
11. De impressionibus aereis.
12. De impressionibus aqueis.
Le quatriesme suivra apres, De humilibus, avec ces chapitres:
 2. De rebus inanimis.
 3. De anima deque corpore et membris animalium.
 4. De generatione, nutritione et incremento animalium.
 5. De sensibus animalium, ac primum de Visu.
 6. De cœteris quattor sensibus Auditu, Olfactu, Gustatu, Tactu.
 7. De mente, seu animo, ejusque sede et actione.
 8. De Appetitu, motu, vigilia, somno, insomniis ac præsensionibus animalium.
 9. De sanitate, morbo, statu, senio, vita et morte animalium.
10. De animorum immortalitate.
Je laisse la partie morale pour une autre fois.

3

Providence and human freedom in Christian Epicureanism: Gassendi on fortune, fate, and divination

Fate is the decree of the divine will, without which nothing at all is done, . . . [and] Fortune is the concourse of events that, although unforeseen by men, nevertheless were foreseen by God.

Pierre Gassendi, Syntagma philosophicum[1]

Having ensured that divine providence played a major role in his mechanical philosophy, Gassendi turned to the question of human freedom in Book III of the "Ethics," the last part of the *Syntagma philosophicum,* entitled "On Liberty, Fortune, Fate, and Divination."[2] In this concluding section of his magnum opus, Gassendi cast his discussion in the form of a debate among the major classical philosophies, particularly Stoicism and Epicureanism. The main issue was freedom – human and divine. While questions about fate, fortune, and divination may, at first glance, appear rather remote from the primary concerns of seventeenth-century natural philosophy, in fact they involve metaphysical issues central to the articulation of the mechanical philosophy: the extent of contingency and necessity

1. Pierre Gassendi, *Syntagma philosophicum,* in Gassendi, *Opera omnia,* 6 vols. (Lyon, 1658; facsimile reprint, Stuttgart-Bad Cannstatt: Friedrich Frommann Verlag, 1964), vol. 2, p. 840.
2. He had originally planned to include this section in the "Physics" as the conclusion of his discussion about God's role in the universe. For the dating of the "Ethics" and for the history of this section, see Louise Tunick Sarasohn, "The Influence of Epicurean Philosophy on Seventeenth Century Ethical and Political Thought: The Moral Philosophy of Pierre Gassendi," Ph.D. dissertation, University of California, Los Angeles, 1979, chap. 5. See also the seventeenth-century English translation of François Bernier's abridgement of Book III of the "Ethics," *Three Discourses of Happiness, Virtue, and, Liberty, collected from the works of the Learn'd Gassendus* (London: Awnsham & John Churchil, 1699).

in the world, the nature of causality, and the role of providence and the extent of human freedom in a mechanical universe.[3] Gassendi's treatment of these issues reflects his underlying voluntarism.

Since classical times, natural philosophers had dealt extensively with questions about fate, fortune, and divination. The concept of fate was central to Stoicism, which had explained the world as governed by a deterministic, rational ordering principle, the Logos. According to Stoic doctrine, fate is the expression of the Logos in the causal nexus of a deterministic universe.[4] This emphasis on causal necessity in the universe had provided foundations for the Stoic belief in astrology and other forms of divination, which were based on the assumption that every part of the universe is connected to every other by the Logos and that events in one realm (say, the heavens) can serve as signs for events taking place elsewhere (say, in human lives).[5] Stoic fate was directly opposed to chance, which the Epicureans had incorporated into the universe by means of the *clinamen* or random swerve of atoms that they introduced to account for the collision of atoms and for free will. While Stoicism was compatible with the idea of providence, it was often interpreted as ruling out human freedom. Epicureanism, which allowed for free will, explicitly denied any kind of providential account of the world.

The evident contradictions between both of these classical philosophies, on the one hand, and the Christian doctrines of divine freedom, providence, and human freedom, on the other, stimulated discussions among early Christian thinkers, among the most influential of whom – on these issues – were St. Augustine, who rejected both Epicureanism and Stoicism,

3. See John Sutton, "Religion and the Failures of Determinism," in *The Uses of Antiquity: The Scientific Revolution and the Classical Tradition*, edited by Stephen Gaukroger (Dordrecht: Kluwer, 1991), pp. 25–51.

4. "Since the entire universe is governed by the divine *logos*, since, indeed, the universe is identical with the divine *logos*, then the universe, by definition, must be reasonable [*sic*]. The *logos* organizes all things according to the rational laws of nature, in which all events are bound by strict rules of cause and effect. Chance and accident have no place in the Stoic system. The causal nexus in the universe is identified with both fate and providence; fate, in turn, is rationalized and identified with the good will of the deity." Marcia L. Colish, *The Stoic Tradition from Antiquity to the Early Middle Ages*, 2 vols. (Leiden: E. J. Brill, 1985), vol. 1, pp. 31–2.

5. A. A. Long, *Hellenistic Philosophy: Stoics, Epicureans, Sceptics,* 2d edition (Berkeley: University of California Press, 1986), pp. 163–70; John M. Rist, *Stoic Philosophy* (Cambridge University Press, 1969), chap. 7; and S. Sambursky, *The Physics of the Stoics* (New York: Macmillan, 1959), pp. 65–71.

and Boethius, who attempted to fuse them with orthodox theology.[6] These early Christian discussions seemed particularly relevant to the Renaissance humanists as they tried to come to grips with the recently recovered texts of the classical philosophers.[7] Although the themes of fate, fortune, and human freedom were ubiquitous in Renaissance writing, one context in which they were particularly relevant was the debate over astrology, which gained prominence following Marsilio Ficino's (1433–99) translation and publication of the Hermetic corpus and other magical-mystical literature.[8] Giovanni Pontano (1427/9–1503) and Giovanni Pico della Mirandola both argued against astrology on the grounds that it limits human freedom in unacceptable ways.[9]

Pomponazzi reasserted the legitimacy of astrology, in *De fato, de libero arbitrio et de praedestinatione,* completed in 1520.[10] Favoring Stoic metaphysics and ethics, he gave an account of contingency as it could be understood in a world ruled by deterministic laws. Contingency, according to Pomponazzi, is not an indication of indifference, "the possibility for an effect to be or not to be." Rather, it refers "only to things which sometimes happen and sometimes do not, such as whether or not it will rain next month. If it does rain, that happens necessarily." Likewise if it does not.[11] Since the human will falls within the "universal hierarchy of natural causes," our intuition of free will is an illusion based on ignorance of the true causes of our actions.[12] Pomponazzi took great pains to show how this kind of Stoic determinism was compatible with moral responsibility. "Everything is therefore subject to the providential order of fate."[13] Given what he construed as a choice between divine providence and human freedom, Pomponazzi opted for the divine.[14] However, in a self-contradiction that reveals the continuing pull of traditional Christian thought, he preserved human freedom by maintaining that "it is in the power of the will to will and to suspend an act – this freedom is pre-

6. Antonio Poppi, "Fate, Fortune, Providence, and Human Freedom," in *The Cambridge History of Renaissance Philosophy,* edited by Charles B. Schmitt, Quentin Skinner, and Eckhard Kessler (Cambridge University Press, 1988), p. 642.
7. Ibid., pp. 644–50.
8. Ibid., pp. 650–1. See also Frances A. Yates, *Giordano Bruno and the Hermetic Tradition* (New York: Vintage, 1969; first published 1964), chaps. 2–4.
9. Poppi, "Fate, Fortune, Providence, and Human Freedom," pp. 651–2.
10. Ibid., p. 654.
11. Ibid., p. 655.
12. Ibid., p. 656.
13. Ibid.
14. Ibid., pp. 656–7.

served."[15] He noted that knowledge of freedom was the product of faith rather than of natural reason.[16]

Human freedom came under further assault at the hands of the Protestant reformers.[17] Luther concluded his debate with Erasmus with the resounding statement that human freedom is incompatible with divine foreknowledge:

> If we believe it to be true that God foreknows and predestines all things, that he can neither be mistaken in his foreknowledge nor hindered in his predestination, and that nothing takes place but as he wills it (as reason itself is forced to admit), then on the testimony of reason itself there cannot be any free choice in man or angel or any other creature.[18]

Calvin's doctrine of election similarly denied human freedom.[19] In his providential relationship to humanity, God determines how people choose and thereby obviates their free will. "God, whenever he wills to make way for his providence, bends and turns men's wills, even in external things; nor are they free to choose that God's will does not rule over their freedom."[20] In order to ensure God's freedom, Calvin was careful to distinguish his doctrine, "that particular events are generally testimonies of the character of God's singular providence," from the "Stoics' dogma of fate:"[21]

---

15. Pietro Pomponazzi, *Libri quinque de fato, de libero arbitrio et de praedestinatione,* edited by R. Le May (Lugano, 1957), 3, 8, 10, as translated by Poppi, "Fate, Fortune, Providence, and Human Freedom," p. 659.

16. Ibid., p. 659. See also Sutton, "Religion and the Failures of Determinism," pp. 31–3.

17. For background on this controversy see Jan Miel, *Pascal and Theology* (Baltimore: Johns Hopkins University Press, 1969), pp. 1–58; Luis de Molina, *On Divine Foreknowledge (Part IV of the* Concordia*),* translated with an introduction and notes by Alfred J. Freddoso (Ithaca, N.Y.: Cornell University Press, 1988), Preface; and Dale van Kley, *The Jansenists and the Expulsion of the Jesuits from France, 1757–1765* (New Haven, Conn.: Yale University Press, 1975), chap. 1.

18. Martin Luther, *De servo arbitrio,* translated by Philip S. Watson and B. Drewery, in *Luther and Erasmus: Free Will and Salvation,* edited by E. Gordon Rupp and Philip S. Watson (Philadelphia: Westminster Press, 1969), p. 332.

19. John Calvin, *Institutes of the Christian Religion,* edited by John T. McNeill and translated by Ford Lewis Battles, 2 vols. (Philadelphia: Westminster Press, 1960), bk. III, chap. xxiv, sec. 5–6.

20. Ibid., bk. II, chap. iv, sec. 7.

21. Ibid., bk. I, chap. xvi, sec. 7.

We do not, with the Stoics, contrive a necessity out of the perpetual connection and intimately related series of causes, which is contained in nature; but we make God the ruler and governor of all things, who in accordance with his wisdom has from the farthest limit of eternity decreed what he was going to do, and now by his might carries out what he has decreed. From this we declare that not only heaven and earth and the inanimate creatures, but also the plans and intentions of men, are so governed by his providence that they are borne by it straight to their appointed end.[22]

In the Catholic world, these controversies came to the fore in the aftermath of the Council of Trent (1545–63), which had addressed the question of formulating doctrine in response to the reformers' challenge.[23] Gassendi's discussion of fate and free will falls clearly within this context. By insisting on human freedom as he did, Gassendi placed himself among the followers of the Jesuit Luis de Molina, whose treatise *Concordia liberi arbitrii cum gratiae donis, divina praescientia, providentia, praedestinatione et reprobatione* (1588) was adopted by his order in its renowned debate with the Dominicans on the relationship between divine grace and human free will.[24] The Spanish Jesuit Francisco Suárez adopted Molina's views, defending and expanding them into a small treatise written in 1594 in response to a request by Pope Clement VIII.[25] The Dominicans had emphasized divine omnipotence to such an extent that they considered God's decree as imposing itself on people, determining their future actions. Although the Dominicans argued that this determination does not

22. Ibid., bk. I, chap. xvi, sec. 8.
23. See Hubert Jedin, *A History of the Council of Trent*, translated by Dom Ernest Graf, 2 vols. (London: Thomas Nelson, and St. Louis: Herder, 1957 and 1961; first published in German, Freiburg im Breisgau: Herder, 1957), vol. 2, chaps. 5–10. For summaries of the issues considered by the Council of Trent, see Steven Ozment, *The Age of Reform, 1250–1550: An Intellectual and Religious History of Late Medieval and Reformation Europe* (New Haven, Conn.: Yale University Press, 1980), pp. 407–9; and Hans J. Hillerbrand, *Men and Ideas in the Sixteenth Century* (Prospect Heights, Ill.: Waveland, 1969), pp. 91–6.
24. Poppi, "Fate, Fortune, Providence, and Human Freedom," p. 667. Sarasohn discusses Molina's influence on Gassendi at length. See Lisa T. Sarasohn, *Freedom in a Deterministic Universe: Gassendi's Ethical Philosophy* (Ithaca, N.Y.: Cornell University Press, forthcoming), chap. 5.
25. William Lane Craig, *The Problem of Divine Foreknowledge and Future Contingents from Aristotle to Suarez* (Leiden: E. J. Brill, 1988), p. 207; Calvin Normore, "Future Contingents," in *The Cambridge History of Later Medieval Philosophy*, edited by Norman Kretzmann, Anthony Kenny, and Jan Pinborg (Cambridge University Press, 1982), pp. 380–1.

destroy free will, the Jesuits rejected their argument, adopting instead the views of Molina that attempted to preserve divine omnipotence without sacrificing human freedom.[26] Whereas the Dominicans had claimed that God's foreknowledge of human actions makes it impossible for those actions to have any other outcome than what God foresees, Molina's account of God's foreknowledge of future contingents seemed to leave more play for human freedom. Molina described three kinds of knowledge that God has of future contingents: (1) his knowledge of naturally necessary states of affairs; (2) his *scientia media*, or knowledge of conditional future contingents (i.e., knowledge of what would follow from any given state of affairs); and (3) knowledge of his own causal contribution to any state of affairs.[27] It was Molina's concept of *scientia media* that enabled him to say that God's foreknowledge does not necessarily determine future human actions: God knows that, given certain circumstances, Peter will deny Christ; but that conditional knowledge does not necessitate Peter's denial.

Gassendi's approach to these questions in Book III of the "Ethics" rests on two important principles: that a proper understanding of the world must include divine freedom, creation, and providence and that the possibility of ethics – moral choice and judgment – requires human freedom. His humanist bent led him to consider these issues in the rhetorical context of a debate among the ancient philosophers, especially the Epicureans and the Stoics.

If divine freedom played a central role in Gassendi's philosophy of nature, human freedom was the cornerstone of his ethics.[28] Moral choice and judgment depend, he argued, on the possibility of those choices being taken freely and deliberately. Actions taken either by accident or by necessity do not merit praise or blame.[29] "Freedom (*libertas*) consists in indifference."[30] That is, the will and intellect are said to be free if they are equally able to choose one or another of possible options and are not in any way determined to one or the other.[31] Real freedom, understood as

---

26. See Molina, *On Divine Foreknowledge,* Preface. See also Rivka Feldhay, "Knowledge and Salvation in Jesuit Culture," *Science in Context, 1* (1987): 204–5.
27. Paraphrased from Molina, *On Divine Foreknowledge,* p. 23.
28. Lisa T. Sarasohn, "The Ethical and Political Philosophy of Pierre Gassendi," *Journal of the History of Philosophy,* 20 (1982): 258–60.
29. Gassendi, *Syntagma philosophicum,* in *Opera omnia,* vol. 2, p. 821.
30. Ibid., p. 823.
31. A distinction of this kind between genuine freedom and necessitated choice has a long history, going back at least to Duns Scotus. See Miel, *Pascal and*

indifference, belongs only to rational beings and differs from what Gassendi called willingness (*libentia*). Willingness characterizes the actions of boys, brutes, and stones, agents lacking the capacity for rational choice.[32] Human freedom is also a concomitant of voluntarist theology; for if any human action were determined necessarily, the universe would contain some element of necessity that would restrict God's power and freedom.

Having articulated his underlying assumptions – namely, a voluntarist theology and a conception of human nature incorporating free will – Gassendi proceeded with his analysis of fortune, fate, and divination. Although he considered them in the context of ancient philosophy, this humanist device was a ploy for discussing some of the most controversial theological and natural philosophical issues of the early seventeenth century. These concepts challenged his Christian, voluntarist, providential view of God, nature, and human nature, by calling for the elimination of either creation, providence, or free will.

Gassendi began his discussion of fortune and chance by adopting a standard, classical definition: Fortune is an unexpected consequence, a cause by accident. To illustrate his meaning, he cited Aristotle's example of a man who discovers treasure while digging in the ground to plant a tree.[33] Finding a treasure is a totally unexpected consequence of his act.

Theology, p. 41. Gassendi's definition of "*libertas*" is very similar to that of Molina, who wrote: "Just as, in order for an act to be a *sin* it is not sufficient that it be spontaneous, but is instead necessary that it be free in such a way that, when the faculty of choice consents to it, it has the power not to consent to it, given all the surrounding circumstances obtaining at that time, so too in order for there to be *merit* or for an act to be *morally* good – indeed, even in order for there to be a free act that is indifferent to moral good and evil – it is necessary that when the act is elicited by the faculty of choice, it be within the faculty's power not to elicit it, given all the circumstances obtaining at that time." Molina, *On Divine Foreknowledge*, disputation 53, pt. 2, sec. 17, pp. 224–5.

32. Gassendi, *Syntagma philosophicum*, in *Opera omnia*, vol. 2, p. 822. Lisa T. Sarasohn discusses Gassendi's concepts of *libertas* and *libentia* in "Motion and Morality: Pierre Gassendi, Thomas Hobbes, and the Mechanical World-View," *Journal of the History of Ideas*, 46 (1985): 371–3; and in "The Ethical and Political Philosophy of Pierre Gassendi," p. 259. See also Sarasohn's extended discussion of these concepts in *Freedom in a Deterministic Universe*, chaps. 3 and 5.

33. Aristotle's example occurs in *Metaphysics*, V, 30 (1025a14–29), translated by W. D. Ross, in *The Complete Works of Aristotle*, edited by Jonathan Barnes, 2 vols. (Princeton, N.J.: Princeton University Press, 1984), vol. 2, p.

While the digging preceded the discovery, it was not the cause of the discovery except accidentally, since the discovery of treasure is not the usual or natural outcome of digging in the ground. Such unexpected consequences are called fortuitous in connection with agents that act freely; they are called chance in connection with inanimate objects.[34] An example of chance would be the occurrence of a storm in the west at sunset. Both events – the storm and the sunset – are the outcome of natural causes, but their coincidence in space and time is both unpredictable and unanticipated. According to Gassendi, fortune and chance are both expressions of contingency in the world. Like fortune, chance is the name given to the kind of contingency that describes an event that may or may not happen in the future. An event that is said to be caused by chance or fortune is one that results from the unexpected concourse of several apparently unrelated causes. In the case of the unexpected discovery of the treasure, there is the concourse of the original burial of the treasure with the present digging in the ground.[35] Each event is the perfectly natural outcome of a series of causes. The two series are unrelated, however, and so their concourse is unexpected: Therein lies the element of chance or fortune, concepts that reflect our ignorance rather than the state of the world. "Fortune [or chance] is truly nothing in itself . . . only the negation of foreknowledge and of the intention of the events."[36]

Certain misunderstandings render the concept of fortune problematic. Those who reify fortune and call it divine – a position that even Epicurus had rejected – are ignorant of the real causes of the events in question.[37] Epicurus had, rather, equated fortune with chance and denied that "there is divine wisdom in the world."[38] He had thus compared life to a game of dice. Here he had erred, according to Gassendi, because he had failed to appreciate that divine providence touches every aspect of nature and of human life.[39]

Fortune and chance, understood as unanticipated outcomes of unexpected concourses of natural causes, can easily be incorporated into an

---

1619. For an extensive discussion of Aristotle's understanding of "coincidence" and "accident," see Richard Sorabji, *Necessity, Cause, and Blame: Perspectives on Aristotle's Theory* (Ithaca, N.Y.: Cornell University Press, 1980).

34. Gassendi, *Syntagma philosophicum*, in *Opera omnia*, vol. 2, p. 828.
35. Ibid.
36. Ibid., p. 829.
37. Ibid.
38. Ibid., p. 830.
39. Ibid., pp. 830–1.

orthodox philosophy of nature by including divine will and providence among the efficient causes operating in the world.[40] Opposing Epicurean materialism and emphasizing the limits of human knowledge, Gassendi thus believed he could reinterpret one of the important components of Epicureanism in a theologically suitable fashion. Accordingly, he defined fortune as "the concourse of events which, although unforeseen by man, were foreseen by God; and they are connected by a series of causes."[41] In other words, events that seem fortuitous to us are nevertheless providential, resulting from God's design, despite our ignorance of the causal sequences producing them.

Whereas fortune and chance raise questions about the nature of contingency, causality, and providence in the world, fate points to the complex problem of free will and determinism. Some interpretations of fate, as Gassendi understood or misunderstood them, seemed to incorporate a kind of natural necessity that would restrict both divine and human freedom. It is this necessitarian interpretation of the notion of fate – whether by ancient philosophers or contemporary theologians – to which Gassendi was primarily opposed.[42] The question of fate and its relation to free will had special relevance in the context of post-Reformation debates about free will. The reformers, especially the Calvinists, seemed to have denied human as well as divine freedom – at least regarding matters of salvation – with their doctrines of predestination and election. The Dominican approach to predestination and divine foreknowledge suffered from similar problems. In working out his own interpretation of the concept of fate in dialogue with the ancient philosophers, Gassendi participated in one of the most heated theological controversies of his own day.

He observed that there are two chief views about fate: that it is something divine and that it is merely natural. Among those who regarded fate as divine, he counted the Platonists and the Stoics. The former group defined fate as "the eternal God or that reason, which disposes all things from eternal time, and thus binds causes to causes."[43] In this sense, Plato (429–347 B.C.) had sometimes considered fate to be part of the soul of the world, sometimes "the eternal reason and law of the universe." The Stoics Zeno (366–c. 264 B.C.) and Chrysippus (280–206 B.C.) had defined fate as "the motive force of matter and the spiritual force and governing

40. Ibid.
41. Ibid., p. 840.
42. See Josiah B. Gould, "The Stoic Conception of Fate," *Journal of the History of Ideas,* 35 (1974): 17–32.
43. Gassendi, *Syntagma philosophicum,* in *Opera omnia,* vol. 2, p. 830.

reason of the order of the universe." Seneca (c. 4 B.C.-65 A.D.) had gone so far as to identify fate with the god Jove.[44]

Despite the apparently theological and providential orientation of the Stoic interpretation of fate, Gassendi found its necessitarianism objectionable: "This necessity seems to be of such a kind that it completely removes the liberty of all human action and leaves nothing within our judgment."[45] Such negation of free will would deprive life of meaning. In a world ruled by fate, there would be no place for plans, prudence, or wisdom, since everything would happen according to fate. "All legislators would be either fools or tyrants, since they would command things [to happen] that were either always to be done or that we absolutely cannot do."[46] Since there would be no freedom of action, no action would be subject to moral judgment. All contingency in the universe would be eliminated. Consequently, all divination, prayer, and sacrifice would be rendered useless.[47]

Among those who considered fate to be something merely natural, Gassendi distinguished those who thought of fate as absolutely binding and those who did not. As Lisa Sarasohn has shown, Gassendi created a dialogue between Democritus (460-c. 356 B.C.) and Epicurus to represent the views of Hobbes in contrast to his own.[48] The hard determinism and materialism endorsed by Democritus in this section of the *Syntagma philosophicum* constitute the nightmare feared by Christian mechanical philosophers. Hobbes in fact held such views. His treatise *Of Liberty and Necessity*, in which he supported a deterministic position on the free-will controversy, was published without his permission in 1654 by the Anglican Bishop John Bramhall, who argued against Hobbes.[49] Hobbes then published *The Questions Concerning Liberty, Necessity, and Chance* (1656) which contained both Bramhall's treatise and his own replies.[50] Although this work was published after Gassendi's death, Gassendi was

44. Ibid.
45. Ibid., p. 831.
46. Ibid., pp. 831–2.
47. Ibid., p. 832.
48. Sarasohn, *Freedom in a Deterministic Universe*, chap. 6.
49. Samuel I. Mintz, *The Hunting of Leviathan: Seventeenth-Century Reactions to the Materialism and Moral Philosophy of Thomas Hobbes* (Cambridge University Press, 1962), p. 110.
50. Thomas Hobbes, *The Questions Concerning Liberty, Necessity, and Chance, Clearly Stated and Debated between Dr. Bramhall, Bishop of Derry, and Thomas Hobbes of Malmesbury*, in *The English Works of Thomas Hobbes of Malmesbury*, edited by Sir William Molesworth, 11 vols. (London, 1839–45; reprinted Aalen: Scientia, 1962), vol. 5.

directly acquainted with Hobbes' views from their interactions in Paris around 1641 or possibly even earlier.[51]

In the dialogue between Democritus and Epicurus in Book III of the "Ethics," Gassendi put Hobbes' opinions in the words of Democritus, the ancient advocate of hard determinism. Democritus had held the view that "every event has a cause and that the same cause is always followed by the same effect" and that the truth of determinism rules out human freedom.[52] He conceived of fate as natural necessity. In Gassendi's presentation, Democritus' view was similar to that of the Stoics, shorn, however, of any remnant of theology: "Democritus taught . . . that Necessity is nothing other than . . . the motion, impact, and rebounding of matter, that is, of atoms, which are the matter of all things. Whence it can be understood that 'Material Necessity' is the cause of all things that happen."[53] Democritus claimed that since everything, including the human soul, is composed of atoms, there is no room for real freedom in the universe.[54] Not only freedom, but also error would be impossible in such a world.[55] And if everything were necessarily determined by the motions and collisions of atoms, there would be no room for divine providence.[56]

In contrast to Democritus, who maintained absolute necessity since nothing can impede a cause from producing its effect,[57] Epicurus believed that the necessity in nature is not absolute. Arguing on logical grounds, he claimed that it is impossible simultaneously to hold that all statements are either true or false and that there is absolute necessity in nature, for to do so would entail giving truth-value to statements about future contingents, something Epicurus regarded as impossible. Following Aristotle in his famous discussion of tomorrow's sea battle,[58] Epicurus "admitted this complex as truth, 'Either Hermachus will be alive tomorrow or he will not be alive.' " But Epicurus could not accept the possbility that either one of the disjuncts – "It is necessary that Hermachus be alive tomorrow" or "It is necessary that Hermachus not be alive tomorrow" – be true; for "There

51. Sarasohn, *Freedom in a Deterministic Universe*, chap. 6.
52. Gassendi, *Syntagma philosophicum*, in *Opera omnia*, vol. 2, pp. 830–2. I take this definition of "hard determinism" from Sutton, "Religion and the Failures of Determinism," p. 27.
53. Gassendi, *Syntagma philosophicum*, in *Opera omnia*, vol. 2, p. 834.
54. Ibid., p. 835.
55. Ibid., p. 834.
56. Ibid., p. 840.
57. Ibid., p. 837.
58. Aristotle, *De interpretatione*, translated by J. L. Ackrill, in *The Complete Works of Aristotle*, edited by Barnes, 18b7–25, vol. 1, p. 29.

is no such necessity in nature."[59] Epicurus thought that necessity could apply only to statements about the past and present: The events described have already occurred or not occurred, and so statements describing them have a determined truth-value. However, statements about the future have an undetermined truth-value, so we cannot reason about them with necessity.[60] Since, according to Epicurus, we cannot have knowledge of future contingents, it follows that there is no such necessity in nature.[61]

As for events that occur by plan or by fortune, these involve human freedom, which Epicurus had tried to preserve by adding the *clinamen*, or random swerve, to the otherwise steady downward motion of the Democritean atoms. According to Gassendi, Epicurus had introduced the swerve with the explicit intention "that it shatter the necessity of fate and thus ensure the liberty of souls."[62] Despite Epicurus' good intentions, Gassendi did not find his solution to the free-will controversy convincing. Epicurus had thought that the unpredictability of the swerve preserves free will.[63] Gassendi did not agree. Events would still always happen by the same chain of necessary consequences. "What always happens by the same necessity would happen by a variety of motions, collisions, rebounds, swerves in a certain external series, like a chain of consequences."[64] Since Epicurus had argued that the soul consists of atoms, its choices would simply be determined by the long causal sequence of the

59. Gassendi, *Syntagma philosophicum*, in *Opera omnia*, vol. 2, p. 837. Cicero discussed Epicurus' ideas about future contingents at some length. See Marcus Tullius Cicero, *De fato*, translated by H. Rackham (Cambridge, Mass: Loeb Classical Library, 1953) pp. 233–5.

60. The concept of necessity employed by Epicurus is not the same as that currently in vogue among twentieth-century philosophers. See Normore, "Future Contingents," pp. 358–81. See also the Introduction in William Ockham, *Predestination, God's Foreknowledge, and Future Contingents*, translated by Marilyn McCord Adams and Norman Kretzmann (New York: Appleton-Century-Crofts, 1969), pp. 1–33.

61. Similar ideas can be found among some of the Stoics. See Rist, *Stoic Philosophy*, p. 122.

62. Gassendi, *Syntagma philosophicum*, in *Opera omnia*, vol. 2, p. 837. Gassendi's analysis of the relationship between the swerve of atoms and free will is borne out by modern scholarship, although the main source for this doctrine appears to have been Lucretius, whom Gassendi cited extensively, rather than Epicurus. See Long, *Hellenistic Philosophy*, pp. 56–61; and Rist, *Epicurus*, pp. 90–9.

63. Gassendi, *Syntagma philosophicum*, in *Opera omnia*, vol. 2, p. 838.

64. Ibid.

material world. Gassendi concluded that the *clinamen* was therefore not a satisfactory explanation of human freedom.

Gassendi found all the traditional accounts of fate to be wanting, primarily for theological and ethical reasons. The deterministic, reductionist atomism of Democritus left room for neither divine providence nor free will:

> Therefore, the opinion of Democritus must be exploded inasmuch as it can by no means stand with the principles of the Sacred Faith (because of having removed from God the care and administration of things), and it is thus manifestly repugnant to the light of nature by which we experience ourselves to be free.[65]

The passion driving Gassendi's attack on Democritus can in part be explained by his identification of "Hobbes as Democritus reincarnated," for Gassendi found his contemporary's hard determinism entirely unacceptable.[66] Epicurus deserved criticism as well, despite his good intentions in attempting to preserve free will, for by denying the possibility of knowing future contingents, he had denied God such knowledge. He "thus supposes that there is no creation of things and no divine providence."[67]

In order to embrace the evident facts of both causal order and contingency within the bounds of his mechanical philosophy, Gassendi undertook a Christian reinterpretation of the concepts of fate, fortune, and chance, providing a providential understanding of these concepts, just as Augustine had done centuries earlier:[68]

> To the extent that Fate can be defended, so can Fortune. If we agree that Fate is the decree of the divine will, without which nothing at all is done, truly Fortune is the concourse of events that, although unforeseen by men, nevertheless was foreseen by God; and they are the connected series of causes or Fate.[69]

Thus, fate is nothing more than God's decree, and fortune and chance are expressions of contingency in the world coupled with human ignorance of the causes of fortuitous events. Fortune, chance, and fate are not autonomous principles running the world. Even if all events have causes and even if God can foresee the unrolling of cause and effect – a foresight not available to humans – both the causes and their effects ultimately depend on divine will. The universe remains a contingent place.

65. Ibid., p. 840.
66. Sarasohn, "Motion and Morality," p. 369.
67. Gassendi, *Syntagma philosophicum*, in *Opera omnia*, vol. 2., p. 840.
68. Vincenzo Cioffari, "Fate, Fortune, and Chance," *Dictionary of the History of Ideas*, edited by Philip P. Wiener, 4 vols. (New York: Scribner, 1973), vol. 2, p. 230; Poppi, "Fate, Fortune, Providence, and Human Freedom," p. 642.
69. Gassendi, *Syntagma philosophicum*, in *Opera omnia*, vol. 2, p. 840.

Gassendi's reinterpretation of fortune, chance, and fate left plenty of room for the exercise of divine freedom and providence. The fact that certain events appear to be fortuitous in no way impairs divine omniscience. They appear fortuitous only because of the limitations of human knowledge:

> The word Fortune . . . indicates two things, the concourse of causes and the previous ignorance of events; Fortune can thus be admitted afterward with respect to man but not God; and on account of this . . . , nothing stands in the way of our saying that Fortune is a part not only of Fate, but also of divine providence, which foresees for man what he cannot foresee [for himself.[70]

Far from imparting a randomness to life's events, fortune itself is an expression of divine foresight and providence. Fortune is an expression of human limitation, but it in no way impugns divine power and freedom.[71]

Reconciling fate with divine providence was not so difficult for Gassendi. A more challenging problem was to reconcile fate with free will. Here again he turned to theology for his solution, and here the connections between Gassendi's discussion and post-Reformation theology become explicit. His discussion was, in effect, a debate with the overly deterministic theologies of Calvin and the Dominicans: "We call Fate, with respect to men, nothing other than that part of Divine Providence that is called Predestination by theologians . . . in order that predestination and thus Fate can be reconciled with liberty."[72] Appealing to the vexatious doctrine of predestination seems an odd way to clarify anything. Indeed, Gassendi acknowledged that the problem of reconciling predestination and divine foreknowledge with human freedom had troubled both philosophers and theologians since antiquity.

---

70. Ibid.
71. This providential interpretation of chance was shared by Gassendi's English contemporary, the Puritan divine William Ames, who argued in his *Medulla theologica* (1623), that the appeal to lots is not an appeal to chance, but to providence: "There is no power of rendering judgement in contingent events themselves and no other fortune judging them than the sure providence of God; so it follows that judgement must be expected in a special way from God's providence. Pure contingency itself cannot be a principal cause in deciding any question, nor can the man for whom the event itself is purely contingent direct it to such an end. Therefore, such direction is rightly to be expected from a superior power." William Ames, *The Marrow of Theology*, translated by John D. Eudsden (Boston: United Church Press, 1968), II, xi, 10–11 (p. 272). See also Margo Todd, "Providence, Chance and the New Science in Early Stuart Cambridge," *Historical Journal*, 29 (1986): 697–711.
72. Gassendi, *Syntagma philosophicum*, in *Opera omnia*, vol. 2, p. 841.

God's foreknowledge of Peter's denial had challenged some philosophers and theologians because they had thought that such foreknowledge entailed a kind of fatalism that denies human freedom.[73] Peter's free will could be saved, but only at the expense of God's omniscience and veracity:

> Either God knew definitely and certainly that Peter would deny Christ, or he did not. It cannot be said that he did not know, because he predicted it and he is not a liar: and unless he knew, he would be neither omniscient nor God. Therefore, he knew it definitely and certainly. Thus, it could not be that Peter would not deny. If God knew and Peter did not deny . . . , it would be argued of God that his foreknowledge was false and that he was a liar. If Peter cannot deny, then he is not free to deny or not deny. Therefore he is without freedom.[74]

In other words, either God's veracity and omniscience or Peter's free will must be denied. Gassendi found both alternatives unacceptable.

In order to resolve this difficulty, Gassendi invoked the Scholastic distinction between absolute necessity and necessity by supposition:

> For example, that double two is four or that yesterday comes before today is absolutely necessary, although that you lay the foundations of your house or leave the city is not necessary: nevertheless if you suppose that you will build your house or that you will be in the country, then for you to lay the foundation or leave the city is, I say, necessary from supposition. Truly it is manifest from this distinction, that absolute necessity hinders that by which a certain action is elicited, however that which is from supposition does not hinder (for he who will lay down a foundation absolutely can *not* lay it, and he who will leave the city can *not* leave).[75]

Molina's theory of *scientia media* provided Gassendi with a way of resolving these difficulties by interpreting the necessity of God's foreknowledge as necessity by supposition:[76]

---

73. Craig, *The Problem of Divine Foreknowledge*, p. 59.

74. Gassendi, *Syntagma philosophicum*, in *Opera omnia*, vol. 2, p. 841.

75. Ibid.; my emphasis. Gassendi's talk about the "absolute necessity" of mathematical truths here should not lead us to conclude that he abandoned his voluntarism. In his debate with Descartes, he defended an empiricist, probabilist account of the epistemological status of mathematics. See Chapter 6. See Pierre Gassendi, *Disquisitio metaphysica, seu dubitationes et instantiae adversus Renati Cartesii metaphysicam et responsa*, edited and translated into French by Bernard Rochot (Paris: J. Vrin, 1962), pp. 468–73; in *Opera omnia*, vol. 3, pp. 374–5.

76. Molina, *On Divine Foreknowledge*, p. 23. Sarasohn argues as well that Gassendi adopted the Molinist approach to the problem of predestination

Peter's future denial was seen by God necessarily, but nevertheless by a necessity from supposition, because of which nothing of liberty is taken away. . . . [T]hus although it was determined from the beginning that [Peter] would deny him, he does it freely in whatever manner he did it; afterward, since he did it, it was necessary.[77]

Gassendi's concept of necessity contains a temporal component. The act of denial does not become necessary, despite being foreseen by God, until Peter commits it. Once the denial has occurred, it is part of the past that cannot be undone:

If indeed when it is said that Peter denied necessarily this necessity is understood, not as something that was truly in Peter antecedently that forced him to act, but only now that it is in this time that is in the past and cannot not be past, thus the thing that is done by him is done . . . and cannot not be done by him.[78]

Necessity of this kind in no way impinges on divine freedom and omniscience, because God could foresee Peter's free choice. "Thus, it can be said that Peter denied not because God foresaw it, but God foresaw since Peter would deny."[79] God's knowledge of future events does not cause those events to happen; but, on account of his omniscience, he has foreknowledge because they will happen. Since some of those events are the acts of free agents, there is no contradiction between God's foreknowledge and human freedom.

Gassendi addressed at some length the question of how to interpret the doctrine of predestination. He rejected the Calvinist view that the members of the elect and the reprobate had been chosen from eternity. He also rejected the Dominican view that God's foreknowledge deprives human agents of their freedom. Instead, he opted for the more liberal, Molinist position. According to Gassendi, God created people with free will as well as the causal order of the world. He knows how an individual will respond in any particular situation, even though that individual will respond freely. In this way, God's foreknowledge in no way restricts the liberty of free agents.[80] Even if everything is included within the domain of divine decree, that inclusion does not eliminate human freedom, for God created free agents as well as determined ones.

and divine foreknowledge. See Sarasohn, *Freedom in a Deterministic Universe*, chap. 5.
77. Gassendi, *Syntagma philosophicum*, in *Opera omnia*, vol. 2, p. 841.
78. Ibid.
79. Ibid., pp. 841–2.
80. Ibid., p. 844.

If discussions about fate and fortune really concerned the roles of contingency and necessity in the universe, divination raised questions about the nature of causality. Divination had played a central role in Stoic thought, where it had been invoked to provide evidence for the causal interconnectedness of the universe.[81] For Gassendi, delineating the boundary between the natural and supernatural was an important part of the task of determining the limits of mechanical causality in the world. This problem was not unique to the mechanical philosophers. All natural philosophers in the period – Aristotelians and natural magicians, as well as mechanical philosophers – faced it in the questions raised by witchcraft, demonology, and other occult pursuits.[82]

The immediate context of Gassendi's concern with divination, as well as that of his contemporaries Mersenne, Naudé, and La Mothe le Vayer, was its notriously naturalistic treatment by Pomponazzi in *De naturalium effectum admirandorum causis sive de incantationibus* (1556) and *De fato, de libero arbitrio et de praedestinatione* (1520). In order to account for many extraordinary effects without appealing to demons, Pomponazzi had sought to explain both natural and human history as determined by natural, astrological, and various occult causes. In so doing, he affirmed an Averroism far more radical than that condemned in 1277. The strong negative reaction to his books was exacerbated by his role in debates about the immortality of the soul.[83]

Whether or not Gassendi had actually read Pomponazzi's works himself, he doubtless knew about them from his friend Mersenne, who went

81. Sambursky, *Physics of the Stoics*, p. 66.
82. See Stuart Clark, "The Scientific Status of Demonology," in *Occult and Scientific Mentalities in the Renaissance*, edited by Brian Vickers (Cambridge University Press, 1984), pp. 351–74.
83. See Brian P. Copenhaver, "Astrology and Magic," p. 273; Poppi, "Fate, Fortune, Providence, and Human Freedom," pp. 653–60; and Eckhard Kessler, "The Intellective Soul," pp. 500–7, all in *The Cambridge History of Renaissance Philosophy*, edited by Schmitt, Skinner, and Kessler. On Pomponazzi's naturalism, see Étienne Gilson, "Autour de Pomponazzi: Problématique de l'immortalité de l'âme en Italie au début du XVIe siècle," *Archives d'histoire doctrinale et littéraire de Moyen Age*, 28 (1961): 163–279. See also Paul Oskar Kristeller, *Eight Philosophers of the Renaissance* (Stanford, Calif.: Stanford University Press, 1964), chap. 5; Jean Céard, "Matérialisme et théorie de l'âme dans la pensée padouane: Le *Traité de l'immortalité de l'âme* de Pomponazzi," *Revue philosophique de France et l'étranger*, 171 (1981): 25–48; and Olivier René Bloch, *La philosophie de Gassendi: nominalisme, matérialisme et métaphysique* (The Hague: Martinus Nijhoff, 1971), pp. 310–11.

to great lengths to refute Renaissance naturalism in *Quaestiones celeberrimae in Genesim* (1623) and *L'impiété des déistes* (1624).[84] Mersenne rejected the Renaissance naturalists as atheists because they "attribute everything to nature alone" and deny God a causal role in the world.[85] In particular, he attacked the naturalists' belief in astrology because it is contrary to the teachings of the church fathers, because it is based on an unacceptable mysticism and a false theory of causation founded on the correspondence between macrocosm and microcosm, and because it is too restrictive of human freedom.[86] Moreover, "only true science, based on an idea of nature submissive to intelligible laws, would permit him to save religion, morality, and science."[87]

While Gassendi rejected Stoic fatalism and the more recent naturalism associated with it, he defended certain forms of divination on theological grounds. In the final chapter of Book III of the "Ethics," entitled "The Meaning of Divination, or the Foreknowledge of Future and Merely Fortuitous Things," Gassendi supported divination in opposition to Epicurus, whose blanket denial of the possibility that knowledge of future contingents might be compatible with human freedom had led him to reject the possibility of any kind of divination. That Epicurus was wrong, Gassendi argued, can be demonstrated straightaway by the fulfillment of the biblical prophecies.[88]

Gassendi began his discussion of divination with the consideration of demons, which some ancient thinkers had invoked as part of a naturalistic way of explaining how divination works. Demons concerned Gassendi, because the question of their existence bore on the deeper question of the causal order of the world and the boundaries between natural and supernatural causation. He rehearsed and rejected various ancient doctrines about demons – that they are particles of the *anima mundi*, that they have a corporeal nature, that they are halfway between humans and gods, that they move the heavenly spheres, that they are of some particular number

---

84. Robert Lenoble, *Mersenne ou la naissance du mécanisme*, 2d edition (Paris: J. Vrin, 1971) chap. 3, esp. pp. 112–21.

85. Marin Mersenne, *Quaestiones celeberrimae in Genesim*, translated by William L. Hine, in "Marin Mersenne: Renaissance Naturalism and Renaissance Magic," in *Occult and Scientific Mentalities in the Renaissance*, edited by Brian Vickers (Cambridge University Press, 1984), p. 167.

86. Lenoble, *Mersenne ol la naissance du mécanisme*, pp. 128–33.

87. Ibid., p. 133; my translation. See also Sutton, "Religion and the Failure of Determinism," pp. 39–41.

88. Gassendi, *Syntagma philosophicum*, in *Opera omnia*, vol. 2, p. 847.

or another.[89] The problem with all of these views is that they remove divine activity from the ordinary workings of the world. "They judged that it was alien to the divine majesty to care for all particulars himself,"[90] thereby impugning divine power by implying that God uses ministers to carry out his will because of some defect in his nature. Gassendi countered with a voluntarist argument: "God uses ministers, not because of disgrace, impotence, or need, but because he wished it for the state of things that is the world. He judged it congruous."[91] Reasserting nature's utter dependence on divine will, he noted that God, "if he had wished to institute another order, he would not have done a disgraceful thing nor would it testify to any impotence or need."[92] Unlike the highest prince in his realm, to whom the philosophers had compared him, God is actually present everywhere in the world, not just to his designated ministers. The philosophers had mistakenly substituted the activity of these demons for both God's general and special providence.[93] Since demons were generally understood to work by natural means, Gassendi sought to maintain a role for the supernatural by defending God's providential activity in the running of the world.

In fact, he believed that various orders of angels and demons do exist as purely spiritual beings, an opinion he based on "sacred scripture and . . . [which was] explained by theologians."[94] But there are also many false superstitions about the activity of these creatures, exploits that are exaggerated by the poets. There are

> many little stories which frequently fill your ears, from which you will often discover something difficult that is true, if you eliminate the fraud of impostors, the tricks of the crafty, the nonsense of old women, the easy credulity of the common people. Something must also be said about this kind of filthy magic, by which the unhappy person thinks himself carried away by he-goats. . . . [And] afterward, put to sleep by narcotic salves, they dream with a most vivid imagination [that they] were present in a most evil assemblage.[95]

Although these and other temptations and possessions actually exist – scripture, the lives of the saints, and the successful practice of exorcism attest to that fact – the point is to attend to our own spiritual and moral state, our relationship to God by virtue of his special providence, rather

89. Ibid., pp. 849–51.
90. Ibid., p. 851.
91. Ibid.
92. Ibid.
93. Ibid., pp. 851–2.
94. Ibid., p. 851.
95. Ibid., pp. 852–3.

than to excuse our sins by blaming evil demons.[96] A proper understanding of demons had not been available to the ancient philosophers who did not possess either the true faith or sacred Scripture, which teach us both of their existence as spiritual beings and of the limits of their powers.

Gassendi had embarked on this long discussion of demons because some ancient advocates of divination had appealed to them in order to explain their practices.[97] Since it is sometimes possible to predict the future, as scripture attests, one must consider "whether the prediction was made by the intervention of demons or the craftiness of the soothsayers or the credulity of those who asked for it."[98] Although there *are* genuine cases of prophecy, many predictions are made of things that have natural causes and are "incapable of impediments, such as eclipses, risings of the stars, and other things of this kind, which depend on the determined disposition and constancy of the motions of the heavenly bodies."[99] In such cases, there is no need to appeal to anything beyond natural causes.

As for genuine divination, Gassendi repeated the traditional Stoic doctrine that there are two kinds. One, like astrology or the ancient interpretations of signs – such as the flight, songs, and feeding of birds or the casting of lots or the interpretation of dreams – depends on art. The other kind does not.[100] In the closest approximation to a joke in the ponderous *Syntagma philosophicum,* Gassendi railed against "geomancers, hydromancers, aeromancers, pyromancers, and others . . . and last those astromancers or astrologers who . . . seek it from the stars," all of them practitioners of "artificial divination."[101] If astrology, which

96. Ibid., p. 852.
97. It should be noted that the most important philosophical account of divination came from the Stoics, whose account was thoroughly materialistic, owing nothing to the personal agency of demons. See Sambursky, *Physics of the Stoics,* pp. 66–71.
98. Gassendi, *Syntagma philosophicum,* in *Opera omnia,* vol. 2, p. 853.
99. Ibid.
100. Sambursky, *Physics of the Stoics,* pp. 66–71. For the Ciceronian roots of this distinction, see *De Divinatione,* translated by William Armistead Falconer (Cambridge, Mass: Loeb Classical Library, 1953), I, 12, 24, 72–92; II, 26.
101. Gassendi, *Syntagma philosophicum,* in *Opera omnia,* vol. 2, p. 854. Apparently this list of various sorts of diviners had a long history. Isidore of Seville wrote as follows in his *Etymologies:* "Varro dicit divinationis quattor esse genera, terram, aquam, aerem et ignem. Hinc geomantiam, hydromantiam, aeromantiam, pyromantiam dictam." Isidore of Seville, *Etymologies,* edited by W. M. Lindsay (Oxford University Press, 1911), Lib. VIII, ix, line

holds the principal place among the arts of divination is "inane and futile, the others ought to be no less inane and futile."[102] Gassendi thus denied that divination by art, the kind the Stoics valued most, is divination at all because it is nothing but the observation of regular sequences of natural events, whether or not we understand the causes of those sequences.

In fact, Gassendi argued, any genuine divination would presume the existence of events that do not have causes.[103] Otherwise, nothing more would be involved in divination than the same kinds of conjectures used in any of the sciences that make predictions about future events. In all conjectural knowledge, we attend to the known causes of events and predict what will likely happen. Such predictions are conjectural, based on reasoning about our observed knowledge of the world. Divining is no different from this kind of conjectural science except that it frequently suffers from a deficit "of ratiocination and consultation."[104] The Stoics had agreed, but in advocating astrology, they had confounded inductive methods with divination by art, which is based on an empirical understanding of the deterministic nexus of the world.[105]

Gassendi's discussion of fortune, fate, and divination in Book III of the "Ethics" reveals his position on the major theological and ethical implications of the mechanical philosophy. His opposition to the hard determinism of Hobbes, the modern Democritus, was drawn from his voluntarist theology, which insisted on freedom, both human and divine. His emphasis on human freedom inclined him toward the more liberal, Molinist interpretation of predestination, probably the single most contentious issue in post-Reformation theology. His views on fortune, fate, and divination clearly situate him in the seventeenth-century debates about the philosophy of nature. He unambiguously advocated a baptized version of

13. I am grateful to Haijo Westra for bringing this point to my attention.

102.   Gassendi, *Syntagma philosophicum*, in *Opera omnia*, vol. 2, p. 854. For a full account of Gassendi's rejection of astrology, see "Physics," sec. II, bk. VI, "De effectibus siderum," in ibid., vol. 1, pp. 713–52. This part of the *Syntagma* was translated into English in the seventeenth century. See Petrus Gassendus, *The Vanity of Judiciary Astrology. Or Divination by the Stars* (London: Humphrey Moseley, 1659). For the context of this polemic, see Jacques E. Halbronn, "The Revealing Process of Translation and Criticism," in *Astrology, Science, and Society: Historical Essays,* edited by Patrick Curry (Suffolk: Boydell Press, 1987), pp. 197–217. Sarasohn thoroughly discusses Gassendi's views on astrology and their relationship to his ethical theory in *Freedom in a Deterministic Universe,* chap. 4.

103.   *Syntagma philosophicum,* in *Opera omnia,* vol. 2, p. 853.

104.   Ibid., p. 855.

105.   Sambursky, *Physics of the Stoics,* p. 67.

Epicurean atomism. By the same token, he clearly rejected the naturalistic Aristotelianism of Pomponazzi, the Stoic cosmological underpinnings of astrology, and the materialism of Hobbes. Gassendi's position on all these issues can be understood as reflecting his underlying theological assumptions, which informed his philosophy of nature at every level.

4

Theology, metaphysics, and epistemology: Gassendi's "science of appearances"

> Nothing would be more beautiful or more desirable for us than to know fully the things that nature has kept in her depths or her farthest recesses; but although we may wish for that, we are being just as absurd as when we yearn to fly like the birds or to stay young forever.
>
> *Pierre Gassendi, Syntagma philosophicum*[1]

Despite the fact that Gassendi's thought developed in a variety of ways during his lifetime, his fundamental principles – a voluntarist theology, an empiricist epistemology, a nominalist or conceptualist account of universals, and an anti-essentialist metaphysics – remained constant throughout his writings. In making this statement, I disagree with Bloch's contention that Gassendi's philosophy does not constitute a system. Bloch regards the *Syntagma philosophicum* as "an assemblage of historical, scientific, and philosophical discussion."[2] He also denies that theology played any important role in Gassendi's thought, claiming instead that it was introduced only after 1641 to dissimulate the materialism of which Gassendi was growing increasingly aware.[3] As I have argued in Chapters 2 and 3, whatever Gassendi's state of belief, the theological views upon which he drew are voluntarist and are reflected – directly and indirectly – throughout his work. Despite changes in his reactions to skepticism, despite his growing interest in natural philosophy, and despite his increasing focus on his Epicurean project, the same theological and philosophical views remained central to his thinking from the early *Exercitationes paradoxicæ adversus Aristoteleos* (1624) to the posthumously published *Syntagma philosoph-*

---

1. Pierre Gassendi, *Syntagma philosophicum*, in Gassendi, *Opera omnia*, 6 vols. (Lyon, 1658; facsimile reprint, Stuttgart-Bad Canstatt: Friedrich Frommann Verlag, 1964), vol. 1, p. 79, in *The Selected Works of Pierre Gassendi*, translated by Craig B. Brush (New York: Johnson Reprint, 1972), pp. 326–7.
2. Olivier René Bloch, *La philosophie de Gassendi: Nominalisme, matérialisme, et métaphysique* (The Hague: Martinus Nijhoff, 1971), p. 3.
3. Ibid., chaps. 9–11.

icum (1658). They guided him in his project to replace the philosophy of Aristotle with a Christianized Epicureanism.

Gassendi's first published work, the *Exercitationes paradoxicæ adversus Aristoteleos,* was an outgrowth of his lectures on Aristotelian philosophy during his previous six years of teaching at the University of Aix.[4] Influenced by his reading of the works of Spanish humanist Juan Luis Vives (1492–1540), the anti-Aristotelian philosopher Gianfrancesco Pico della Mirandola (1469–1533),[5] the logician and educational reformer Peter Ramus (1515–1572), and Montaigne's disciple in skepticism Pierre Charron (1541–1603), Gassendi became very critical – not to say impatient – with Aristotelian philosophy:

> And that is why, having been charged to teach Philosophy and especially that of Aristotle during six whole years at the Academy at Aix, in truth I always made sure that my auditors could defend Aristotle perfectly; but nevertheless, in the form of appendices, I also presented them the principles by which the teachings of Aristotle could be completely destroyed.[6]

Although he published only Book I of the *Exercitationes* in 1624 and then suppressed Book II in 1625 until it finally saw the light of day in 1649, he had originally planned to write seven books, systematically refuting every aspect of Aristotelian philosophy.[7] The extant Books I and II indicate just how thorough he intended his destruction of the Aristotelian

---

4. Pierre Gassendi, *Dissertations en forme de paradoxes contres les aristotéliciens* (*Exercitationes paradoxicæ adversus Aristoteleos*), bks. I and II, translated into French by Bernard Rochot (Paris: J. Vrin, 1959), pp. 6–7; in Gassendi, *Opera omnia,* vol. 3, pp. 99–100.

5. Although Rochot identifies Gassendi's reference to "Mirandulanus" in the Preface of the *Exercitationes* as Gianfrancesco's more famous uncle, the Renaissance humanist and cabbalist Giovanni Pico della Mirandola (1463–94), Schmitt makes a compelling argument that the reference is to the nephew Gianfrancesco, who had made an "extensive critique of the philosophy of Aristotle using a sceptical approach" in his *Examen vanitatis doctrinae gentium* (1520). Basing his argument on a close comparison of Pico's *Examen vanitatis* and Gassendi's *Exercitationes,* Schmitt demonstrates a resemblance so close that it is convincing of filiation. See Gassendi, *Exercitationes,* pp. 6–7 n11 (in *Opera omnia,* vol. 3, p. 100); and Charles B. Schmitt, *Gianfrancesco Pico della Mirandola (1469–1533) and His Critique of Aristotle* (The Hague: Martinus Nijhoff, 1967), pp. 175–8. Gregory also notes the relationship between the two texts. See Tullio Gregory, *Scetticismo ed empirismo: Studio su Gassendi* (Bari: Laterza, 1961), pp. 24–5, 33, 40–1.

6. Gassendi, *Exercitationes,* pp. 6-7 (*Opera omnia,* vol. 3, pp. 99–100).

7. Ibid., pp. 12–15 (pp. 102–3). On Gassendi's reasons for suppressing further publication of this work, see Joy, *Gassendi the Atomist,* pp. 33–8.

philosophy to be. In Book I, "Against the Whole Doctrine of the Aristotelians," Gassendi argued against the very style of Aristotelian philosophy. Considering each of Aristotle's major works in turn, he demonstrated them to be full of "omissions, superfluities, errors, and contradictions."[8] In Book II, Gassendi scrutinized the dialectic of Aristotle, showing it to be "neither necessary nor useful."[9] Arguing at length against the Aristotelian understanding of universals, categories, and propositions, he concluded this section by using skeptical arguments to prove that scientific demonstration of the kind Aristotle had described in the *Posterior Analytics* is impossible.[10]

Gassendi not only found Aristotelian philosophy to be full of contradictions and empty of meaning, he also found it useless as a method for natural philosophy. He repeatedly criticized Aristotelianism for providing no insight into the structure or function of things in the natural world. Mocking Aristotelian science, which was based on the concepts of matter, form, and privation, he asked what those concepts taught about the real world known by observation – things like the minute organs of the mite, which the recently invented microscope revealed.[11] As his correspondence with Peiresc and other natural philosophers demonstrates, Gassendi was deeply interested in the science of his day. The ability of a philosophy to describe a method suitable for the pursuit of that science was of critical importance to him.

In his sketch of projected but unwritten books of the *Exercitationes*, Gassendi described his intention to replace Aristotelianism with a philosophy more amenable to the new science. The Aristotelian physics of forms and natural motion would be replaced with a physics incorporating the void and a non-Aristotelian definition of time. He would attack Aristotle's theory on simple corporeal substances, his theory of the elements, his theory of mixed bodies, and his psychology. He would argue against Aristotelian metaphysics, affirming that "it is only the orthodox Faith to which one attributes all knowledge that one has of purely intelligent beings and of the thrice powerful God, while the vanity of arguments by means of which one is accustomed to philosophize by natural light about separate substances is demonstrated." Finally, he intended to replace Aristotelian ethics with "the Epicurean doctrine of pleasure." Recognizing the monolithic character of Aristotelianism, Gassendi concluded his summary of intentions by stating that it is not necessary to refute every detail of

8. Gassendi, *Exercitationes*, pp. 12–13 (*Opera omnia*, vol. 3, p. 102).
9. Ibid.
10. Ibid., pp. 234–519 (pp. 149–210).
11. Ibid., pp. 488–91 (pp. 203–4).

Aristotelian philosophy, because once the foundations are removed, the whole structure will crumble.[12] Although Gassendi never completed the *Exercitationes* as outlined in 1624, the Epicurean project on which he embarked full tilt by the late 1620s ultimately fulfilled many of these goals.

In the posthumous *Syntagma philosophicum,* he worked out a complete philosophy, a Christianized Epicureanism, to replace the logic, physics, and ethics of Aristotle. Among the continuities between the *Exercitationes* and the *Syntagma philosophicum* are his epistemology and his metaphysics, both of which are closely tied to his voluntarist theology. In the *Exercitationes,* Gassendi used the methods of the skeptics, especially the arguments of Sextus Empiricus, to prove that science in the Aristotelian sense is impossible. In the course of this demonstration, Gassendi enunciated a new definition of "science," laid the foundations of his empiricist epistemology, denied the independent existence of universals, and articulated his antiessentialist metaphysics. According to Gassendi, the Aristotelians had regarded science as consisting of certain a0d evident knowledge, obtained by means of syllogistic demonstrations about necessary causes. He argued that syllogisms alone do not generate knowledge of the world. They yield true conclusions only if their premises are true. The premises, then, must be known on some independent basis. Aristotle himself had maintained that the principles on which demonstration is based must be tested against "sensation, . . . a kind of tribunal before which one makes appeal."[13]

Gassendi questioned whether sensory knowledge could serve as the basis of demonstrable, certain science. "I want only to observe that Aristotelian demonstration, being based on the senses and the senses being very deceptive and uncertain, how much certainty can there be in the demonstration and in science?"[14] Gassendi's skeptical critique of the senses in the *Exercitationes* followed Sextus Empiricus quite closely.[15] Not satisfied with the suspension of judgment advocated by the ancient skeptics. Gassendi sought a middle way, which Popkin has happily called "mitigated skepticism":[16]

12. Ibid., pp. 12–15 (p. 102).
13. Ibid., pp. 388–9 (p. 182).
14. Ibid.
15. See Sextus Empiricus, *Outlines of Pyrrhonism,* bk. I, chap. XIV, translated by R. G. Bury, 4 vols. (Cambridge, Mass.: Harvard University Press, 1976), vol. 1, pp. 25–94. See Gassendi, *Exercitationes,* pp. 388–93 (*Opera omnia,* vol. 3, pp. 182–3).
16. For the revival of the skeptical texts and Gassendi's use of the skeptical

We would do best to holdosome middle way between the Skeptics . . . and the dogmatics. For the dogmatics do not really know everything they believe they know, nor do they have the appropriate criteria to determine it; but neither does everything that the Skeptics turn into the subject of debate seem to be so completely unknown that no criteria can be found for determining it.[17]

In contrast to Descartes, who responded to skepticism by seeking a completely reliable criterion of true knowledge, Gassendi, like Mersenne, changed the criterion of knowledge itself. Mersenne agreed with the skeptics that there is no knowledge of which we can be certain, but he believed that the sciences provide useful knowledge that psychologically we cannot doubt. He devoted most of *La vérité des sciences* to examples of such knowledge in the sciences.[18] He acknowledged that this was not a philosophical solution to the skeptical crisis, but thought it constituted a pragmatic resolution.[19] Gassendi shared Mersenne's constructive approach to the problem. Accepting the force of the skeptical arguments but not content with skeptical conclusions, he redefined the epistemic goal of science so that certainty is no longer its necessary characteristic. Instead, "knowledge," for Gassendi, consists of probable statements based on our experience of the phenomena. Probability is the most we can attain, according to his phenomenalist epistemology.[20] He did not consider accepting probability to be a terrible compromise. Rather, it is an acceptance of our own limitations.

arguments see Richard H. Popkin, *The History of Scepticism from Erasmus to Spinoza* (Berkeley: University of California Press, 1979), pp. 18–41, 101–9, 129–50; and Charles B. Schmitt, *Cicero Scepticus: A Study of the Influence of the* Academica *in the Renaissance* (The Hague: Martinus Nijhoff, 1972). While Popkin emphasizes the influence of Sextus Empiricus and Pyrrhonian skepticism, Schmitt emphasizes the influence of Cicero and Academic skepticism in the Renaissance. See also Henri Berr, *Du scepticisme de Gassendi*, translated by Bernard Rochot (Paris: Albin Michel, 1960; first published in Latin in 1898); Gregory, *Scetticismo ed empirismo;* and Robert Walker, "Gassendi and Skepticism," in *The Skeptical Tradition*, edited by Miles Burnyeat (Berkeley and Los Angeles: University of California Press, 1983), pp. 319–36.

17. Gassendi, *Syntagma philosophicum*, in *Opera omnia*, vol. 1, p. 79 (translated by Brush, *Selected Works of Pierre Gassendi*, pp. 326–7).

18. Richard H. Popkin, "Preface," in Henry van Leeuwen, *The Problem of Certainty in English Thought, 1630–1680* (The Hague: Martinus Nijhoff, 1963), pp. vii–viii; and Popkin, *History of Scepticism*, pp. 129–41.

19. Popkin, *History of Scepticism*, p. 139.

20. Ibid. Bloch, *La philosophie de Gassendi*, p. 26.

In settling for probability rather than certainty as the epistemic goal of natural philosophy, Gassendi was rejecting the traditional Aristotelian and Scholastic conception of *scientia* or demonstrative knowledge. "Probability," in Scholastic discourse, is an attribute of *opinio*, "beliefs or doctrines not gotten by demonstration. It may also cover propositions which, not being universal, cannot (according to Aquinas) be demonstrated."[21] Gassendi was not alluding to the mathematical concept of probability, which was apparently not articulated until the decade around 1660.[22] He used the term in much the same sense as the seventeenth-century English intellectuals who concerned themselves with degrees of belief and evidence.[23]

On what epistemological foundation did Gassendi think such probable knowledge could be justified? Despite the cogency of the skeptical critiques of the senses, Gassendi somewhat paradoxically thought that sensation provides a reliable basis for knowledge. Gassendi's theory of knowledge began from a classic statement of empiricism. "All the ideas that are contained in the mind derive their origin from the senses. . . . The intellect or mind is a *tabula rasa* in which nothing is engraved [prior to

---

21. Ian Hacking, *The Emergence of Probability: A Philosophical Study o Early Ideas about Probability, Induction and Statistical Inference* (Cambridge University Press, 1975), pp. 21–2.
22. Ibid., p. 11. On the development of the mathematical theory of probability, see Lorraine Daston, *Classical Probability in the Enlightenment* (Princeton, N.J.: Princeton University Press, 1988), chap. 1.
23. "As the natural scientists became more empirical and more concerned with matters of fact, and as those in other fields became more sensitive to issues relating to evidence and proof, knowledge in all fact-related fields was seen to fall along a continuum. The lower reaches of this continuum were characterized as 'fiction,' 'mere opinion,' and 'conjecture'; its middle and high ranges as 'probable' and 'highly probable'; and its apex as 'morally certain'. 'Knowledge' was no longer reserved for the logically demonstrable products of mathematical and syllogistic 'science'. The morally certain was also a form of knowledge, and the highly probable came close to being another." Barbara J. Shapiro, *Probability and Certainty in Seventeenth-Century England: A Study of the Relationships between Natural Science, Religion, History, Law, and Literature* (Princeton, N.J.: Princeton University Press, 1983), p. 4. Franklin argues that probabilistic reasoning also had roots in legal thought. See James Franklin, "The Ancient Legal Sources of Seventeenth-Century Probability," in *The Uses of Antiquity: The Scientific Revolution and the Classical Tradition*, edited by Stephen Gaukroger (Dordrecht: Kluwer, 1991), pp. 123–44.

sensation]."[24] Ideas may come into the mind directly from sensation, or they may be the product of the action of the mind on those directly received. The mind forms this second kind of idea by the processes of conjoining, enlarging, diminishing, transferring, adapting, analogizing, and comparing.[25] However complex the process by which the mind transforms the ideas coming directly from sensation, the fact remains that sensateon – directly or indirectly – is the only source of ideas in the mind.

Gassendi further argued that, understood properly, sense never fails. That is to say, the content of a sensation, considered in itself without reference to anything else, is what it is; and in that sense it is free from error. Drawing on Epicurus, Gassendi distinguished between "truths of existence" and "truths of judgment":[26]

> Two sorts of truth can be appropriately distinguished; the one called truths of 'existence' or of 'being' and the other truths of 'judgment' or 'statement'. Now in the first place, although anything regarded in itself is just what it actually is, and nothing more, yet for the sake of a somewhat greater explanation it is customary to apply the attribute 'true' to it. And so we habitually say that a thing exists truly, or is according to itself, for example 'true' gold, a 'true' man, and the like, although it cannot be called 'false' in the same sense, since it is always something true according to itself – even fool's gold is not false gold but true fool's gold, and a painting of a man is not a false man but a true image of a man.[27]

24. Gassendi, *Syntagma philosophicum*, in *Opera omnia*, vol. 1, p. 92. For the history of earlier views of this statement of the empiricist credo, see Paul F. Cranefield, "On the Origin of the Phrase '*Nihil est in intellectu quod non prius fuerit in sensu*,'" *Journal of the History of Medicine*, 1970, 25: 77–80.
25. Gassendi, *Syntagma philosophicum*, in *Opera omnia*, vol. 1, p. 93.
26. For Epicurus' account of sensation, see J. M. Rist, *Epicurus: An Introduction* (Cambridge University Press, 1972), pp. 17–25; and Elizabeth Asmis, *Epicurus' Scientific Method* (Ithaca, N.Y.: Cornell University Press, 1984), pp. 141–59. For the relationship between Gassendi's theory and that of Epicurus, see Antonina Alberti, "La Canonica di Epicuro nell' interpretazione di Gassendi," *Annali dell' Instituto di filosofia dell' Università di Firenze*, 2 (1980): 151–94. For the influence of Gassendi's theory on Locke and Berkeley, see Thomas M. Lennon, "The Epicurean New Way of Ideas: Gassendi, Locke, and Berkeley," in *Atoms, Pneuma, and Tranquillity: Epicurean and Stoic Themes in European Thought*, edited by Margaret J. Osler (Cambridge University Press, 1991), pp. 259–71. For a more comprehensive discussion of his influence on philosophy, see Thomas M. Lennon, *The Battle of the Gods and Giants: The Legacies of Descartes and Gassendi, 1655–1715* (Princeton, N.J.: Princeton University Press, 1993).
27. Gassendi, *Syntagma philosophicum*, in *Opera omnia*, vol 1, p. 67 (translated by Brush, *Selected Works of Gassendi*, p. 286).

"True," in this first sense, refers to the genuineness of things, not to the truth or falsity of propositions.

The second kind of truth, "truth of judgment," applies to propositions. Truth and falsity, in this sense, apply to judgments about the external referents of our sensations:

> Secondly, . . . there, is a certain truth which consists in the conformity of the judgment and statement with the thing judged and reported in the statement; and it is this truth for which there is in fact a falsehood opposed to it, consisting obviously in the discrepancy between the judgment and statement and the thing judged and reported in the statement.[28]

It is only in this second sense of "truth" that it makes sense to speak about error. Error arises when we make mistaken judgments about the referents of our sensations.

It is on the basis of this distinction between "truths of being," on the one hand, and "truths of judgment," on the other, that Gassendi claimed the infallibility of the senses:

> It is not the senses themselves but the intellect which makes the error; and when it makes a mistake, it is not the fault of the senses but of the intellect whose responsibility it is as the higher and dominant faculty before it pronounced what a thing is like to inquire which of the different appearances produced in the senses (each one of them is the result of a necessity that produces them as they are) is in conformity with the thing.[29]

Not only the infallibility of the senses, but also an answer to the skeptics followed from Gassendi's distinction between "truths of being" and "truths of judgment":

> If this advice were duly observed, many things would become more certain and indubitable about which it would be possible to offer a true judgment: for instance, the tower would be square when you came close to it, the stick would be really straight when taken out of the water and held entirely in the air, and things like that.[30]

Skeptical arguments about round towers, bent oars, and the varying experiences of different individuals and kinds of animals in different circumstances are meaningful when applied to the judgments we make on the basis of our sensations.[31] They do not apply to sensations taken in themselves. These sensations, which Gassendi called the "appearances," provide the basis for our knowledge of the world, a knowledge that cannot penetrate to the inner natures of things precisely because it is knowledge

28. Ibid., p. 68 (p. 287).
29. Ibid., p. 85 (pp. 345–6).
30. Ibid.
31. Gassendi, *Exercitationes*, pp. 486–7 (*Opera omnia*, vol. 3, p. 203).

of how they appear to us. Gassendi's middle way led him to maintain that even if we cannot have science in the Aristotelian sense of demonstrative knowledge about real essences, we can achieve a science of appearances. Given his view that sense itself cannot fail – that is to say, that we have reliable knowledge of the appearances, if not of the intimate natures of things – he claimed that "the conditions for science exist, but always an experimental science . . . based on appearances. . . . [A]ll that we are denying is that one can penetrate to the intimate natures of things."[32] On the basis of the appearances, however, it is possible to seek causal explanations, with the understanding that such reasoning is always conjectural, to be judged by how well it explains other effects too.[33] This science of appearances can never achieve certainty.[34] It can, however, attain a measure of probability that is not an unhappy compromise:

> As it is certain that probability is neighbor enough of truth, the danger of error . . . is the same when, in seeking the truth you turn away from probability as it is for him who, on his way from Paris to Holland, takes the road which leads to Marseilles.[35]

Gassendi thus redefined the goal of natural philosophy, replacing the traditional search for demonstrative knowledge of real essences with probable knowledge of the appearances, departing not only from Aristotle and the Scholastics, but also from two of his contemporaries, Bacon and Descartes, both of whom continued to maintain that certainty is the attainable goal of scientific inquiry.[36] This change in the ideal of knowledge

32. Ibid., pp. 504–5 (p. 207).
33. Gassendi, *Syntagma philosophicum,* in *Opera omnia,* vol. 1, p. 207.
34. Gassendi, *Exercitationes,* pp. 498–501 (*Opera omnia,* vol. 3, p. 206).
35. Gassendi, *Disquisitio metaphysica seu dubitationes et instantiae adversus Renati Cartesii metaphysicam et responsa,* edited and translated into French by Bernard Rochot (Paris: J. Vrin, 1962), pp. 54–5 (in *Opera omnia,* vol. 3, p. 283).
36. Bacon wrote in the *New Organon,* "Now what the sciences stand in need of is a form of induction which shall analyse experience and take it to pieces, and by a due process of exclusion lead to *an inevitable conclusion." The Works of Francis Bacon, Baron of Verulam, Viscount St. Alban, and Lord High Chancellor of England,* collected and edited by James Spedding, Robert Leslie Ellis, and Douglas Denon Heath, 14 vols. (London, 1857–74), vol. 4, p. 25; my italics. Like Descartes, Bacon based his epistemology of certitude on an essentialist metaphysics: "Whosoever is acquainted with Forms, embraces the unity of nature in substances the most unlike; and is able therefore to detect and bring to light things never yet done, and such as neither the vicissitudes of nature, nor industry in experimenting, nor accident itself, would ever have brought into act, and which would never have occurred in the thought of man. From the discovery of Forms therefore

was a major step in the history of philosophy, influencing the tradition that later came to be known as British Empiricism.[37]

Gassendi applied his empiricism to all areas of investigation. He adopted an approach to natural philosophy that, while fully acknowledging the probabilistic status of scientific knowledge, described empirical methods for ferreting out information about hidden things. In the "Proemium" to the "Physics," he wrote, "Physics cannot contemplate how things are [in themselves] except insofar as it can observe them carefully and thus uncover them."[38] Although we must settle for a science of appearances, ways exist to discover less evident facts about the world, such as reasoning on the basis of signs, appearances that indicate the existence of other things not directly observed. For example, smoke is a sign of fire, and lactation is a sign of pregnancy. "[I]f the truth in question is hidden, lying concealed beneath appearances; we must then inquire, since its nature is not open to us, whether we have a criterion by which we may recognize the sign and judge what the thing really is."[39] Gassendi argued that "we may distinguish two criteria in ourselves: one by which we perceive the sign, namely the senses, and the second by which we understand something hidden by means of reasoning, namely the mind, intellect, or reason."[40] The senses are limited, and judgment is fallible; so reasoning from signs may not be entirely reliable. "Still reason, which is superior to the senses, can correct the perception of the senses so that it will not accept a sign from the senses unless it has been corrected and then

results truth in speculation and freedom in operation" (p. 120). On Descartes' rather complex conception of certainty, see Margaret Morrison, "Hypotheses and Certainty in Cartesian Science," in *An Intimate Relation: Studies in History and Philosophy of Science Presented to Robert E. Butts,* edited by James R. Brown and Jürgen Mittelstrass (Dordrecht: Kluwer, 1989), pp. 43–64.

37. See Margaret J. Osler, "John Locke and the Changing Ideal of Scientific Knowledge," *Journal of the History of Ideas,* 31 (1970): 1–16; also David Fate Norton, "The Myth of 'British Empiricism,'" *American Philosophical Quarterly,* 1 (1981): 331–44. See also Gaston Coirault, "Gassendi et non Locke créateur de la doctrine sensualiste moderne sur la génération des idées," in *Actes du Congrès du tricentenaire de Pierre Gassendi (1655–1955),* edited by Comité du Tricentenaire de Gassendi (Digne: CNRS, 1955), pp. 69–94; Richard W. F. Kroll, "The Question of Locke's Relation to Gassendi," *Journal of the History of Ideas,* 45 (1984): 339–60; Lennon, *The Battle of the Gods and Giants;* and Lennon, "The Epicurean Way of Ideas."
38. Gassendi, *Syntagma philosophicum,* in *Opera omnia,* vol. 1, p. 126.
39. Ibid., p. 80 (translated by Brush, *Selected Works of Gassendi,* p. 329).
40. Ibid., p. 81 (p. 333).

at last it deliberates, or reaches a judgment of the thing."[41] It is by reasoning upon observable signs that Gassendi thought his science of appearances could be constructed. For example, it is possible to argue that there are unobservable pores in the skin on the basis of the observation of sweating. The inference to pores rests, inter alia, on the assumption that it is impossible for two bodies to occupy the same place at the same time. This assumption itself, Gassendi thought, is subject to proof on the basis of experience and observation.[42]

He noted that our knowledge of observable signs and thus our knowledge of the world can be improved with new technologies of observation such as the microscope:

> With the passage of time helpful appliances are being found that will make them visible to the senses. For example, take the little animal the mite, which is born under the skin; the senses perceive it as a certain unitary little point without parts; but since, however the senses saw that it moved by itself, reason has deduced from this motion as from a perceptible sign that this little body was an animal and because its forward motion was somewhat like a turtle's, reason added that it must get about by the use of certain tiny legs and feet. And although this truth would have been hidden to the senses, which never perceived these limbs, the microscope was recently invented by which sight could perceive that matters were actually as predicted.[43]

In this example, signs can be observed only with the aid of an instrument that enhances the powers of vision. But the reasoning from observed signs to knowledge that the mite has limbs just like larger animals is epistemologically on a par with reasoning from signs about macroscopic things. Similarly, telescopic observations have confirmed Democritus' speculation that the Milky Way is composed of individual stars.[44] Although these advances in scientific instrumentation have improved our ability to confirm the conjectures to which sensible signs lead us, they are all examples of reasoning upon observable signs and do not change the fact that, at bottom, natural philosophy consists of probabilistic reasoning about the appearances. Experiment, which Gassendi described as "experience weighed in the ballance of Reason,"[45] was the method he recommended

41. Ibid.
42. Ibid., p. 85 (pp. 346–7).
43. Ibid., p. 82 (p. 334).
44. Ibid.
45. Pierre Gassendi, *The Vanity of Judiciary Astrology. Or Divination by the Stars. Lately written in Latin, by that Great Schollar and Mathematician, the Illustrious PETRUS GASSENDUS; Mathematical Professor to the King of*

for justifying the explanations of the sensible signs. Gassendi thought that the entire edifice of natural philosophy could be constructed in this way on the basis of empirical knowledge. Such a natural philosophy is not without utility, even if it fails to achieve the penetration and certainty sought by the Aristotelians.

In addition to his empiricist and probabilist account of knowledge, Gassendi denied independent existence to both universals and essences:[46]

> These famous universals are nothing other than what Grammarians call appellative nouns, for example, "man" or "horse," for which each is attributed to several objects, the same as individuals are nothing other than proper nouns, such as "Plato," "Bucephalus," and all those that are only given to simple things.[47]

He not only articulated a nominalist position – in the course of writing the *Exercitationes* he mentioned such nominalists as Gregory of Rimini, the Scotists, and Pierre d'Ailly – he also clearly identified himself as sharing that position: "What? Do you say, do you admit this foolish opinion of the Nominalists who do not recognize any other universality than that of concepts or names? It is thus; I admit it, but in pretending to admit only a perfectly reasonable opinion."[48] At this point Gassendi's understanding of universals seems to be sliding from an unambiguous nominalism, which equates universals with names, into a conceptualism, which identifies universals with concepts.[49] According to both nominalism and conceptualism, only particulars exist in the world.

Whether conceptualist or nominalist, Gassendi was consistent in denying the ontological independence of universals. "God is the most singular of beings, and all his creatures are particulars: this angel, this man, this Sun, this rock, finally nothing can be met which is not this particular

France, translated into English by a Person of Quality (London: Humphrey Moseley, 1659), p. 130.

46. Gassendi's views are summarized by Bloch, *La philosophie de Gassendi,* pp. 113–17.

47. Gassendi, *Excercitationes,* pp. 280–1 (in *Opera omnia,* vol. 3, p. 159.

48. Ibid.

49. A. D. Woozley, "Universals," in *The Encyclopedia of Philosophy,* edited by Paul Edwards, 8 vols. (New York: Macmillan and Free Press, 1967), vol. 8, p. 199. On various philosophical approaches to the problem of universals, see D. M. Armstrong, *Nominalism and Realism: Universals and Scientific Realism,* 2 vols. (Cambridge University Press, 1978), vol. 1, pp. 11–87. See also David H. Degrood, *Philosophies of Essence: An Examination of the Category of Essence,* 2d edition (Amsterdam: B. R. Grüner, 1976).

thing."[50] Only particulars exist. In the *Syntagma philosophicum,* he enunciated the same view, further explaining how the mind creates universal concepts by the processes of generalization and abstraction from singular ideas of sense:

> Every idea that is transmitted through sensation is singular; it is the mind, however, that makes general ideas from the combination of singular ones, for as with all things that are in the world and can impinge on the senses, they are singular, as Socrates, Bucephalus, this stone, that plant, and the other things demonstrable by pointing.[51]

Since only particulars exist in the world, no forms, whether Platonic or Aristotelian, have real existence outside of the mind. If universals did have independent existence, that fact would raise unanswerable questions: "In which corner of the universe do you wish the residence of universals to be seen? You say, for example, that there exists a human nature that is universal. But where is this universal nature seen?"[52]

Universal natures do not exist. The concept of such natures does not even make sense. Rather, each individual possesses its own nature. Corporeal objects each have their own atomic structure. Individual human beings each have their particular body united with their particular soul. "For me, I see the human nature of Plato, that of Socrates, but these are all singular natures."[53] Beyond these singular natures, what else is there that would be a universal nature? Where could it be seen?

> If you are better than a Lynx, where would you say you see another nature that is universal? . . . You have your body, your Soul, your organs, and your own qualities; and I too have all those of my own. What would be this natural substance that would be at the same time in me and in you?[54]

---

50. Gassendi, *Exercitationes,* pp. 280–1 (*Opera omnia,* vol. 3, p. 159).
51. Gassendi, *Syntagma philosophicum,* in *Opera omnia,* vol. 1, p. 93. Gassendi gave a standard empiricist account of the formation of concepts of universals. Ibid., pp. 93–5.
52. Gassendi, *Exercitationes,* pp. 280–1 (*Opera omnia,* vol. 3, p. 159).
53. Ibid.
54. Ibid. In fact, the Scholastics had a far more complex idea of universal than Gassendi discusses: "St. Thomas thus admits (i) the *universale ante rem,* while insisting that it is not a subsistent thing, either apart from things (Plato) or in things (early mediaval ultra-realists), for it is God considered as perceiving His Essence as imitable *ad extra* in a certain type of creature; (ii) the *universale in re,* which is the concrete individual essence alike in the members of the species; and (iii) the *universale post rem,* which is the abstract universal concept." Frederick Copleston, *A History of Philosophy,* 9 vols. (Garden City, N.Y.: Image Books, 1962; first published 1950), vol. 2, pt. 1, p. 176.

Discourse about "individual natures" from a philosopher who has just denied the existence of universal natures may seem puzzling until the fact is recognized that Gassendi tacitly redefined the concept "nature." For Thomas Aquinas and the other Aristotelians, universal natures are internal to things, necessarily causing them to possess essential properties and to behave in natural ways. Aristotelian science appealed to these universal natures as ultimate terms of explanation. It is these universal natures to which Gassendi objected so vigorously. When speaking of individual natures, he spoke as a mechanical philosopher and was thinking of the physical structures of individual objects, structures composed of atoms. The existence of both the individual and its nature is contingent. Each individual stands alone, ontologically distinct from every other individual.[55] Another way of making this point is to say that no reified laws of nature exist in Gassendi's world. There may be groups of similar individuals that behave similarly because of their similar individual natures.[56] But there do not exist additional laws or relations that determine the behavior of these individuals in any necessary way. What really is at stake here is what counts as an explanation. For Gassendi, the observed properties of an individual are explained by its own, particular atomic structure, not by the possession of some universal form or essence. It was the identification of universals and essences that gave Aristotelian science the basis for demonstrative knowledge of the natures of things. Gassendi's nominalism undermined the foundations of Aristotelian essentialism.

Gassendi's nominalism, like Ockham's, was closely tied to his voluntarism. In Gassendi's day, realism about universals took one of two traditional forms, Platonic or Aristotelian. According to the Platonic view, universals exist apart from the particulars that instantiate them. Two presuppositions of Platonism conflict with the voluntarists' insistence on God's absolute freedom: (1) that the universals are permanent and real; and (2) that the ideal reality of these universals can be comprehended without appeal to observation and experience.[57] The permanent existence of universals in the Platonic sense would imply that they exist independent of God's creative act. The possibility of knowing them by purely rational means would imply that a necessary relation exists between the human understanding and the ideal world of universals, a necessity that impedes

---

55. On this point, see Messeri, *Causa e spiegazione*, p. 20.
56. One of the standard objections to nominalism springs to mind here: If universals do not exist, by what criteria do we judge the similarity of two things? Gassendi neither discussed this objection nor suggested a way to respond to it.
57. Degrood, *Philosophies of Essence*, p. 5.

God in the free exercise of his will. For both of these reasons, a Platonic understanding of universals is totally unacceptable to voluntarist theology.

The Aristotelian theory of universals is likewise incompatible with voluntarist principles. Aristotle's theory of universals is connected to his theory of essences or forms. Universals are affirmations that are "predicated of many things."[58] Aristotle's theory of predication is based on the ontology of forms.[59] Always correlative to matter, forms provide the underpinning for his method of reasoning from essences and causes, a method that he thought ensured that scientific knowledge is necessary knowledge.[60] Once again, the necessary relations between universals and the properties of the substances that possess them and the necessary relations between universals and the causal knowledge of nature that we can possess with certainty are unacceptable to a voluntarist understanding of divine power.

Rejecting the reality of universals, Gassendi described a world consisting of particulars that are contingent on the free exercise of divine will. His empiricist epistemology was inseparable from his nominalism. Both forms of realism – Platonic and Aristotelian – contradicted his epistemological and theological views. On the Platonic account, knowledge of universals was thought to be purely rational, in direct contradiction with Gassendi's phenomenalism. On Aristotelian assumptions, knowledge of universals is knowledge of forms – knowledge of "intimate natures" – which Gassendi's science of appearances ruled out. A priori knowledge of any kind about the world would entail the existence of some kind of necessity: It would be necessary, at the very least, that the world correspond to our a priori conceptions. Gassendi's nominalism had already eliminated this kind of necessity by insisting that God and all his creatures are particulars and that there can be no necessary relations between them.

For Gassendi, scientific knowledge was confined to the phenomenal level. We cannot know with certainty the inner mechanisms by which the observed phenomena are produced on the basis of the appearances (although, given Gassendi's extensive arguments in the *Syntagma philosoph-*

58. Aristotle, *De interpretatione,* translated by J. L Ackrill, in *The Complete Works of Aristotle,* edited by Jonathan Barnes, 2 vols. (Princeton, N.J.: Princeton University Press, 1984), 17a37–40, vol. 1, p. 27.

59. G. E. R. Lloyd, *Aristotle: The Growth and Structure of His Thought* (Cambridge University Press, 1968), p. 114; W. D. Ross, *Aristotle: A Complete Exposition of His Works and Thought,* 5th edition (New York: Meridian, 1959), pp. 27–30.

60. Degrood, *Philosophies of Essence,* pp. 17–18.

icum and elsewhere, we can be sure that these mechanisms consist of atoms colliding in void space). The mechanisms that produce the appearances are ultimately unknowable. "What we know of things does not pertain, in any case, to their intimate nature; for the knowledge we have of it is not more necessary than that of accidents."[61] This limitation to our knowledge is not merely a practical limitation resulting from the fact that the appearances that impinge on our senses necessarily come from the external parts of bodies. Rather,

> it is thus that the all-Powerful Good Lord established the creation and has left it to our use. It is in fact all that it is necessary for us to know . . . , he has revealed it to us, giving properties to things, permitting us to recognize them and diverse senses by which we may apprehend them and an interior faculty permitting us to judge them. But for that which is the interior nature, . . . as that is the thing which is not necessary to know, he wished that it would be hidden from us; and when we wish presumptuously to give an air of knowing it, we bear the pain of our fault in measure.[62]

To the extent that we seek to know the secrets of nature – and Gassendi devoted hundreds of pages in his *Syntagma philosophicum* to unearthing those secrets – the results of our research must be regarded as provisional and probable, never as fully established, certain truths. We can know no necessary connections between individual natures and the observed qualities of things of the kind that Aristotle sought.

Gassendi's theory of scientific knowledge can be understood as his response to the skeptical crisis, tempered by his humanist use of Epicurean philosophy and the antiessentialism that was closely linked to his voluntarist theology. The empiricist epistemology that precipitated out of Gassendi's learned mix characterized his whole approach to philosophy and natural philosophy. Empirical methods, conjectural reasoning, and probable conclusions marked his philosophy of science. In all these respects he differed from his contemporary Descartes, who continued to search for certain knowledge of the real essences of things.

61. Gassendi, *Disquisitio Metaphysica*, pp. 186–9 (*Opera omnia*, vol. 3, p. 312).
62. Ibid.

5

Eternal truths and the laws of nature: The theological foundations of Descartes' philosophy of nature

All philosophy is like a tree, of which the roots are metaphysics, the trunk is physics, and the branches coming from this trunk are all the other sciences.

René Descartes, Principes de la philosophie[1]

Like Gassendi, Descartes belonged to the early-seventeenth-century community of natural philosophers committed to Galilean science and the mechanical philosophy. Educated by the Jesuits at La Flèche – where Mersenne was also a student from 1604 to about 1609[2] – Descartes qualified as a lawyer at the University of Poitiers in 1616, after which he served as an officer in the army of Prince Maurice of the Netherlands and then spent several years traveling in Europe.[3] He had a significant encounter with Isaac Beeckman in 1619 and, like Gassendi, found contact with the Dutch schoolmaster extremely stimulating.[4] Shortly after meeting Beeckman, Descartes dedicated himself to the pursuit of natural philosophy. He settled permanently in the Netherlands in 1629, remaining there

1. René Descartes, *Principes de la philosophie,* in *Oeuvres de Descartes,* edited by Charles Adam and Paul Tannery (hereafter AT), 11 vols. (Paris: J. Vrin, 1897–1983), vol. 9-2, p. 14.
2. Peter Dear, *Mersenne and the Learning of the Schools* (Ithaca, N.Y.: Cornell University Press, 1988), pp. 12–13.
3. William R. Shea, *The Magic of Numbers and Motion: René Descartes' Scientific Career* (Canton, Mass: Science History Publications, 1991), pp. 8–9. For a critical discussion of Descartes' life, see Geneviève Rodis-Lewis, "Descartes' Life and the Development of His Philosophy," in *The Cambridge Companion to Descartes,* edited by John Cottingham (Cambridge University Press, 1992), pp. 1–20.
4. Descartes to Beeckman, 23 April 1619, AT, vol. 10, pp. 162–3.

until his final, fatal move to Sweden in the last year of his life.[5] He corresponded frequently with Mersenne and many other natural philosophers, including Beeckman, Constantijn Huygens (father of the physicist Christiaan), William Cavendish, the brothers Jacques and Pierre DuPuy, Jean-Baptiste Morin, Gassendi, Hobbes, and Henry More. His letters contain a wealth of information about key issues in seventeenth-century natural philosophy.

Descartes published a number of important and widely read works in natural philosophy. In 1637 he published the *Discours de la méthode pour bien conduire la raison, & chercher la vérité dans les sciences,* in which he announced his method, which was, among other things, an answer to the skeptical challenge. Like Mersenne's *La vérité des sciences,* the essays published together with the *Discours* – *La dioptrique, Les météores,* and *La géométrie* – were intended to illustrate the fruitfulness of his new method for natural philosophy. In 1641, Descartes published his *Meditationes de prima philosophia,* in which he demonstrated the metaphysical and epistemological foundations of the new mechanical philosophy, as he understood them. He worked out his mechanical philosophy in detail in his *Principia philosophiae* (1644), which he hoped the Jesuits would adopt as a physics textbook for use in their schools to replace the traditional Aristotelian textbooks still in use.[6] His attempt to replace the Thomist theory of transubstantiation with a mechanical explanation of real presence in the Eucharist[7] led to the denunciation of *The Principles of Philosophy* by Father Thomas Compton Carlton, a mathematician and theologian at the College of Liège, the condemnation of the book by the theological faculty of the University of Louvain in 1662, and ultimately its being placed on the Index in 1663.[8]

Descartes' natural philosophy, like Gassendi's, was deeply informed by his theological presuppositions. Descartes, however, came to the consideration of theological matters by a very different route from the one taken by Gassendi. Instead of using a different ancient model to replace Aristotelianism, Descartes wanted to provide metaphysical and epistemological

5. Shea, *The Magic of Numbers and Motion,* p. 165.
6. See Henri Gouhier, *La pensée religieuse de Descartes,* 2d edition (Paris: J. Vrin, 1972), chap. 4; and "La crise de la théologie au temps de Descartes," *Revue de théologie et de philosophie,* 4, 3d ser. (1954): 54.
7. J.-R. Armogathe, *Theologia cartesiana: L'explication physique de l'eucharistie chez Descartes et Dom Desgabets* (The Hague: Martinus Nijhoff, 1977).
8. Pietro Redondi, *Galileo Heretic,* translated by Raymond Rosenthal (Princeton, N.J.: Princeton University Press, 1987), pp. 285–6.

foundations for his new philosophy of nature.[9] Where Gassendi responded to the skeptical crisis by embracing a "mitigated skepticism," which redefined the traditional goal of *scientia* from demonstrative truth about essences to probable conclusions about appearances, Descartes deployed the skeptical arguments as instruments for establishing an epistemological warrant, an indubitable foundation upon which to erect a system of certain and demonstrative knowledge.[10] In the course of this search, Descartes appealed to God as the guarantor of human knowledge.[11] In so doing, he addressed questions about how God created man and the world in such a way that – the skeptics notwithstanding – certainty in human knowledge is attainable. It is in this context that theological presuppositions played a key role in his thinking.[12] Descartes' quest for epistemological foundations led him to articulate theological assumptions that – as in Gassendi's case – bore the imprint of the medieval dialectic about the absolute and ordained powers of God. But in contrast to Gassendi, who espoused a voluntarist theology, Descartes adopted a kind of intellectualism. Briefly, Descartes believed that after God's initial exercise of absolute power in creating the world, he was bound by the necessity he had freely introduced into the created order. This necessity in the natural world provided Descartes with the warrant he sought for

9. See Daniel Garber, *Descartes' Metaphysical Physics* (Chicago: University of Chicago Press, 1992), chap. 1; John Cottingham, "A New Start? Cartesian Metaphysics and the Emergence of Modern Philosophy," in *The Rise of Modern Philosophy: The Tension Between the New and Traditional Philosophies from Machiavelli to Leibniz*, edited by Tom Sorell (Oxford University Press, 1993), pp. 145–66.

10. For an account that interprets Descartes's skepticism in the context of his search for foundations for science, see Margaret Dauler Wilson, *Descartes* (London: Routledge & Kegan Paul, 1978), chap. 1. See also Martial Gueroult, *Descartes' Philosophy Interpreted According to the Order of Reasons*, translated by Roger Ariew, 2 vols. (Minneapolis: University of Minnesota Press, 1984–5; first published 1952); and E. M. Curley, *Descartes Against the Skeptics* (Cambridge, Mass.: Harvard University Press, 1978). Curley is more concerned with evaluating the validity of Descartes' arguments than with placing them in the context of his thought more generally.

11. The Third Meditation is the *locus classicus* in which he made this appeal. See *The Philosophical Writings of Descartes* (hereafter *PWD*), translated by John Cottingham, Robert Stoothoff, Dugald Murdoch, and Anthony Kenny, 3 vols. (Cambridge University Press, 1984, 1985, 1991), vol. 2, pp. 24–36 (AT, vol. 7, pp. 34–52).

12. On the relationship between Descartes' search for foundations for philosophy and medieval theology, see Jean-Luc Marion, *Sur la théologie blanche de Descartes: Analogie, création des vérités éternelles et fondement* (Paris: Presses Universitaires de France, 1981).

claiming certainty for physics, the first principles of which could be known a priori. It marks the central difference between his theological presuppositions and those of Gassendi.

What connected Descartes' theology to his conception of the sciences? More particularly, what were the methodological implications of Descartes' theological assumptions? The answers to these questions lie in his theory of the creation of the eternal truths and, consequently, the epistemological and ontological status of the laws of nature. Descartes' statements on the status of eternal truths, initially and most fully elaborated in several letters to Mersenne in April and May 1630, have been the subject of a considerable amount of scholarly attention.[13] Although most

13. On Descartes' theory of the creation of the eternal truths, see Étienne Gilson, *La liberté chez Descartes et la théologie* (Paris: Félix Alcan, 1913); and *Études sur le rôle de la pensée médiévale dans la formation du système cartésien,* 2d edition (Paris: J. Vrin, 1951); Émile Boutroux, *Des vérités éternelles chez Descartes,* translated by M. Canguilhem (Paris: Félix Alcan, 1927; reprinted J. Vrin, 1989); Arnold Reymond, "Le problème cartésien des vérités éternelles et la situation présente," *Études philosophiques,* n.s. 8 (1953): 155–70; Pierre Garin, *Thèses cartésiennes et thèses thomistes* (Paris: Desclée de Brouwer, 1931); Leonard G. Miller, "Descartes, Mathematics, and God," *Philosophical Review,* 66 (1957), 451–65; Norman J. Wells, "Descartes and the Scholastics Briefly Revisited," *New Scholasticism 35* (1961): 172–90; Émile Bréhier, "The Creation of the Eternal Truths in Descartes' System," in *Descartes: A Collection of Critical Essays,* edited by Willis Doney (Notre Dame, Ind.: University of Notre Dame Press, 1968), pp. 192–208, first published as "La création des vérités éternelles," in *Revue philosophique de la France et de l'étranger, 113* (May-August 1937): 15–29; Geneviève Rodis-Lewis, *L'oeuvre de Descartes,* 2 vols. (Paris: J. Vrin, 1971); Amos Funkenstein, "Descartes, Eternal Truths, and the Divine Omnipotence," *Studies in History and Philosophy of Science,* 6 (1975): 185–99; Harry Frankfurt, "Descartes on the Creation of the Eternal Truths," *Philosophical Review, 86* (1977): 36–57; and Marion, *Sur la théologie blanche de Descartes.* See also Jean-Marie Beyssade, "Création des vérités éternelles et doute métaphysique," Geneviève Rodis-Lewis, "Polémiques sur la création des possible et sur l'impossible dans l'école cartésienne," and Gérard Simon, "Les vérités éternelles de Descartes, évidences ontologiques," all in *Studia cartesiana 2* (Amsterdam: Quadratures, 1981), pp. 124–36; Norman J. Wells, "Descartes' Uncreated Eternal Truths," *New Scholasticism, 56* (1982): 185–99; E. M. Curley, "Descartes on the Creation of the Eternal Truths," *Philosophical Review, 93* (1984): 569–97; and Gary Hatfield, "Reason, Nature, and God in Descartes," *Science in Context, 3,* no. 1 (Spring 1989): 175–201, revised version in *Essays on the Philosophy and Science of René Descartes,* edited by Stephen Voss (New York: Oxford University Press, 1993), pp. 259–87.

of these studies are based on close analysis of the Cartesian texts and some on a consideration of the Scholastic background against which they developed, few of them deal with the relationship between Descartes' theological presuppositions and his ideas about natural philosophy and its method.[14] It is this relationship that I pursue in the present chapter.

The actual nature of Descartes' religious beliefs is irrelevant to this discussion and analysis. The point is, rather, that the substance and style of his philosophizing were tempered by certain theological presuppositions drawn from traditions to which he had been exposed, particularly as a student. Even if he introduced theological considerations into his philosophy simply to placate the ecclesiastical authorities – his reaction to the condemnation of Galileo reveals genuine anxiety on his part – the theological tradition that he tapped is significant for its influence on his philosophical and scientific ideas.[15] The Jesuit Aristotelianism to which Descartes was exposed as a student at La Flèche continued to influence his conception of human knowledge and the laws of nature throughout his philosophical life.[16] In this respect, Descartes' intellectual development resembles that of Mersenne, whose philosophical agenda was deeply influenced by his Jesuit education at La Flèche.[17]

In order to establish this point, I consider Descartes' theory about the

14. Important exceptions to this generalization are Hatfield, "Reason, Nature, and God"; Gary C. Hatfield, "Force (God) in Descartes' Physics," *Studies in History and Philosophy of Science, 10* (1979): 113–40; and Garber, *Descartes' Metaphysical Physics.*

15. The fact that Descartes' early treatise *Le monde* – which he suppressed in response to news of Galileo's trial – contains virtually none of the theology present in the *Meditations* and *The Principles of Philosophy* is often taken as prima facie evidence for his lack of religious piety. This feeling is reinforced by some of his statements in response to the Galileo affair: "But, since I would not for anything in the world want a discourse to issue from me that contained the least word which the Church would disapprove of, so I would prefer to suppress it than to have it appear crippled." Descartes to Mersenne, end of November 1633, in AT, vol. 1, 271. See also Descartes to Mersenne, February 1634, in AT, vol. 1, pp. 281–2, in which he spoke with considerable alarm of "suppressing" *Le monde* and thereby losing "almost all my work of four years in order to make an entire obedience to the Church." In perceiving a continuing influence of medieval theology on the development of Descartes' metaphysics and philosophy of nature, I am in agreement with Marion, *Sur la théologie blanche de Descartes,* chap. 1.

16. Roger Ariew, "Descartes and Scholasticism: The Intellectual Background to Descartes' Thought," in *The Cambridge Companion to Descartes,* edited by Cottingham, pp. 58–90.

17. Dear, *Mersenne and the Learning of the Schools.*

creation of the eternal truths in the context of medieval discussions about the absolute and ordained powers of God.[18] I then demonstrate the close relationship between Descartes' position on these theological questions and his formulation of a method suitable for natural philosophy. In Chapter 9, I discuss the further ramifications of these theological and methodological views in his detailed articulation of the mechanical philosophy.

The creation of the eternal truths

Despite the fact that Descartes presented his philosophy as his own creation, independent of any historical tradition and radically severed from the Scholastic philosophy in which he had been educated, medieval theological concepts continued to play a formative role in his philosophy and in his understanding of scientific knowledge.[19] Descartes' acquaintance with Scholastic philosophy and theology stemmed from his early youth when he studied at La Flèche.[20] According to the founding documents of the Jesuit order, "The teaching of Aristotle conformed to the *true philosophy* . . . , and the only authorized interpreter of the Aristotelian doctrine was that of the Angelic Doctor [Thomas Aquinas]."[21] By basing their

18. Funkenstein discusses Descartes's theory of the eternal truths in this same context, but his concern – which is to understand Descartes' place in the history of the distinction between physical and logical necessity – is different from mine – to show the influence of ideas about divine power on Descartes' ideas about scientific method. See Amos Funkenstein, *Theology and the Scientific Imagination from the Middle Ages to the Seventeenth Century* (Princeton, N.J.: Princeton University Press, 1986), pp. 179–92. This section of Funkenstein's book is basically a republication of his article "Descartes, Eternal Truths, and the Divine Omnipotence."
19. On the influence of medieval philosophy on Descartes' thought, see Étienne Gilson, *Études sur le rôle de la pensée médiévale dans la formation du système cartésien*, 2d edition (Paris: J. Vrin, 1951), and *Index scolastico-cartésien*, 2d edition (Paris, J. Vrin, 1979). See also R. Dalbiez, "Les sources scolastiques de la théorie cartésienne de l'être objectif: À propos du 'Descartes' de M. Gilson," *Revue d'histoire de la philosophie*, 3 (1929): 464–72; Norman J. Wells, "Objective Being: Descartes and His Sources," *Modern Schoolman*, 45 (1967): 49–61; and J. R. Armogathe, "Les sources scolastiques du temps cartésien: Éléments d'un débat," *Revue international de philosophie*, 37 (1983): 326–36.
20. Garber, *Descartes Metaphysical Physics*, pp. 5–9.
21. P. Camille De Rochemonteix, *Un collège des Jésuites aux XVIIe et XVIIIe siècles: Le Collège Henri IV de La Flèche*, 4 vols. (Le Mans: Leguicheux, 1889), vol. 4, p. 8; my translation.

curriculum on Aristotle and Aquinas, the Jesuits hoped to fulfill their mission of providing "an effective antidote to Protestantism."[22] The philosophy course at La Flèche – consisting of logic, physics, metaphysics, and ethics – lasted for three years and was based largely on the works of Aristotle and his Jesuit commentators.[23] In addition to Aristotelian philosophy, a year of the program was devoted to the study of mathematics.[24] It is important to bear in mind the fact that Descartes' knowledge of medieval philosophy and theology in general and Thomism in particular was filtered through the lens of the Jesuit curriculum at La Flèche.[25] These facts explain some of the subtleties of Descartes' theory of the creation of the eternal truths and the scholarly debates surrounding it.[26]

In addition to the Jesuit Aristotelianism of La Flèche, another possible influence on Descartes was the Augustinian theology of Pierre Bérulle

22. Steven Ozment, *The Age of Reform, 1250–1550: An Intellectual and Religious History of Late Medieval and Reformation Europe* (New Haven, Conn.: Yale University Press, 1980), p. 414.

23. See Charles H. Lohr, "Jesuit Aristotelianism and Sixteenth-Century Metaphysics," in ΠΑΡΑΔΟΣΙΣ *(Paradosis): Studies in Memory of Edwin A. Quain,* edited by Harry George Fletcher III and Mary Beatrice Schulte (New York: Fordham University Press, 1976), pp. 203–20.

24. Rochemonteix, *Un collège des Jésuites,* pp. 21–36. Aristotle's works dominated the teaching of philosophy during the seventeenth century in the universities and other non-Jesuit institutions as well. See L. W. B. Brockliss, *French Higher Education in the Seventeenth and Eighteenth Centuries: A Cultural History* (Oxford University Press, 1987), pp. 331–3, 337–50. See also L. W. B. Brockliss, "Philosophy Teaching in France, 1600–1740," *History of Universities,* 1 (1981): 131–68.

25. In this connection, see Garin, *Thèses cartésiennes et thèses thomistes,* Introduction. For a discussion of the curriculum at La Flèche, see Ariew, "Descartes and Scholasticism," pp. 60–1; and Dear, *Mersenne and the Learning of the Schools,* pp. 12–15. See also Marion, *Sur la théologie blanche de Descartes,* p. 15.

26. Vendler finds close stylistic and substantive parallels between Loyola's *Spiritual Exercises* and Descartes' *Meditations,* suggesting a very strong influence of Jesuit spirituality on Cartesian philosophy. See Zeno Vendler, "Descartes' Exercises," *Canadian Journal of Philosophy,* 19 (1989): 193–224. See also Bradley Rubidge, "Descartes' *Meditations* and Devotional Meditations," *Journal of the History of Ideas,* 51 (1990): 27–49. Rubidge denies that the *Meditations* resembles Loyola's *Spiritual Exercises* or any other devotional literature of the period. Rather, Rubidge argues, Descartes "invoked the tradition of devotional exercises in order to advertise the orthodoxy of his intentions as he extended methodological doubting to theological matters" (p. 28).

(1575–1629) and the Oratorians.[27] Descartes knew Bérulle at least as early as 1626 and chose him to be his spiritual advisor. Gilson has argued that evidence of Bérulle's influence on Descartes is strong.[28] Bérulle had abandoned scholastic dialectic and preferred to write in French rather than Latin. His theology emphasized God's goodness and, significantly, his unity. Divine unity became the centerpiece of Descartes' understanding of divine will. Many of Descartes' contemporaries and followers – including Mersenne,[29] Antoine Arnauld (1612–94),[30] Claude Clerselier (1614–85), Ambrosius Victor (aka André Martin (1621–95)), Jacques Rohault (1620–72), Louis de la Forge (1605?-79?), Gérauld de Cordemoy (1620–84), Nicholas Malebranche (1638–1715), Bernard Lamy (1640–1715), Pierre-Daniel Huet (1630–1721) – saw a resemblance between certain Cartesian ideas and those of Augustine.[31] Gouhier has challenged Gilson's emphasis on the Oratorians as the source of the Augustinian strands in Descartes' philosophy. Arguing on the basis of chronology, Gouhier claims that by the time Descartes became involved with Bérulle in the late 1620s, he had already formed his basic philosophical outlook. He thinks that Descartes' attraction to the Oratorians can be explained more adequately as a mutual recognition of like minds (on certain issues). It is a mistake, Gouhier says, to look to the Oratorians every time we find an Augustinian thesis in Descartes. On precise points, when the documents are supportive, we can indicate undeniable filiation, as Gilson has done with regard to divine liberty and innatism. In other cases, direct borrowing cannot be documented.[32] Augustinian ideas and attitudes were accessible in many books readily available in the seven-

27. See Paul Cochois, *Bérulle et l'École française* (Paris: Éditions du Seuil, 1963).
28. Gilson, *La liberté chez Descartes,* pp. 161–2.
29. Mersenne mentions the resemblance between Descartes' *cogito* and Augustine's *City of God,* bk. 11, chap. 26, in letters he wrote to Descartes in June 1637, 15 November 1638, and December 1640. *Correspondance du P. Marin Mersenne,* edited by Cornelius de Waard and Armand Beaulieu, 15 vols. (Paris: CNRS, 1933–83), vol. 6, p. 277; vol. 10, pp. 199–200, 331.
30. Arnauld wrote in "Objections IV," referring to the *cogito,* "The first thing I find remarkable is that our distinguished author has laid down as the basis for his entire philosophy exactly the same principle as that laid down by St. Augustine." *PWD,* vol. 2, p. 139 (AT, vol. 7, pp. 198–9).
31. Henri Gouhier, *La pensée religieuse de Descartes,* 2d edition (Paris: J. Vrin, 1972), p. 257.
32. Ibid., pp. 261–2. Gouhier pursues these themes at greater length in *Cartésianisme et augustinisme au XVIIe siècle* (Paris: J. Vrin, 1978).

teenth century, and it is probably not possible to identify the precise source of Augustinian themes in Descartes' thought.[33]

Given this background, what motivated Descartes to address the question of the eternal truths? Mersenne's half of the correspondence with Descartes in the spring of 1630 no longer survives, leaving obscure the questions that propelled Descartes into his elaborate discussion of divine power and the eternal truths. One possibility is that Descartes, who was at that very time in the midst of composing *Le monde*, found himself faced with certain theological difficulties that he wished to avoid. In particular, he might have recognized that his arguments against the void in *Le monde* were reminiscent of similar Aristotelian arguments condemned in 1277.[34] Possibly he formulated his doctrine of the creation of the eternal truths in order to avoid the kind of problems that had arisen in the thirteenth century.[35] Since there seems to be no documentary evidence either for or against this explanation, one can only acknowledge it as a plausible speculation.[36]

A more likely possibility is that Descartes was struck by the apparent necessity of mathematical truths, which he then tried to incorporate into a suitable theological framework. Indeed, his first epistolary discussion with Mersenne about the question of the eternal truths began by his identifying them with mathematical truths. "The mathematical truths, which you call eternal, were established by God and totally depend on him just like all the other creatures."[37] His concern to prove that the truths of

33. Gouhier, *Cartésianisme et augustinisme*, pp. 173–8.

34. Edward Grant, *Much Ado About Nothing: Theories of Space and Vacuum from the Middle Ages to the Scientific Revolution* (Cambridge University Press, 1981), pp. 103–15; and Pierre Duhem, *Medieval Cosmology: Theories of Infinity, Place, Time, Void, and the Plurality of Worlds*, edited and translated by Roger Ariew (Chicago: University of Chicago Press, 1985), chap. 9.

35. Daniel Garber suggests this explanation of Descartes' motives. See Garber, *Descartes' Metaphysical Physics*, pp. 148–55.

36. Edward B. Davis has suggested a possible link between Descartes and William Ames, the English Puritan, who was professor of theology at the University of Frankener at the same time that Descartes was there in 1629. While a possible encounter with Ames may have stimulated Descartes to think about the status of the eternal truths, I find Davis' speculation unconvincing, particularly because he interprets Descartes' theory as voluntarist. Davis fails to consider the Scholastic context of Descartes' philosophy. See Edward B. Davis, "God, Man and Nature: The Problem of Creation in Cartesian Thought," *Scottish Journal of Theology*, 44 (1991): 325–48.

37. Descartes to Mersenne, 15 April 1630, AT, vol. 1, p. 145.

mathematics were created by God was probably reinforced, if not initially stimulated, by his acquaintance with the works of Johannes Kepler (1571–1630) and Galileo Galilei (1564–1642),[38] both of whom considered mathematics central to the natural order. For Kepler, the mathematical truths are eternal and exist independently of God.[39] Galileo had stated that the book of nature "is written in the language of mathematics, and its characters are triangles, circles, and other geometric figures."[40] By means of this mathematical language, human knowledge can attain the same truths that God knows, the only difference between human and divine knowledge being the mode by which these truths are apprehended.[41] Like Kepler, Galileo maintained that the truths of mathematics possess an absolute nature that is independent of God.

Whatever the sources of his interest in the status of the eternal truths, the significant facts remain that Descartes' discussions drew on the language found in medieval discussions of divine power and that his doctrine of the creation of the eternal truths played an important role in his accounts of epistemology and scientific method. Descartes alluded to and explicitly rejected as blasphemous the view "Si Deus non esset, nihilominus istae veritates essent verae" (If God did not exist, nonetheless these truths would be true).[42] In Scholastic terms, this objectionable view stated that although God does create individual existents, he does not create their essences, which exist independently of the actual individual existents they inform.[43] More than a little ink and midnight oil have been

38. Descartes knew Galileo's work directly as evidenced by his frequent mention of it in his correspondence with Mersenne. See, e.g., Descartes to Mersenne, 11 October 1638, AT, vol. 2, p. 380; Descartes to Mersenne, 15 November 1638, AT, vol. 2, p. 433; Descartes to Mersenne, 23 February 1642, AT, vol. 3, p. 634. For evidence of Descartes's acquaintance with Kepler's work, see Marion, *Sur la théologie blanche de Descartes,* chap. 10.

39. Johannes Kepler, *Harmonies of the World,* translated by Charles Glenn Wallis, in *Ptolemy, Copernicus, Kepler* (Great Books of the Western World, vol. 16) (Chicago: Encyclopedia Britannica, 1952), bk. 5, chap. 3, p. 1017. Marion argues that Kepler was a likely target of Descartes' concern. See Marion, *Sur la théologie blanche de Descartes,* p. 196.

40. Galileo Galilei, *The Assayer,* in *Discoveries and Opinions of Galileo,* edited by Stillman Drake (Garden City, N.Y.: Doubleday Anchor, 1957), p. 238.

41. Galileo Galilei, *Dialogue Concerning the Two Chief World Systems – Ptolemaic and Copernican,* translated by Stillman Drake (Berkeley and Los Angeles: University of California Press, 1953), pp. 103–4.

42. Descartes to Mersenne, 6 May 1630, AT, vol. 1, pp. 149–50 (*PWD,* vol. 3, p. 24).

43. Wells, "Descartes and the Scholastics Briefly Revisited," p. 174.

consumed in attempting to identify who held this view, which Descartes rejected so vehemently.[44] Thomas Aquinas is not a likely candidate for two reasons: First, Descartes was probably not acquainted with Aquinas' position directly but rather with a Jesuit rendition of it, and second, the doctrine of uncreated essences is not to be found in Aquinas' writings.[45] Such a view can, however, be found in Duns Scotus.[46] Moreover, some of Thomas Aquinas' followers espoused this doctrine, notably Henry of Ghent (d. 1293), Paulus Barbus Soncinas (d. 1494), Thomas Cajetan (1468–1534), Sylvester of Ferrara (1474–1528), Chrysostomus Javellus (1470–1538), John Capreolus (d. 1444), and the Jesuit Aristotelians Francisco Suárez (1548–1617) and Gabriel Vasquez (1549–1604).[47] Several scholars have argued that Suárez is the most likely candidate for Descartes' unnamed adversary.[48] However, because Suárez was critical of

44. On this issue, see Gilson, *La liberté chez Descartes;* T. J. Cronin, "Eternal Truths in the Thought of Suarez and of Descartes," *Modern Schoolman, 38* (May 1961): 269–88, and *39* (November 1961): 23–38; Wells, "Descartes and the Scholastics Briefly Revisited"; John P. Doyle, "Suarez on the Reality of the Possibles," *Modern Schoolman, 45* (1967): 29–48; James C. Doig, "Suarez, Descartes, and the Objective Reality of Ideas," *New Scholasticism, 51* (1977): 350–71; Norman J. Wells, "Old Bottles and New Wine: A Rejoinder to J. C. Doig," *New Scholasticism, 53* (1979): 515–23; Norman J. Wells, "Suarez on the Eternal Truths," *Modern Schoolman, 58* (1981): 73–104, 159–74. For a useful discussion of some of the issues separating these scholars, see Dear, *Mersenne and the Learning of the Schools,* pp. 54–62.

45. Originally Gilson had argued that Thomas Aquinas was Descartes' adversary, although he has also considered the possibility that Descartes' adversary was Suárez. See Gilson, *La liberté chez Descartes,* chap. 2. Pierre Garin countered that Descartes was not exposed to Thomas Aquinas directly at La Flèche and that his acquaintance with Thomism was mediated by the Jesuit Aristotelians, particularly Suárez. See Garin, *Thèses cartésiennes et thèses thomistes,* chap. 3.

46. "Si poneretur, per impossibile, quod Deus non esset, et quod triangulus esset, adhuc habere tres angulos resolveretur ut in naturam trianguli," Duns Scotus, *Reportata parisiensia,* prologus, III, qu. 4, as cited in Gilson, *Index scolastico-cartésien,* p. 357.

47. Wells, "Descartes and the Scholastics Briefly Revisited," pp. 178–81.

48. A strong argument in favor of Suárez as Descartes' adversary is to be found in Marion, *Sur la théologie blanche de Descartes,* chaps. 3–7, esp. p. 28. See also Cronin, "Eternal Truths in the Thought of Suarez and Descartes." For brief expositions of Suárez' philosophy, see John A. Trentman, "Scholasticism in the Seventeenth Century," in *The Cambridge History of Later Medieval Philosophy,* edited by Norman Kretzmann, Anthony Kenny, and Jan Pinborg (Cambridge University Press, 1982), pp. 822–7; Frederick

certain aspects of the doctrine of uncreated essences, the suggestion has also been made that Soncinas and Cajetan are more suitable candidates for this position.[49] An additional and closer target for Descartes' attack may have been Bérulle, who is known to have encouraged Descartes in his intellectual pursuits.[50] Bérulle, following Augustine, who had accepted the Neoplatonic view that the forms and immutable essences of things reside in the mind of God, believed that the eternal truths, if not actually independent of God, were uncreated because they are eternally included in his Word.[51] He claimed that human knowledge of these entities are the eternal truths that can be explained by the theory of divine illumination.[52]

Given this complicated filiation of the problem, what did Descartes himself have to say about the status of the eternal truths? Descartes began by asserting that the eternal truths, like everything else in the world, have been created by God. "The mathematical truths which you call eternal have been laid down by God and depend on Him entirely no less than the rest of his creatures."[53] Like Gassendi's laws of nature, they can be understood as the products of God's absolute power. "Please do not hesitate to assert and proclaim everywhere that it is God who has laid down these laws in nature just as a king lays down laws in his kingdom."[54] To think of the eternal truths as existing independently of God would be a blasphemous affront to divine omnipotence. Unlike Plato's Demiurge[55] or the pagan gods, God is not limited by anything in the world. "Indeed to say that these truths are independent of God is to talk of Him as if He were Jupiter or Saturn and to subject him to the Styx and the Fates."[56] From the standpoint of his absolute power, God is subject to nothing. His freedom is complete.

   Copleston, *A History of Philosophy*, 9 vols. (Garden City, N.Y.: Doubleday, 1963), vol. 3, pp. 22, 23.
49. Wells, "Descartes and the Scholastics Briefly Revisited," pp. 178–90. For Suárez' criticisms of Soncinas and Sylvester of Ferrara, see ibid., pp. 162–4.
50. Gouhier, *La pensée religieuse de Descartes*, p. 333.
51. Marion, *Sur la théologie blanche de Descartes*, p. 140. See also Henri Busson, *La pensée religieuse française de Charron à Pascal* (Paris: J. Vrin, 1933).
52. Copleston, *A History of Philosophy*, vol. 2, pt. 1, p. 75; and R. A. Markus, "Marius Victorinus and Augustine," in *The Cambridge History of Later Greek and Early Medieval Philosophy*, edited by A. H. Armstrong, corrected edition (Cambridge University Press, 1970), pp. 365–6.
53. AT, vol. 1, p. 145 (*PWD*, vol. 3, p. 23).
54. Ibid.
55. Plato, *Timaeus*, 29D–30C, in Francis MacDonald Cornford, *Plato's Cosmology: The Timaeus of Plato* (Indianapolis, Ind.: Bobbs-Merrill, n.d.), p. 38.
56. AT, vol. 1, p. 145 (*PWD*, vol. 3, p. 23).

Given Descartes' emphasis in these passages on God's free creation of the eternal truths and the laws of nature, it is easy to understand why some commentators have intepreted Descartes as a voluntarist. After all, there is a substantial difference between Descartes' views here and the Platonic and Scholastic notions that there are some truths that exist independently of God's creation.[57] Commentators who claim that Descartes is a voluntarist based on these passages alone fail to take account of what comes next. Descartes proceeded to argue that God has created us in such a way that we can have a priori knowledge of the eternal truths and that his own nature prevents him from changing what he once created freely. In other words, God's ability to intervene in the created world is limited by his initial act of creation. According to Descartes, God's ordained power does not reach to the full extent of his absolute power. In this sense, Descartes differed from the voluntarists. It is with regard to his ordained power that Descartes was an intellectualist: He accepted the existence of some necessity in the world, something that the voluntarists could never accept because of their emphasis on the utter contingency of the world.

Descartes' first difference from the voluntarists lay in his claim that we have innate knowledge of the eternal truths. God not only created the eternal truths, he also created our minds in such a way that we possess an innate capacity to understand them. Here Descartes introduced a major element of necessity into the world, the connection between the eternal truths and our minds. From rather traditional talk about the powers of God, he produced the metaphysical groundwork for what would become the basis of his revolution in philosophy: "There is no single one [of the eternal truths] that we cannot understand if our mind turns to consider it. They are all *inborn in our minds* just as a king would imprint his laws on the hearts of all his subjects if he had enough power to do so."[58]

If the eternal truths are somehow inborn in our minds, how can we be assured that God will not change them, that we will not be deceived in believing them to be true?

> It will be said that if God had established these truths He could change them as a king changes his laws. To this the answer is: "Yes he

---

57. This point is clearly stated in John Cottingham, *A Descartes Dictionary* (Oxford: Blackwell, 1993), p. 58.
58. AT, vol. 1, p. 145 (*PWD*, vol. 3, p. 23). It is this important point that Steven Nadler overlooks when he claims that there is a conflict between Descartes' theory of the creation of the eternal truths and his claim that the laws of nature can be derived with certainty from the divine attributes. See Steven M. Nadler, "Scientific Certainty and the Creation of the Eternal Truths: A Problem in Descartes," *Southern Journal of Philosophy*, 25 (1987): 186–7.

can, if his will can change." – "But I understand them to be eternal and unchangeable." – "I make the same judgment about God." – "But His will is free." – "Yes, but his power is incomprehensible."[59]
The immutability of God's will is entailed by the unity of his will and his understanding. If an act of willing (say, to create the eternal truths) is simultaneously an act of understanding (of these same truths), then a change of his will would entail a change of his understanding. But any change in the divine understanding would entail some imperfection in God: His understanding would at some time – before or after the change – have been mistaken or incomplete. That is to say, something he knew at one time he would not know at some other time. It would then not be possible to say that he has perfect knowledge. Such imperfection would be incompatible with divine perfection. Consequently, it is not possible to assert without contradiction that God's will can change. The immutability of God's understanding and the unity of his will and intellect jointly entail the immutability of his will.[60] Similarly, divine immutability provides Descartes justification for the necessity of the eternal truths that God created freely.

Writing to Princess Elizabeth on 6 October 1645, Descartes' reiterated his insistence that divine immutability precludes any change in God's will:

> When Your Highness speaks of the particular providence of God as being the foundation of theology, I do not think that you have in mind some change in God's decrees occasioned by actions that depend on our free will. No such change is theologically tenable; and when we are told to pray to God, that is not so that we should inform him of our needs, or that we should try to get him to change anything in the order established from all eternity by his providence – either of these aims would be blameworthy – but simply to obtain whatever he has, from all eternity, willed to be obtained by our prayers.[61]

In other words, God's design for the world, maintained by his ordained power, is not subject to further change, even by his absolute power. This position clearly separates Descartes from Gassendi and the voluntarists, who insisted that God is always free to change any of his decrees.

Descartes' argument reflects the traditional discourse about the absolute and ordained power of God. By his absolute power, God freely created the eternal truths, just as he freely created the other creatures. In this act of creation, God's power and freedom were absolute, unconstrained by anything else in the universe. Quite simply, nothing in the universe

59. AT, vol. 1, pp. 145–6 (*PWD*, vol. 3, p. 23).
60. On the interconnection between God's will and understanding, see Gilson, *La liberté chez Descartes*, pp. 34–48.
61. AT, vol. 4, 315–16 (*PWD*, vol. 3, p. 273).

exists independent of God, even the laws of mathematics. The eternal truths are true because God knows them. It is not the case that he knows them because they are true, for in that case both the truths and an absolute standard of truth would exist independent of God's creative act:

> As for the eternal truths, I say once more that they are all true or possible only because God knows them as true or possible. They are not known as true by God in any way which would imply that they are true independently of Him. If men really understood the sense of their words they could never say without blasphemy that the truth of anything is prior to the knowledge which God has of it.[62]

His knowledge of the eternal truths follows from his willing their creation:

> In God willing and knowing are a single thing in such a way that by the very fact of willing something he knows it and it is only for this reason that such a thing is true. So we must not say that if God did not exist nonetheless these truths would be true; for the existence of God is the first and most eternal of all possible truths and the one from which alone all others derive.[63]

Note that the existence of God is an *uncreated* eternal truth, but obviously not one that in any way limits divine freedom.[64]

In making this argument, Descartes' similarity to Thomas Aquinas is striking. Aquinas had written:

> Everything eternal is necessary. Now, that God should will some effect to be is eternal, . . . His willing is measured by eternity, and is therefore necessary. But it is not necessary considered absolutely, because

62. Descartes to Mersenne, 6 May 1630, AT, vol. 1, p. 149 (*PWD*, vol. 3, p. 24).
63. Descartes to Mersenne, 6 May 1630, AT, vol. 1, pp. 149–50 (*PWD*, vol. 3, p. 24). Harry Frankfurt points out that by uniting God's will and intellect, Descartes rejected the Scholastic view that possible essences are "dependent on God's understanding but independent of His will." By identifying the divine will and intellect, Descartes was led to regard freedom as indifference. "Since there *are* no truths prior to God's creation of them, His creative will cannot be determined or even moved by any considerations of value or rationality." See Frankfurt, "Descartes and the Creation of the Eternal Truths," pp. 40–1. Wells criticizes Frankfurt's argument that the eternal truths are "mind dependent" on the grounds that these uncreated eternal truths have nothing to do with the human mind. See Norman J. Wells, "Descartes' Uncreated Eternal Truths," pp. 185–99. See also Descartes, "Sixièmes Responses," AT, vol. 9–1, pp. 232–4.
64. Wells, "Descartes' Uncreated Eternal Truths," pp. 185–99.

the will of God does not have a necessary relation to this willed
object. Therefore it is necessary by supposition.[65]

For Aquinas, God is free in his choice to will something into eternal
existence; but once having been created, that thing is both eternal and
necessary. This similarity between Aquinas and Descartes underscores the
resonance of the dialectic of God's absolute and ordained powers in
Descartes' thinking.

Descartes' claim of the unity of God's will and intellect acquires broader
significance in the context of the Counter-Reformation controversy between the Dominicans and Jesuits on the question of predestination. In
this debate, the Dominicans believed that "the divine decree embodies
both God's will and his foreknowledge of men's future acts" and that "the
necessity of predestination derives from the absolute character of God's
foreknowledge and will."[66] The fact that the Jesuit Molina separated
God's will and foreknowledge would seem to indicate that Descartes was
closer to the Dominicans on this issue.[67] His rejection of the Molinist view
may have its roots in the anti-Molinist understanding of the Oratorian
Guillaume Gibieuf (1591–1650).[68] Although Descartes was deeply influenced by his Jesuit education, his rejection of the Molinist position on the
relationship between the divine attributes underscores the fact that he was
not a blind follower of Jesuit teachings in all matters.

To recapitulate, according to Descartes, God freely willed to create the
eternal truths and the laws of nature. "You ask also what necessitated
God to create these truths; and I reply that just as He was free not to create
the world, so He was no less free to make it untrue that all the lines drawn
from the centre of a circle to its circumference are equal."[69] In this statement, Descartes directly contradicted Aquinas, who had stated:

65. Thomas Aquinas, *Summa contra Gentiles,* translated by Charles J. O'Neil, 4
     vols. (Notre Dame, Ind.: University of Notre Dame Press, 1957), bk. I, chap.
     83, ¶3, vol. 1, p. 263.
66. Rivka Feldhay, "Knowledge and Salvation in Jesuit Culture," *Science in
     Context, 1* (1987): 204–5.
67. Ibid., 205. See also Lisa T. Sarasohn, *Freedom in a Deterministic Universe:
     Gassendi's Ethical Philosophy* (Ithaca, N.Y.: Cornell University Press,
     forthcoming).
68. Gilson, *La liberté chez Descartes,* pp. 184–92. See also Francis Ferrier, *Un
     Oratorien ami de Descartes: Guillaume Gibieuf et sa philosophie de la
     liberté* (Paris: J. Vrin, 1980).
69. Descartes to Mersenne, 27 May 1630, AT, vol. 1, p. 152 (*PWD,* vol. 3, p.
     25).

Since the principles of certain sciences – of logic, geometry, and arith-
metic, for instance – are derived exclusively from the formal principles of things, upon which their essence depends, it follows that God
cannot make the contraries of those principles; He cannot make the
genus not to be predicable of the species, nor lines drawn from a
circle's center to its circumference not to be equal, nor the three angles
of a rectilinear triangle not to be equal to two right angles.[70]

At this point, Aquinas' intellectualism is more extreme than Descartes',
since the principles of these sciences lie outside God's creative act, thereby
limiting his subsequent freedom of action. Descartes, on the contrary,
asserted that God created everything, even the principles of these sciences.
This act of creation – a product of God's absolute power – established the
order of nature, which, because of the immutability of his will and the
unity of his intellect and will, he maintains and conserves in the manner in
which he first created it. The ensuing necessity of these truths and laws is a
necessity relative to this natural order and to our understanding of it. Like
Aquinas' necessity of supposition, it can thus be understood in terms of
God's ordained power. "And even if God has willed that some truths
should be necessary, this does not mean that he willed them necessarily;
for it is one thing to will that they be necessary, and quite another to will
them necessarily, or be necessitated to will them."[71] Again, note the striking parallel with Thomas Aquinas:

Now, the fact that God is said to have produced things voluntarily,
and not of necessity, does not preclude His having willed certain
things to be which are of necessity and others which are contingently,
so that there may be an ordered diversity in things. Therefore, nothing prevents certain things that are produced by the divine will from
being necessary.[72]

According to Descartes, the eternal truths are necessary, even though
God created them freely and their existence is entirely dependent on him.
Our ability to know them a priori, moreover, entails the existence of a
necessary relation between at least some of the ideas in our minds and
their external referents. Descartes' position, like Aquinas', is an intellectualism with regard to the ordained, not the absolute, power of God. The
differences between this position and the voluntarism of Ockham, the

70. Aquinas, *Summa contra Gentiles*, bk. II, chap. 25, ¶ 14, pp. 74–5. See also
    Cottingham, *A Descartes Dictionary*, pp. 57–9.
71. Descartes to Mesland, 2 May 1644, in AT, vol. 4, pp. 118–19 (*PWD*, vol. 3,
    pp. 234–5).
72. Aquinas, *Summa contra Gentiles*, bk. II, chap. 30, ¶ 4, vol. 2, p. 86.

nominalists, and Gassendi are subtle. Although both groups denied the existence of anything that God did not create, they differed as to the status of the order that he did create. In essence, the voluntarists equated God's absolute and ordained power so that his ordained power was absorbed into his absolute power, leaving nothing beyond his direct control. In contrast to the voluntarists' insistence on the utter contingency of the creation, Descartes, like Aquinas and his followers, believed in the existence of at least some elements of necessity in the created world. The relative necessity of Descartes' eternal truths closely resembles Aquinas' idea of necessity of supposition.[73] God created this necessity, and it delineates an area in which God's absolute and ordained powers do not overlap. Whatever we think of the coherence of this concept – how necessary is relative necessity? – the fact remains that it is a mode that no voluntarist would have countenanced.

Is Descartes' position consistent? Why does God's freedom in creating the eternal truths not defeat his immutability, and conversely, why does his immutability not interfere with the free exercise of his will in the act of creation? The resolution of this conundrum lies in the distinction between the absolute and ordained power of God. On the one hand, God willed the eternal truths freely, and it follows that there are no external constraints or principles of rationality to restrict God's absolute, creative act in any way. God was absolutely free to create circles with unequal radii. On the other hand, once having exercised his will with absolute freedom, he cannot alter the eternal truths. This element of necessity follows from the relationship between the divine attributes. It does not impede the freedom of his creative act.

Descartes' theory of the creation of the eternal truths served as the metaphysical foundation for his philosophy. On this foundation he erected his theory of knowledge – his response to the skeptics, the *cogito*, his proof for the existence of God, and the criterion of clear and distinct ideas. He thought that he had thereby answered the challenge of the skeptics by providing an indubitable premise upon which to build his philosophical structure. The eternal truths lay at the very base of this structure, providing the foundations for his a priori demonstration of the first principles of philosophy. Descartes' foundationalist approach to philosophy has its roots in the long history of the dialectic between the absolute and ordained powers of God.

---

73. See, e.g., ibid., bk. II, chap. 30, ¶1, vol. 2, p. 85. This point is discussed at length in Chapter 2.

The status of the laws of nature

God's creation of the eternal truths provided the metaphysical and epistemological foundations for the first principles of Descartes' natural philosophy.[74] Descartes believed that, in creating this world, God established
certain fundamental principles that are necessarily true and the truth of
which we can prove a priori. Descartes' claim to have a priori, certain
knowledge constituted his response to the skeptical challenge. Rather
than working within a general acceptance of the skeptical arguments, as
both Mersenne and Gassendi did, Descartes wanted to defeat skepticism
by finding an indubitable kernel of certainty from which to construct his
deductive system. He found the ultimate justification for this certainty in
his doctrine of the eternal truths.

Descartes' arguments in the *Meditations* – systematic doubt, the *cogito*,
and the proofs of the existence of God, the soul, and matter – are well
known and need not be reiterated here. What may be less well known is
the connection between these arguments and his theory of the creation of
the eternal truths. Having employed systematic doubt to eliminate any
dubitable claims, Descartes discovered the certainty he sought in the proposition "Cogito ergo sum" (I think, therefore I am). In order to proceed
from the *cogito* to the existence of the physical world, he needed a criterion by which to judge the reliability of his knowledge. He found that
criterion in the principle that "everything that we clearly and distinctly
understand is true in a way that corresponds exactly to our understanding
of it."[75] Lying at the heart of both the *Meditations* and *The Principles of
Philosophy*,[76] this principle follows from the fact that if what we perceive
clearly and distinctly should turn out to be false, God would be a deceiver,
something incompatible with divine perfection.[77] The truth of the clear
and distinct principle follows from the divine nature, and it functions as
the necessary connection between the contents of our minds and the natural order God created by his absolute power. The criterion of clear and

74. Marion argues convincingly that the letters of 1630 on the status of the
     eternal truths determine the point of departure for the *Meditations*. See
     Marion, *Sur la théologie blanche de Descartes*, chap. 14.
75. Descartes, *Meditations*, in *PWD*, vol. 2, p. 9 (AT, vol. 7, p. 13).
76. Descartes, *Principia philosophiae*, in AT, vol. 8–1, pp. 20–3, and vol. 9–2,
     pp. 42–50; vol. 8–1, p. 16, and vol. 9–2, p. 38.
77. Ibid., vol. 8–1, p. 16, and vol. 9–2, p. 38. Note that this solution was not
     open to Ockham because of his conception of divine liberty. See Marilyn
     McCord Adams, "Intuitive Cognition, Certainty, and Scepticism in William
     of Ockham," *Traditio*, 26 (1970): 397

distinct ideas served as the main underpinning for Descartes' claim to have a priori knowledge of the nature of matter as extension, its infinite divisibility, and the doctrine of primary and secondary qualities – the fundamental components of his mechanical philosophy.

Descartes asserted that the principles from which the rest of science could be derived

> should have two conditions attached to them: the one that the human mind not doubt their truth when it applies itself with attention to consider them; the other, that it would be on them that the knowledge of other things depends, so that they [the principles] can be known without them [the other things], but not reciprocally.[78]

He thought that the laws of nature, which are the first principles of his natural philosophy, possess precisely this epistemological status, a status that is closely related to his understanding of the eternal truths.

Consider his first law of nature: "That God is the first cause of motion, and that he always conserves an equal quantity of it in the universe. . . . [I]t is evident that there is none other than God who by his omnipotence created matter with motion and who now conserves in the universe by his ordinary concourse (*concursum ordinarium*) as much motion and rest as he placed in it creating it."[79] Descartes' language here reflects the discourse about the two powers of God. By his omnipotence, his absolute power, God created matter and motion; by his ordinary concourse, his ordained power, he conserves them.[80] That is, he relies on the natures and

78. Descartes, *Principes de la philosophie,* in AT, vol. 9–2, p. 2.
79. Descartes, *Principia philosophiae,* in AT, vol. 8–1, p. 61, and vol. 9–2, p. 83. For a thorough discussion of the meaning and implications of this law, see Garber, *Descartes' Metaphysical Physics,* pp. 203–9.
80. Freddoso explains the concept of God's ordinary concourse as follows: "When God brings about a created effect by Himself, He acts as a *particular* cause, since His causal contribution by itself determines the specific nature of the effect. However, the medieval Aristotelians maintain, in opposition to occasionalists, that all *creatures* have genuine causal power, too – though . . . in order for them to exercise this power God must also act to produce the relevant effect. When He thus cooperates with secondary causes, He acts as a *general* or *universal* cause of the effect, and His causal contribution is called His *general concurrence* or *concourse* (*concursus generalis*). The nomenclature is indicative of the fact that in such a case the particular nature of the effect is traceable not to God's causal contribution, necessary though it is in order for any effect to be produced at all, but rather to the natures and causal contributions of the relevant *secondary* causes, which act as *particular* causes of the effect." Luis de Molina, *On Divine Foreknowledge (Part IV of the* Concordia*)*, translated with an introduction and notes by Alfred J. Freddoso (Ithaca, N.Y.: Cornell University Press, 1988), p. 17.

second causes he created to bring about particular effects.[81] Descartes'
justification for this statement rests, once again, on one of the divine
attributes, in this case immutability:

> We also know that it is a perfection in God, not only that he is
> immutable in his nature, but also that it is the question of a mode
> which he never changes: such that besides the changes which we see
> . . . *in the world,* and those which we believe because God has re-
> vealed them and which we know happen . . . or *have happened in*
> *nature,* without any change on the part of the Creator, we ought not
> to suppose it of other of his works, for fear of attributing inconstancy
> to him. From which it follows that . . . since he moved the different
> parts of matter in several different ways when he created them, and
> since he always maintains them in the same way and with the same
> laws which *he made them obey* at their creation, he ceaselessly con-
> serves in this matter an equal quantity of motion.[82]

Note that Descartes committed a non sequitur here: Just because God
conserves the laws, it doesn't follow that they must be conservation
laws.[83] Nevertheless, the laws of nature, as Descartes described them in
The Principles of Philosophy, have the same epistemological status as the
eternal truths. Part of the initial creation, they possess the same kind of
necessity as the eternal truths, a necessity following from divine immutability. Descartes' laws of nature were a far cry from Gassendi's, which
God is free to violate at will.

Similarly, Descartes appealed to God's immutability in justifying his
second and third laws of nature, the principle of inertia[84] and his fundamental law of impact.[85] He believed that these laws provided the founda-

81. Theories about the nature of divine concurrence were directly related to the
controversy about freedom and divine foreknowledge. See J. A. van Ruler,
"New Philosophy to Old Standards: Voetius' Vindication of Divine Concurrence and Secondary Causality," *Nederlands archeif voor kerkgeschiedenis/*
Dutch Review of Church History, 71 (1991): 67–79.
82. Descartes, *Principia philosophiae,* in AT, vol. 8-1, pp. 61–2 and vol. 9-2,
pp. 83–4. For a different approach to the argument from divine immutability, see Garber, *Descartes' Metaphysical Physics,* pp. 279–91.
83. I am grateful to J. J. MacIntosh for noting this point. Garber makes a similar
point. See *Descartes' Metaphysical Physics,* p. 281. See also Steven M.
Nadler, "Deduction, Confirmation, and the Laws of Nature in Descartes'
Principia philosophiae," *Journal of the History of Philosophy,* 28 (1990):
366.
84. On Descartes' understanding of the persistence of motion, see Garber,
Descartes' Metaphysical Physics, pp. 209–24.
85. On Descartes' laws of impact, see ibid., chap. 8. See also Alan Gabbey,
"Force and Inertia in the Seventeenth Century: Descartes and Newton," in

tion from which everything in the universe as we know it can be explained.[86] Because these laws follow directly from God's immutability, they are necessarily true, and consequently we can know them in an a priori, demonstrative manner:

> I showed what the laws of nature were, and without basing my arguments on any principle other than the infinite perfection of God, I tried to demonstrate all those laws about which we could have any doubt, and to show that they are such that, even if God created many worlds, there could not be any in which they failed to be observed.[87]

Natural light and our knowledge of God's attributes enable us to have a priori knowledge of the laws of nature.[88] These laws, however, are very general, so general that they would apply to any world that God might have created. This conclusion follows from the fact that not only the existence of the laws, but also their content flow from the divine attributes.[89] Any of the worlds that God might create would be governed by the same laws of nature because the nature and existence of God are *uncreated* eternal truths and therefore universally applicable.[90] It follows that, if we are to have knowledge of *this particular world,* knowledge of the laws of nature alone is not enough:

> The principles which I have explicated above are so ample that one can deduce from them more things than we see in the world. . . . That

Descartes: Philosophy, Mathematics, and Physics, edited by Stephen Gaukroger (Sussex: Harvester, 1980), pp. 230–320.

86. For Descartes' overall contribution to mechanics in the seventeenth century, see Richard S. Westfall, *Force in Newton's Physics: The Science of Dynamics in the Seventeenth Century* (New York: American Elsevier, 1971), chap. 2.

87. Descartes, *Discourse on Method,* in *PWD*, vol. 1, p. 132 (AT, vol. 6, p. 43). The necessity Descartes described in this passage seems to be less relative and more absolute than that expressed in the letters he wrote to Mersenne and Mesland.

88. The concept of the natural light is one that Descartes used extensively in his philosophy and that had a complex history in medieval philosophy. Descartes' source was probably Thomas Aquinas, despite certain key differences between their conceptions of the natural light. See John Morris, "Descartes' Natural Light," *Journal of the History of Philosophy,* 11 (1973): 169–88. The Platonic notion of inner light probably entered medieval thought with Augustine's illuminationist theology. See Joseph Owens, "Faith, Ideas, Illumination, and Experience," in *The Cambridge History of Later Medieval Philosophy,* edited by Kretzmann, Kenny, and Pinborg, pp. 442–4.

89. See Nadler, "Deduction, Confirmation, and the Laws of Nature in Descartes' *Principia philosophiae,*" pp. 359–83, for a full discussion of this point.

90. Wells, "Descartes' Uncreated Eternal Truths," pp. 185–99.

is why I will make a brief description of the principal phenomena, of
which I aspire to seek the causes, not in order to draw the reasons
which serve to prove what I have to say hereafter: for I aim to explain
the effects by their causes and not the causes by their effects; but in
order that we choose amongst an infinity of effects which would be
deduced from the same causes, those which we ought principally to
try to deduce from them.[91]

Thus, Descartes considered his laws of nature as necessary but not sufficient conditions for explaining the phenomena of nature. Not only are
they too broad to deal with the specific phenomena of this world exclusively, but more detailed information – empirical information – is
needed if we are to know and to explain the phenomena of this world.

Two factors prevent us from deriving a priori the whole body of science
from these general principles alone. First, without observing, we do not
know which of many possible phenomena are actually in this world.
Second, any given phenomenon can be explained by several possible
mechanisms, each of which is compatible with the same set of general
principles. Despite the necessity of the eternal truths and the laws of
nature, many aspects of the world are in fact contingent. The particular
implementation of the laws of nature in the creation is contingent. "I
agree that a simple effect can be explained in several possible ways; but I
do not think that the possibility of things in general can be explained
except in one way, which is the true one."[92] In order to know which of
many possible phenomena are in the world and which of many possible
mechanisms God in fact used to produce the phenomena in question, one
must appeal to observation and experiment.[93]

91. Descartes, *Principia philosophiae,* in AT, vol. 8-1, pp. 80–2, and vol. 9-2,
     pp. 104–5.
92. Descartes to Mersenne, 28 October 1640, AT, vol. 3, p. 212 (PWD, vol. 3, p.
     154). Descartes' statement here that a single phenomenon can be explained
     in several different ways is the hypothetical physics to which Newton ob-
     jected in his statement "Non fingo hypotheses."
93. On the role of experiments in Descartes' science, see Daniel Garber, "Science
     and Certainty in Descartes," in *Descartes: Critical and Interpretive Essays,*
     edited by Michael Hooker (Baltimore: Johns Hopkins University Press,
     1978), pp. 114–51; James Collins, "Descartes' Philosophy of Nature,"
     American Philosophical Quarterly, Monograph Series, no. 5 (Oxford:
     1971); Paul J. Olscamp, "Introduction," in Descartes, *Discourse on Method
     for Rightly Directing One's Reason and Searching for Truth in the Sciences,*
     translated by Paul J. Olscamp (Indianapolis, Ind.: Bobbs-Merrill, 1965).
     Desmond M. Clarke, *Descartes' Philosophy of Science* (University Park:
     Pennsylvania State University Press, 1982), discusses in great detail the
     meanings of the words "experience," "experiment," and "reason" in

Descartes advocated this methodological strategy in the *Discourse on Method* (1637):

> When I sought to descend to more particular things, I encountered such a variety that I did not think the human mind could possibly distinguish the forms or species of bodies that are on the earth from an infinity of others that might be there if it had been God's will to put them there. Consequently I thought the only way of making these bodies useful to us was to progress to the causes by way of the effects and to make use of many special observations. And now, reviewing in my mind all the objects that have ever been present to my senses, I venture to say that I have never noticed anything in them which I could not explain quite easily by the principles I had discovered. But I must also admit that the power of nature is so ample and so vast, and these principles so general, that I notice hardly any particular effect of which I do not know at once that it can be deduced from the principles in many different ways; and my greatest difficulty is usually to discover in which of these ways it depend on them. I know no other means to discover this than by seeking further observations whose outcomes vary according to which of these ways provides the correct explanation.[94]

The fecundity of the laws of nature is such that particular effects can be derived from them in more than one way, just like a problem in geometry that can be solved by different paths of reasoning, all equally correct. Observation is needed in order to determine which derivation is the right one.

Descartes followed this method in the scientific treatises published in conjunction with the *Discourse on Method* as examples illustrating the power of the methods he was recommending. Consider, for example, the opening pages of his *Dioptrique*. Seeking to explain the observed properties of light, Descartes elaborated three different mechanical models – he called them *comparaisons* – which enabled him to explain the phenomena of colors, the transmission of light through solid, transparent matter, and the laws of reflection and refraction. He made plausible the perception of colors by purely mechanical means by comparing it to a blind man's use of a stick to perceive the various objects in his environment. He compared the transmission of light through solid, transparent matter to juice flowing through a vat full of grapes and passing out of a hole in the bottom of the vat; and he compared the reflection and refraction of light rays to a tennis ball, which, in the case of reflection, bounces off a hard surface,

Descartes' scientific works. See also Bernard Williams, *Descartes; The Project of Pure Enquiry* (Harmondsworth: Penguin Books, 1978), chap. 9.

94. Descartes, *Discourse on Method*, in *PWD*, vol. 1, p. 144 (AT, vol. 6, p. 64).

and, in the case of refraction, has its motion altered in passing through a "soft" material, "as when it strikes linen sheets, or sand, or mud."[95]

What epistemological status did Descartes ascribe to these analogies?

> Now since my only reason for speaking of light here is to explain how its rays enter into the eye, and how they may be deflected by the various bodies they encounter, I need not attempt to say what is its true nature. It will, I think, suffice if I use two or three comparisons in order to facilitate that conception of light which seems most suitable for explaining all those of its properties that we know through experience and then for deducing all the others that we cannot observe so easily. In this I am imitating the astronomers, whose assumptions are almost all false or uncertain, but who nevertheless draw many very true and certain consequences from them because they are related to various observations they have made.[96]

The models Descartes constructed were no more arbitrary than those of the astronomers. Because of the underdetermination of theories by facts, he considered it possible to generate true observations from theoretical premises of undetermined truth-value. Like the astronomers in the instrumentalist, pre-Copernican tradition who assumed that their models must be constructed from various combinations of uniform circular motion, Descartes assumed that his models must be couched in terms of the fundamental laws of nature, matter, and motion. But Descartes believed that he could do better than the astronomers had. He thought he could move beyond instrumentalism and actually establish the truth of his mechanical models by a complex process of experiment and observation.[97] Such mechanical analogies, once proven to account adequately for the observed phenomena, could then be deduced from first principles, or so Descartes thought:

> We have noted above that all the bodies which compose the universe are made of the same matter, that it is divided into several parts which are moved diversely and whose motions are in some manner circular; and that there is always an equal quantity of these motions in the world; but we have not been able to determine in the same way how great are the parts into which this matter is divided nor what is the speed with which they move, nor what circles they describe. For these things having been ordained by God in an infinity of diverse ways, it is only by experience, *and not by the force of reasoning* that one can know which of these he has chosen.[98]

95. Descartes, *Optics,* in *PWD,* vol. 1, p. 155 (AT, vol. 6, p. 88).
96. Ibid., pp. 152–3 (AT, vol. 6, p. 83).
97. See Garber, "Science and Certainty in Descartes."
98. Descartes, *Principia philosophiae,* in AT, vol. 8-1, pp. 100–1, and vol. 9-2, p. 124.

That is to say, God, by his absolute power, has created this particular world. God's freedom of will in that act of creation is reflected by the fact that we do not know a priori which particular phenomena, of the infinite number consistent with the laws of nature, he actually created.[99] That determination demands the use of empirical methods. Experiments and observations are required to decide which consequences, from a potentially infinite set deducible from the laws of nature, are the actual ones that describe the created world. According to the Cartesian method, empirical evidence serves to eliminate possibilities. If it were possible to enumerate all possible conclusions from first principles and to use experiments and observations to eliminate all possible phenomena and explanatory mechanisms but one, then the demonstrative character of a completed natural philosophy would be guaranteed. In this way, despite the fact that experiments and observations are genuinely empirical, Descartes believed that they serve to ensure the certainty of his natural philosophy.[100]

In this sense, my interpretation is consonant with Nadler's argument that although Descartes was able to derive the laws of nature a priori from knowledge of the divine attributes, he needed to appeal to observations and experiments in order to confirm the application of the laws to particular phenomena in this world.[101] Whether or not Descartes considered these empirical methods necessary for *confirming* the laws of nature, as Nadler argues,[102] he certainly invoked them to show how the very general laws apply to particular phenomena.[103] My analysis goes beyond Nadler's, however, by showing how Descartes' use of a priori and empirical methods can be understood in relation to his interpretation of God's *potentia absoluta* and *potentia ordinata*. The parts of natural philosophy

99.   See Garber, *Descartes' Metaphysical Physics*, pp. 109–10.
100. See Garber, "Science and Certainty in Descartes," p. 136. I think that Garber overstates his claim that Descartes came close to abandoning the deductivist ideal for science because of difficulty in generating appropriate deductions in the later sections of the *Principles of Philosophy*. See also Morrison, "Hypotheses and Certainty in Cartesian Science," in *An Intimate Relation,* edited by Brown and Mittelstrass, pp. 43–64. Morrison ignores the theological and metaphysical context of Descartes' methodological concerns and sees him as struggling simply to formulate an adequate approach to science.
101. Nadler, "Deduction, Confirmation, and the Laws of Nature in Descartes' *Principia philosophiae.*"
102. Ibid., pp. 272–4.
103. See Daniel Garber, "Descartes and Experiment in the *Discourse* and *Essays,*" in *Essays on the Philosophy and Science of René Descartes,* edited by Stephen Voss (New York: Oxford University Press, 1993), pp. 288–310.

that can be known a priori are those aspects of the created order that God created to be necessary or that follow necessarily from the divine attributes. They are the things on which his absolute power can no longer act. Those aspects of the creation that can be known only by appeal to empirical methods are the things still subject to God's absolute power and, accordingly, contingent on divine will.

In order to explain the observed phenomena, Descartes proposed mechanisms that were subject to further experimental test. In Part III of *The Principles of Philosophy*, he discussed at length the role of what he called "suppositions" in science. In order to explain the formation of this world, he supposed the existence of certain initial conditions from which the present effects could have been formed in accordance with the laws of nature:

> Indeed, in order to explain natural things better, I may even seek their causes here higher than any that are visible. . . . just as for an understanding of the natures of plants or men it is better by far to consider how they can gradually grow from seeds than how they were created by God at the very beginning of the world. Thus if we can invent some principles which are very simple and easily known, from which we can demonstrate that the stars and the Earth, and indeed everything which we perceive in this visible world, could have originated from certain seeds, although we know that that was not their origin, we shall in that way explain their natures much better than if we were to describe them as they are now.[104]

Descartes proceeded to make a number of suppositions about the ways in which God divided matter at the beginning of the world, the sizes of the parts of matter, and the amount of motion with which he endowed it. "The falsity of these suppositions does not prevent us from deducing what is true and certain from them." From just a few of them, we can explain everything in the world according to the laws of nature.[105] In other words, given that the world is as we know it to be, suppose certain initial conditions from which it could have developed according to the laws of nature.

The method of supposition (*ex suppositione*) that Descartes introduced without further explanation or justification in the middle of *The Principles of Philosophy* had a long history in medieval commentaries on Aristotle and had occupied a prominent place in Thomas Aquinas' discussions of scientific method. It was a method designed to validate reasoning from causes to effects despite the contingency of natural phenomena.[106] No

104. Descartes, *Principia philosophiae*, in AT, vol. 8-1, pp. 99–100.
105. AT, vol. 8-1, p. 101, vol. 9-2, p. 125.
106. William A. Wallace, "Galileo and Reasoning *Ex Suppositione*," in *Prelude to Galileo: Essays on Medieval and Sixteenth-Century Sources of Galileo's*

doubt, Descartes became acquainted with this method through his contact with Jesuit Aristotelian thought at La Flèche.[107]

The method of suppositions was an important component of Descartes' attempt to explain the effects he discovered in the world. Couched in the mechanical terms stipulated by the general principles on which his physics rested and developed to explain the observed phenomena, these suppositions provide causal explanations of specific phenomena:

> It hardly matters in what way I suppose that matter has been disposed at the beginning, since its disposition must afterwards have changed following the laws of nature, and one barely knows how to imagine from what one could prove that by these laws, *it continually changes until it finally composes a world entirely similar to this* (although it would take longer to deduce from one supposition than from another): for the laws being the cause that matter must successively take all the forms of which it is capable, if one considers by the order of all these forms, one could finally arrive at that which happens presently in this world.[108]

In addition to alluding to the method of supposition, this passage echoes the discourse of the absolute and ordained powers of God, shorn of its theological language. The possible variety of forms that matter can assume is a consequence of the absolute power by which God disposed things in one of an infinite number of ways. We make suppositions about which possible configurations of matter are the ones that actually exist. Ultimately the choice among rival suppositions must be based on how well they agree with experience. Once experience tells us which phenomena need explaining and which explanations best conform to reality, then Descartes believed, it would be possible to deduce both the phenomena and their mechanical explanations from first principles, which embody the necessity with which God endowed them at creation and which he maintains by his ordained power.

Like many writers on the methodology of science, Descartes did not always practice the methods that he preached. Most of the mechanical models he described in *Le monde* and later in *The Principles of Philoso-*

Thought, edited by William A. Wallace (Dordrecht: Reidel, 1981), pp. 132–3.

107. Feldhay, "Knowledge and Salvation in Jesuit Culture," pp. 197–213.
108. Descartes, *Principia philosophiae,* in AT, vol. 8–1, p. 103. In this passage, there seems to be some resonance from the principle of plenitude, which Lovejoy finds in Descartes' astronomical hypotheses as well. See Arthur O. Lovejoy, *The Great Chain of Being: The Study of the History of an Idea* (Cambridge, Mass.: Harvard University Press, 1936), p. 123. I am grateful to B. J. T. Dobbs for noting this point.

phy were highly speculative and were not grounded in the kind of experimental practice he called for in these methodological remarks. Nevertheless, I think his dicta still apply if we regard *Le monde* and *The Principles of Philosophy* as prescriptive works, stating what natural philosophy should look like. In this light, they can be understood as establishing the terms in which explanations and models ought to be couched, rather than standing as the finished products of completed research. Accordingly, the speculative models that Descartes used to explain various phenomena serve as exemplars of mechanical explanations rather than actual explanations that have survived the rigors of experimental test.

Was Descartes really an intellectualist?

Twentieth-century Anglo-American philosophers in the analytic tradition have frequently called Descartes a voluntarist, citing his theory of the creation of the eternal truths as clear evidence that he considered God's will to be primary. In so doing, they fail to come to grips with subtleties of the medieval dialectic between the absolute and ordained powers of God and the influence of that discourse on Descartes' theory. Consequently, they do not distinguish between God's exercise of his absolute power in creating the world, something about which virtually all Christian thinkers agreed, and the question of God's relationship to the natural order that he thereby created. It is in the latter context that we should look for the differences between voluntarists and intellectualists.

For example, in a frequently cited article Frankfurt claims that Descartes denied that rationality is in any way essential to God and that, rather than reason, unconstrained will or power is at the source of the universe.[109] There are two arguments against Frankfurt's interpretation: First, The point is not whether rationality is essential to God, but whether it is essential to the world. It is essential to the world, for Descartes, because God has given us the means for having insight into the necessities of the creation. That is to say, it is the rational necessity God imparted to aspects of the creation that underwrites our capacity for a priori knowledge of the eternal truths and the laws of nature. Descartes' understanding of God's power and his relationship to the creation is highlighted by concentrating on the role of contingency and necessity in the creation rather than on the divine attributes. A second problem with Frankfurt's argument is that God's will and understanding are united for Descartes; therefore, will alone cannot be at the source of the universe.

109. Frankfurt, "Descartes and the Creation of the Eternal Truths," p. 54.

In an important study of the development of Descartes' thought, Schuster presents Descartes as a thoroughgoing voluntarist.[110] Schuster sees Descartes' alleged voluntarism as separate from and independent of his justificatory scheme of metaphysics. He speaks of a "marriage between Voluntarist theology and mechanical philosophy in the seventeenth century"[111] and describes voluntarists as "trying to avoid the conclusion that God's intellect may be necessitated by truths existing in some sense independently of Himself."[112] Schuster then argues that Descartes' voluntarism is evident in God's creation of the eternal truths and laws of nature, a theory that forecloses the possibility that "true essences can somehow subsist independently of God."[113] While Platonic realism is one kind of metaphysics antithetical to voluntarism, it is not the only kind. Understanding how Cartesian rationalism differs from voluntarism depends on understanding the distinction between God's absolute and ordained powers, a distinction that Schuster does not adequately consider. True, Descartes (and Thomas Aquinas and Ockham and Gassendi and Boyle – and virtually all Christian natural philosophers) would agree that nothing exists apart from what God created freely by his absolute power. What distinguishes Descartes and Aquinas from the voluntarists is the fact that the former thought that God created the eternal truths and laws to be necessary, whereas Gassendi and the voluntarists did not. By overlooking the distinction between the two powers of God, Schuster fails to see the differences between Descartes' intellectualism and rationalism, on the one hand, and Gassendi's and Boyle's voluntarism, on the other.

Underlying Descartes' belief that the laws of nature can be known a priori was his belief in the stability of the natural order. "He [God] always maintains them [the different parts of matter] in the same way and with the same laws which *he made them obey* at their creation."[114] Indeed in his outspoken, but unpublished, treatise *Le monde,* Descartes made such a strong claim for the regularity and stability of nature that he denied the occurrence of miracles in his mechanical universe, a startlingly heterodox claim. This denial of miracles occurs at the climax of his discussion of the laws of nature and the possibility of having a priori, demonstrative knowledge of "everything which could be produced in this new world":

110. John Andrew Schuster, "Descartes and the Scientific Revolution, 1618–1634," Ph.D. dissertation, Princeton University, 1977, pp. 622–47.
111. Ibid., p. 624.
112. Ibid.
113. Ibid., pp. 625–7.
114. Descartes, *Principia philosophiae,* in AT, vol. 8–1, p. 62, and vol. 9–2, p. 84.

And in order that there would be no exceptions which would prevent us from it [i.e., a priori knowledge], we will add, if you please, to our suppositions that God will never produce any miracles and that the Intelligences or reasonable souls, which we could suppose in it hereafter, would not in any way disturb the ordinary concourse of nature.[115]

Descartes' concern to establish the possibility of an a priori science known with certainty moved him to deny divine intervention in the world. It is noteworthy that this denial of miracles is a much stronger claim for necessity than his understanding of divine omnipotence demands. Descartes' extreme emphasis on regularity is a far cry from the voluntarists' insistence on the possibility of divine intervention and the resultant contingency of the creation and uncertainty of human knowledge. Even though voluntarist natural philosophers believed in the regularity of the natural order, they understood that regularity to be grounded in God's free choice rather than in any necessity embedded in the creation.[116] In Descartes' world, even the omnipotent God would not disturb the ordinary concourse of nature lest he prevent us from having a priori, demonstrative knowledge of the world. God established this stability by creating necessary laws, which restricted his further freedom of action.

Descartes' emphasis on the regularity of the created world might seem to be contradicted by his well-known statements about continuous creation.[117] He enunciated this doctrine in a famous passage in the Third Meditation:

A lifespan can be divided into countless parts, each completely independent of the others, so that it does not follow from the fact that I existed a little while ago that I must exist now, unless there is some cause which as it were creates me afresh at this moment – that is, which preserves me. For it is quite clear to anyone who attentively considers the nature of time that the same power and action are needed to preserve anything at each individual moment of its duration as would be required to create that thing anew if it were not yet in existence. Hence the distinction between preservation and creation is only a conceptual one, and this is one of the things that are evident by the natural light.[118]

115. Descartes, *Le monde, ou traité de la lumière*, in AT, vol. 11, pp. 47–8.
116. Francis Oakley, *Omnipotence, Covenant, and Order: An Excursion in the History of Ideas from Abelard to Leibniz* (Ithaca, N.Y.: Cornell University Press, 1984), p. 62.
117. Descartes' theory of continuous creation is discussed at length in Garber, *Descartes' Metaphysical Physics*, pp. 265–72.
118. Descartes, *PWD*, vol. 2, p. 33 (AT, vol. 7, pp. 48–9).

If this passage is interpreted to mean that God actually re-creates the world at each instant, what is to prevent him from creating radically different worlds at each instant, that is, to annihilate any regularity that guarantees our a priori knowledge of the laws of nature? There are two ways to solve this problem. One is to deny that the passage actually states that time is discontinuous, the product of successive, independent acts of creation.[119] The second solution avoids the necessity of dealing with technical questions about the continuity or discontinuity of time; it relies, once again, on Descartes' conception of the absolute and ordained power of God and the resultant status of the laws of nature. As Descartes claimed in Part VI of the *Discourse on Method,* because the necessity of the laws of nature flows from God's immutability, the laws must hold in all possible, created worlds.[120] Even if the occasionalist interpretation of Descartes' statements about continuous creation is correct (an interpretation I am not inclined to accept), the worlds created in successive moments must all be governed by the same laws of nature. Consequently, the possibility of an a priori, demonstrative knowledge of the world remains.

Descartes' contemporaries well understood the distinction between voluntarism and intellectualism, even if they did not use these terms to describe their positions. Some seventeenth-century natural philosophers criticized Descartes for being insufficiently voluntarist, and they understood the issues in terms very similar to the ones I have been developing. Not only Gassendi, whose criticisms occupy much of the next chapter, but also Robert Boyle and Isaac Newton criticized Descartes on the grounds that his system placed unacceptable limitations on divine freedom.

It is likely that Boyle actually read Descartes' letters of 1630 in which the question of the eternal truths was central. Boyle was thoroughly acquainted with Descartes' major writings, references to which are sprinkled throughout his works. He made at least two direct references to the published volumes of Descartes' letters,[121] the publication of which both

---

119. Richard T. W. Arthur, "Continuous Creation, Continuous Time: A Refutation of the Alleged Discontinuity of Cartesian Time," *Journal of the History of Philosophy,* 26 (1988): 349–75. See also Daniel Garber, "How God Causes Motion: Descartes, Divine Sustenance, and Occasionalism," *Journal of Philosophy,* 84 (1987): 567–80.

120. Descartes, *Discourse on Method,* in *PWD,* vol. 1, pp. 141–51 (AT, vol. 6, pp. 60–78).

121. Robert Boyle, *The Excellency of Theology, Compared with Natural Philosophy . . .* (1674), in *The Works of the Honourable Robert Boyle,* edited by Thomas Birch, 6 vols. (London, 1772; reprinted, Hildesheim: Georg Olms, 1965), vol. 4, pp. 21–2; and *Some Considerations About the Reconcileableness of Reason and Religion* (1675), in *Works,* vol. 4, p. 163.

Samuel Hartlib (c. 1600–62) and Henry Oldenburg (1617?-77) mentioned in letters to him.[122]

At several places in his works, Boyle took Descartes to task for not sufficiently emphasizing God's absolute power and freedom of will:

> I have often wished, that learned gentleman [Descartes] had ascribed to the divine Author of nature a more particular and immediate efficiency and guidance, in contriving the parts of the universal matter into that great engine we call the world; and . . . I am still of the opinion that he might have ascribed more than he has to the supreme cause, in the first and original production of things corporeal without the least injury to truth, and without much, if any, prejudice to his own philosophy.[123]

Boyle implicitly criticized Descartes' rationalism as constraining God's creativity by restricting it to the capacities of human understanding:

> There is no necessity, that intelligibility to a human understanding should be necessary to the truth or existence of a thing, any more than that visibility to a human eye should be necessary to the existence of an atom, or of a corpuscle of air, or of the effluviums of a loadstone, or of the fragrant exhalations of a rose.[124]

And he objected to Descartes' proof of the law of the conservation of motion on the basis of divine immutability. That proof, according to Boyle, unacceptably limited God's power and was inconsistent with God's ongoing creative relationship with the world:

> It seems not clear, why God may not as well be immutable, though he should sometimes vary the quantity of motion, that He has put into the world, as He is, though, according to the opinion of most of the

---

122.   Quoting Oldenburg, Hartlib wrote: "I suppose you have heard, that there is another volume of *Des Cartes* letters under the press, which will shortly be finished. . . . They are advanced by an admirer of his, called Monsieur *Clerselier.*" Hartlib to Boyle, 10 May 1659, in *Works*, vol. 6, p. 124. Oldenburg wrote to Boyle at another time, saying, "My philosophical friend tells me, that another volume of *Des Cartes*' letters is printed, promising to send me a copy by the first opportunity." Oldenburg to Boyle, 25 September 1666, in *Works*, vol. 6, p. 231. Oldenburg was referring here to the first edition of the *Letters of Descartes*, edited by Claude Clerselier, which appeared in three volumes published in 1657, 1659, and 1667. Descartes' letters from April and May 1630 were published in the first two volumes. See AT, vol. 1, intro.

123.   Boyle, *An Hydrostatical Discourse, Occasioned by the Objections of the Learned Dr. Henry More* . . . (1672), in *Works*, vol. 3, p. 597.

124.   Boyle, *The Christian Virtuoso: Shewing, That, by being addicted to the Experimental Philosophy, a man is rather assisted than indisposed to be a good Christian* (1690), in *Works*, vol. 6, p. 694.

> Cartesians themselves, He does daily create multitudes of rational
> souls to unite these to human bodies.[125]

In other words, Boyle criticized Descartes for not adhering to voluntarist principles.

Voluntarism was also central to Newton's thinking. As Dobbs has demonstrated, he was devoted to establishing the reality of divine activity in the created world.[126] Newton believed that God is capable of doing anything that is not logically contradictory and that "the world might have been otherwise than it is" because its creation was the voluntary act of an omnipotent God.[127] Although God created the laws of nature, he is not bound by them: "It may also be allow'd that God is able . . . to vary the Laws of Nature, and make Worlds of several sorts in several Parts of the Universe. At least, I see nothing of Contradiction in all this."[128]

Like Boyle, Newton found Descartes' philosophy of nature insufficiently voluntarist. Criticizing the identification of matter with extension, which Descartes considered eternal and immutable "because" in Newton's words, "it is the emanent [*sic*] effect of an eternal and immutable being,"[129] Newton insisted that the existence of matter and its observed properites are utterly dependent on divine will. Matter, he asserted,

> does not exist necessarily but by divine will, because it is hardly given
> to us to know the limits of divine power, that is to say whether matter
> could be created in one way only, or whether there are several ways by
> which different beings similar to bodies could be produced.[130]

In a passage reminiscent of the nominalists' insistence that God can do directly whatever he usually does by second causes, Newton asserted that

125. Boyle, *A Disquisition About the Final Causes of Natural Things* . . . (1688), in *Works*, vol. 5, p. 397. See Eugene M. Klaaren, *Religious Origins of Modern Science: Belief in Creation in Seventeenth-Century Thought* (Grand Rapids, Mich.: Eerdmans, 1977), p. 120.

126. B. J. T. Dobbs, *The Janus Faces of Genius: The Role of Alchemy in Newton's Thought* (Cambridge University Press, 1991).

127. Isaac Newton, "Of natures obvious laws & processes in vegetation," Dibner Collection MSS 1031 B (1, n. 30), f.4v, as quoted by Dobbs, *Janus Faces of Genius*, p. 266.

128. Isaac Newton, *Opticks, or A Treatise of the Reflections, Refractions, Inflections & Colours of Light*, with a forward by Albert Einstein, intro. by Sir Edmund Whittaker, and preface by I. Bernard Cohen. Based on the 4th edition, London, 1730 (New York: Dover, 1952), pp. 403–4.

129. Isaac Newton, "De Gravitatione et Aequipondio Fluidorum" (Cambridge MS. Add. 4003), in *Unpublished Scientific Papers of Isaac Newton*, edited by A. Rupert Hall and Marie Boas Hall (Cambridge University Press, 1962), p. 137.

130. Ibid., p. 138.

God could produce the observed properties of matter directly, without the mediation of any substantial creation:

> If he should exercise this power, and cause some space projecting above the Earth, like a mountain or any other body, to be impervious to bodies and thus stop or reflect light and all impinging things, it seems impossible that we should not consider this space to be truly body from the evidence of our senses (which constitute our sole judges in this matter); for it will be tangible on account of its impenetrability, and visible, opaque and coloured on account of the reflection of light, and it will resonate when struck because the adjacent air will be moved by the blow.[131]

Newton's voluntarism lay behind his fear that Descartes' equation of matter and extension would lead to atheism,

> both because extension is not created but has existed eternally, and because we have an absolute idea of it without any relationship to God, and so in some circumstances it would be possible for us to conceive of extension while imagining the non-existence of God.[132]

He thus perceived Descartes' idea of extension as contradicting a voluntarist understanding of God's relationship to the creation. While Newton did not address the question of the status of eternal truths in criticizing the theological implications of Descartes' philosophy of nature, it is nonetheless significant that he objected to the intellectualist strands of his predecessor's thought.

It was indeed these intellectualist strands that gave Descartes' system its defining characteristics – the rationalist elements of his theory of knowledge, his belief that the first principles of natural philosophy could be known a priori, and his essentialist metaphysics. Based, ultimately, on his theory of the creation of the eternal truths, they spell out the consequences of a world containing some elements of necessity. To his voluntarist contemporaries, most significantly Gassendi, it was these very features of his thought that seemed particularly deserving of criticism. It is in this light that I examine the heated controversy between Gassendi and Descartes that followed the publication of the *Meditations* in 1641.

131. Ibid., p. 139.
132. Ibid., p. 143. See Westfall, *Force in Newton's Physics*, p. 341.

6

Gassendi and Descartes in conflict

When . . . I imagine a triangle, even if perhaps no such figure exists, or has ever existed, anywhere outside ·my thought, there is still a determinate nature, or essence, or form of the triangle which is immutable and eternal, and not invented by me or dependent on my mind.
René Descartes, Meditations[1]

It is hard to agree that there exists some *immutable and eternal nature* other than [that of] omnipotent God.
Pierre Gassendi, Disquisitio metaphysica[2]

Prior to the publication of the *Meditations* in 1641, Descartes asked his friend and correspondent Marin Mersenne to circulate the manuscript for comment and criticism among a number of philosophers, including his fellow mechanical philosophers, Hobbes and Gassendi.[3] These comments and Descartes' subsequent responses, were published as a lengthy appendix to the *Meditations,* the "Objections and Replies." Despite the fact that Gassendi and Descartes shared a strong anti-Aristotelian sentiment, a

1. René Descartes, *Meditations on First Philosophy,* in *The Philosophical Writings of Descartes* (hereafter *PWD*), translated by John Cottingham, Robert Stoothoff, Dugald Murdoch, and Anthony Kenny, 3 vols (Cambridge University Press, 1984, 1985, 1991), vol. 2, pp. 44–5; René Descartes, *Oeuvres de Descartes,* edited by Charles Adam and Paul Tannery (hereafter AT), 11 vols. (Paris: J. Vrin, 1897–1974), vol. 7, p. 64.
2. Gassendi, *Disquisitio metaphysica seu dubitationes et instantiae adversus Renati Cartesii metaphysicam et responsa,* edited and translated into French by Bernard Rochot (Paris: J. Vrin, 1962), pp. 468–9; in Pierre Gassendi, *Opera omnia,* 6 vols. (Lyon, 1658; facsmile reprint, Stuttgart-Bad Cannstatt: Friedrich Frommann Verlag, 1964), vol. 3, p. 374.
3. For a brief account of the circumstances that led to the "Objections and Replies" to Descartes' *Meditations,* see *PWD,* vol. 2, pp. 63–5.

consuming interest in natural philosophy, and a dedication to the establishment of a new, mechanical philosophy of nature, their arguments in the "Objections and Replies," later amplified in Gassendi's *Disquisitio metaphysica* (1644), were heated, to say the least. This controversy signaled a major rift between the two founding fathers of the mechanical philosophy. What was really at stake between them?

Their conflict can be understood as yet one more episode in the ongoing confrontation between the Judeo-Christian belief in divine omnipotence and the classical Greek preoccupation with the intelligibility of nature. Erupting again in the profound disagreement between these two mechanical philosophers, the controversy still bore the mark of ideas that had been forged in medieval debates about the relationship between God's absolute and ordained powers. In the seventeenth century, when interest in natural philosophy was rising, these medieval ideas were transformed into views about the epistemological status of knowledge about the world. The underlying controversy persisted. Seventy-five years later, it was one of the major points of contention between Leibniz and Clarke in their debate about how to understand providence in a mechanical world.[4]

Gassendi and Descartes held different views about God's absolute and ordained powers. In this debate, Gassendi's emphasis on divine will and the contingency of the creation directly conflicted with Descartes' acceptance of necessity in the created world. This controversy not only illuminates the divergent philosophical consequences of their theological assumptions; it also helps to explain the differences in their respective versions of the mechanical philosophy.

In this chapter, I consider some of the particular philosophical differences that arose in their exchange and show how they are tied to their respective theological presuppositions. Previous discussions of this episode have not adequately taken into account the role that theology played in their dispute. Pintard interpreted the controversy in terms of Gassendi's association with the *libertins érudits,* an association in which he saw far more significance than I do, given Gassendi's serious concern with theological issues.[5] Rochot emphasized certain parallels between the text of the *Disquisitio metaphysica* and the sections of the *Syntagma philosophicum* treating logic, although he did discuss theological aspects

4. See *The Leibniz-Clarke Correspondence,* edited by H. G. Alexander (Manchester: Manchester University Press, 1956).

5. René Pintard, "Descartes et Gassendi," in *Travaux du IXe Congrès de Philosophie – Congrès Descartes II, Etudes Cartésiennes* (Paris: Hermann, 1937).

of the dispute briefly.[6] Bloch emphasizes Gassendi's rejection of Descartes' "intuitive mode" of reasoning without explaining its theological matrix.[7] Sarasohn interprets the controversy in the context of Gassendi's concern with Christianizing Epicurean ethical theory, paying particular attention to questions about human freedom.[8] Marjorie Grene focuses on purely philosophical issues, such as Gassendi's and Descartes' theories of knowledge, universals, and philosophical methods.[9] I will argue that all of the issues dividing Gassendi and Descartes were closely tied to the differences in their understandings of God's relationship to the creation.

Despite their differences, Gassendi and Descartes were well acquainted with each other's work, and before the publication of the *Meditations* in 1641, the two men had regarded each other with mutual respect. In his correspondence, Descartes frequently expressed admiration for Gassendi's scientific knowledge; and the attention Gassendi devoted to Descartes' thought – not only in the "Objections" and in the *Disquisitio metaphysica,* but also in scattered comments in the *Syntagma philosophicum* – indicates his recognition of the importance of the Cartesian philosophy.

Gassendi's lengthy "Objections" followed the order of topics in the *Meditations,* and Descartes replied to them point by point. In the *Disquisitio metaphysica,* Gassendi responded in tedious detail to each of Descartes' replies. From the outset, the two philosophers considered the overt issue between them to be methodological and epistemological. Gassendi explained that he had raised his objections "not against the things you have undertaken to demonstrate, but against the method and kind of proof [which you use] to demonstrate [them]."[10] Similarly, Descartes objected to Gassendi's approach, counting him among "those whose mind is so drowned by the senses that they cannot endure metaphysical thoughts. . . . That is why I do not respond to you as a very subtle philosopher but as one of those made of flesh."[11] From this point on, the two

6.   Bernard Rochot, "Gassendi et la 'Logique' de Descartes," *Revue philosophique de la France et de l'étranger, 80* (1955): 300–308; Bernard Rochot, "Les véritées éternelles dans la querelle entre Descartes et Gassendi," *Revue philosophique de France et l'étranger, 141* (1951): 288–98.

7.   Olivier René Bloch, "Gassendi critique de Descartes," *Revue philosophique de la France et de l'étranger, 91* (1966): 217–36.

8.   Lisa T. Sarasohn, *Freedom in a Deterministic Universe: Gassendi's Ethical Philosophy* (Ithaca, N.Y.: Cornell University Press, forthcoming), chap. 5.

9.   Marjorie Grene, *Descartes* (Minneapolis: University of Minnesota Press, 1985), chap. 6.

10.   Gassendi, *Disquisitio metaphysica,* pp. 10–11 (*Opera omnia,* vol. 3, p. 273).

11.   Ibid., pp. 12–13 (p. 274).

philosophers addressed each other as "O Mind" and "O Flesh." These epithets – exaggerated though they were – epitomized their respective positions. The fundamental epistemological difference between them concerned how deeply reason, unaided by the senses, can penetrate the essences of things and how adequately any knowledge based on the senses can represent reality. These differences bore the imprint of the philosophers' underlying theological differences.

Descartes' emphasis on the a priori aspects of scientific knowledge was based on in his distrust of observation as a reliable source of knowledge. The *Meditations* began with a skeptical critique of empirical knowledge. "From time to time I have found that the senses deceive, and it is prudent never to trust completely those who have deceived us even once."[12] Descartes made this statement in the context of his systematic use of the skeptical arguments to eliminate any dubitable statements in his search for a foundation for certain knowledge. His distrust of the senses was no mere rhetorical ploy: It accurately expressed his attitude toward claims of empirical knowledge. In his arguments for the existence and nature of the human mind, God, and the material world, he constantly appealed to a priori methods and rejected the use of empirical ones.

Gassendi regarded the skeptical arguments as both a weaker and a stronger epistemological challenge than did Descartes. He did not share Descartes' need to doubt everything he had formerly believed. Descartes voiced this need in the opening of the *Meditations:*

> Some years ago I was struck by the large number of falsehoods that I had accepted as true in my childhood, and by the highly doubtful nature of the whole edifice that I had subsequently based on them, I realized that it was necessary, once in the course of my life, to demolish everything completely and start again from the foundations if I wanted to establish anything at all in the sciences that was stable and likely to last.[13]

Gassendi did not see the need for such a radical attack on all previous knowledge claims in order to secure sound foundations for the new science. Indeed, his chapter "On Method" in the *Syntagma philosophicum* simply describes the proper way to go about acquiring scientific knowledge. It does not attempt to justify these methods by providing epistemological foundations for them.[14] Gassendi refused to play the philosophical

12. Descartes, *Meditations*, in *PWD*, vol. 1, p. 12 (AT, vol. 7, p. 18).
13. Ibid., p. 12 (p. 17).
14. Pierre Gassendi, *Syntagma philosophicum*, in *Opera omnia*, vol. 1, pp. 120–4. For an English translation, see *The Selected Works of Pierre Gassendi*, edited by Craig B. Brush (New York: Johnson Reprint, 1972), pp. 366–79.

game that lay at the heart of the Cartesian revolution in philosophy.[15] It is Gassendi's refusal to join Descartes in his search for absolute and indubitable foundations that explains why the controversy between them often seems to be conducted at cross purposes.

Their epistemological differences hinged on their respective assessments of the power and reliability of human reason. Descartes saw reason as reliable and, if properly used, free from error.[16] According to Gassendi, the appearances, free from any judgment imposed by the intellect, are reliable and true. Error, he thought, arises from the misuse of the faculty of reason.[17] This disagreement about the reliability of reason as a source of knowledge was a constantly repeated theme in their exchange. I have argued in earlier chapters that the differences in their epistemological views reflect differences in their views about God's relationship to the creation. In their direct controversy, this relationship becomes explicit.

In each of the specific points about which they disagreed, the argument between them assumed a similar form. Descartes employed a rationalist approach, taking intelligibility as a criterion for real existence. Using his rule that "everything that we clearly and distinctly understand is true in a way which corresponds exactly to our understanding of it,"[18] he repeatedly argued from the conceptions in his mind to the natures of things in the world. He also accepted an essentialist metaphysics, claiming that his method of arguing led him to certain knowledge of the essences of things. Gassendi, on the contrary, consistently used empiricist and nominalist arguments to criticize Descartes and to defend his own position. Gassendi found Descartes' use of the criterion of clear and distinct ideas as a rule for reasoning from ideas in his mind to truths about the world entirely unacceptable.[19] Since this method lay at the very heart of Descartes' project, Gassendi's criticism of it was central to his attack on Cartesian philosophy more generally. Gassendi argued that we have no assurance of the adequacy of the clear and distinct rule because the skeptical arguments undermine our epistemological confidence: We may be deceived even when we perceive something clearly and distinctly.[20] Descartes' only response was

15. Pintard, "Descartes et Gassendi," p. 117.
16. Descartes, *Meditations*, pp. 39–41 (AT, vol. 7, pp. 55–9).
17. Gassendi, *Disquisitio metaphysica*, pp. 416–17, 428–9 (*Opera omnia*, vol. 3, pp. 363, 366).
18. Descartes, *Meditations*, in *PWD*, vol. 2, p. 9 (AT, vol. 7, p. 13).
19. Olivier René Bloch, *La philosophie de Gassendi: Nominalisme, matérialisme, et métaphysique* (The Hague: Martinus Nijhoff, 1971), p. 21.
20. Gassendi, *Disquisitio metaphysica*, pp. 198–9, 458–9 (*Opera omnia*, vol. 3, pp. 314, 372).

to say that Gassendi had not taken sufficient pains to understand his reasoning.[21]

The two philosophers disagreed about the origin of our ideas. Descartes believed that at least some of our ideas are innate.[22] Gassendi disagreed, stating that all our ideas originate in the senses. He explained that abstract ideas such as Thing, Truth, and Thought are formed by the mind by a process of abstracting from the ideas it has received from sensation of many particular things (or true statements or thoughts).[23] Gassendi justified his claim that all ideas have a sensory origin by citing the ideas "which are absent among the blind and the deaf."[24] Descartes replied that Gassendi was begging the question, for he had no way of knowing what kinds of ideas a blind or deaf person has.[25] Gassendi, as one might expect, proceeded to cite empirical evidence in support of his claim.[26]

In Descartes' eyes, use of the clear and distinct rule provided a warrant for reasoning from the ideas of things to knowledge of their essences. The fabric of the *Meditations* is woven from arguments of this form: the *cogito*, from which Descartes concluded that the essence of the human mind is a thinking thing; his proof of the existence of God; and his argument that the essence of matter is extension. Gassendi rejected Descartes' essentialism and rationalism by appealing to his own nominalism and empiricism. He denied that we can have any knowledge of the intimate natures or essences of things, reiterating his claim that we are only acquainted with their appearances. Knowledge of their attributes or properties is not the same as knowledge of their essences:

> All that we can know is that, if such-and-such properties belong to such a substance or such a nature, that appears to observation and becomes evident to our experience; and one does not penetrate to the substance or intimate nature; thus when we see water spout from a source, although we know water comes from that source, that does not permit us to plunge to the interior and subterranean source of that spouting.[27]

Gassendi's epistemology confined human knowledge to the external appearances of things. Such knowledge of the appearances is sufficient for

21. Ibid., pp. 460–1 (p. 372). Grene, *Descartes*, chap. 6, documents Gassendi's perceptive undertanding of Descartes.
22. Descartes, *Meditations*, in *PWD*, vol. 2, p. 26 (AT, vol. 7, pp. 37–8).
23. Gassendi, *Disquisitio metaphysica*, pp. 214–15 (*Opera omnia*, vol. 3, p. 318).
24. Ibid., pp. 224–5 (p. 320).
25. Ibid., pp. 226–9 (p. 321).
26. Ibid., pp. 238–41 (pp. 323–4).
27. Ibid., pp. 186–9 (p. 312).

our needs.[28] It is illusory to imagine that we can proceed beyond the appearances to knowledge of inner natures. "[Not] that it does not appear beautiful and desirable to know that," Gassendi continued, "but that is the same manner in which it is good and desirable for us . . . to have wings or to remain forever young.59.[29]

Descartes disagreed, asserting that the idea of substance is logically prior to that of the accidents. "It is the accidents which are conceived by means of the substances . . . In fact, no reality . . . can be attributed to the accidents which was not borrowed from the idea of substance."[30] For that reason, he thought, it makes no sense to talk about knowing the accidents without first having an idea of the substance to which they belong. As with almost every other issue dividing them, Descartes ard Gassendi viewed the question of the knowledge of essences and substances from opposing standpoints. Descartes, the essentialist and rationalist, regarded the abstract and intelligible as primary, whereas Gassendi, the nominalist and empiricist, presumed the primacy of the individual, the observable, and the concrete.

By far the longest section of the *Disquisitio metaphysica* concerns Descartes' proof of the existence of God in the Third Meditation. Descartes had tried to prove the existence of God using an a priori argument, proceeding from his innate idea of God as "a substance that is infinite, <eternal, immutable,> independent, supremely intelligent, supremely powerful, and which created both myself and everything else . . . that exists." From rational analysis of the contents of this idea, he argued "that it must be concluded that God necessarily exists."[31] Descartes pro-

---

28. Ibid.
29. Ibid.
30. Ibid., pp. 240–3 (p. 324). This Aristotelian analysis of the question is very similar to statements by Eustachius a Sancto Paulo, in *Summa philosophica quadripartita, de regus dialecticis, moralibus, physicis et metaphysicis, authore, Fr. Eustachio a Sancto Paulo, a congregatione Fuliensi*, 2 vols. (Paris, 1609), as cited by Étienne Gilson, *Index scolastico-cartésien*, 2d edition (Paris: J. Vrin, 1979), pp. 203–4. Descartes wrote to Mersenne on 11 November 1640 that he had purchased a copy of Eustachius a Sancto Paulo's *Summa philosophica*, which he read to prepare himself to reply to the "Objections." AT, vol. 3, p. 233 (*PWD*, vol. 3, pp. 156–7). See Daniel Garber, *Descartes' metaphysical Physics* (Chicago: University of Chicago Press, 1992), pp. 24–5.
31. Descartes, *Meditations*, in *PWD*, vol. 2, p. 31 (AT, vol. 7, p. 45). (The material in angle brackets appeared in the French translation by Louis-Charles d'Albert, Duc de Luynes, published with Descartes' approval in 1647; ibid., pp. 1–2). For the historical development of the ontological

ceeded, as usual, from idea to existence; and it was on this inference that Gassendi focused his attack.

He objected to Descartes' claim that the idea of God is innate. It is quite plausible, he argued, to suppose that we have acquired this idea of God either from revelation or from other people's accounts of their revelations. In either case, our idea of God comes from outside of our minds: It is not innate. Originally,

> our first knowledge of God came from a revelation made by God himself, who manifested himself to the first men he created; then that knowledge was spread to all men, and by consequence to you and to me. . . . If we receive the idea of God from divine revelation, . . . that proves that God exists, but that also proves that the idea of him is not innate as you pretend, but acquired and introduced to us by revelation.[32]

Revelation is thus a form of empirical knowledge.[33] Furthermore, if the idea of God were innate, all people would share it. The fact that certain primitives in the New World do not possess any idea of God is an argument against its innateness, as is the fact that not all believers share the same idea of God.[34]

Gassendi criticized the logic of Descartes' proof in great detail. His fundamental objection was that one cannot validly argue from the idea of

argument from the Middle Ages to the seventeenth century, see Amos Funkenstein, "The Body of God in the Seventeenth Century Theology and Science," in *Millenarianism and Messianism in English Literature and Thought, 1650–1800,* edited by Richard H. Popkin (Leiden: E. J. Brill, 1988), pp. 149–75.

32. Gassendi, *Disquisitio metaphysica,* pp. 250–1 (*Opera omnia,* vol. 3, p. 326).
33. Robert Boyle also held the view that revelation is a form of empirical knowledge: "the knowledge we have of any matter of fact, which, without owing it to ratiocination, either we acquire by the immediate testimony of our own senses and other faculties, or accrues to us by the communicated testimony of others . . . I see not why that, which I call theological experience, may not be admitted: since the revelations, that God makes concerning what he has done, or purposes to do, so far forth as they are merely revelations, cannot be known by reasoning, but by testimony: whose being divine, and relating to theological subjects, does not alter its nature, though it give it a peculiar and supereminent authority." *The Christian Virtuouso,* in *The Works of the Honourable Robert Boyle,* edited by Thomas Birch, new edition, 6 vols. (London, 1772; reprinted, Hildesheim: George Olms, 1965), vol. 5, p. 525.
34. Gassendi, *Disquisitio metaphysica,* pp. 390–3 (*Opera omnia,* vol. 3, p. 357). Note that this statement contradicts his statement in the *Syntagma philosophicum* that belief in God·is universal (*Opera omnia,* vol. 1, p. 290).

a thing to its existence.[35] In rejecting Descartes' a priori proof for the existence of God, Gassendi made it clear that he rejected neither the existence of God nor the possibility of proving his existence. He affirmed the existence of God and supported his belief with an empirical argument, the argument from design, which he described as

> the royal road, smooth and easy to follow, by which one comes to recognize the existence of God, his power, his wisdom, his goodness, and his other attributes, which is nothing other than the marvelous working of the universe, which proclaims its author by its grandeur, its divisions, its variety, its order, its beauty, its constancy, and its other particularities.[36]

Gassendi's appeal to the argument from design was doubly consistent with his empiricism: First, the argument from design is based on empirical evidence; second, it encourages the practice of science in the name of religion.[37] Furthermore, according to Gassendi, the argument from design has the added advantage of supporting the existence of a God who is a creator, a conclusion he thought that Descartes' argument does not necessarily establish.[38]

Their disagreement about how to prove the existence of God led to a discussion of the use of final causes in physics. In elaborating the argument from design, Gassendi had given many examples of what he considered to be final causes observable in nature.[39] Descartes objected:

> Everything you bring to bear in favor of the final cause is to report on the efficient cause. Thus, from . . . the parts of plants, animals, etc., it is right to admire God who made them, and from inspection of his works to recognize and glorify the workman, but not to seek to know for what end he created each thing.[40]

Furthermore, all divine ends are "hidden in the inscrutable abyss of wisdom."[41] The trouble with the argument from design, Descartes con-

---

35. Gassendi, *Disquisitio metaphysica*, pp. 326–7 (*Opera omnia*, vol. 3, p. 343).
36. Ibid., pp. 302–3 (p. 337).
37. The English virtuosi likewise saw a close relationship between the practice of science and religious worship. See Richard S. Westfall, *Science and Religion in Seventeenth-Century England* (New Haven, Conn.: Yale University Press, 1958), chap. 2. See also James E. Force, "Hume and the Relation of Science to Religion Among Certain Members of the Royal Society," *Journal of the History of Ideas*, 45 (1984): 517–36.
38. Gassendi, *Disquisitio metaphysica*, pp. 338–9 (*Opera omnia*, vol. 3, pp. 345–6).
39. Ibid., pp. 396–9 (p. 359).
40. Ibid. See also Descartes, *Principia philosophiae*, in AT, vol. 8–1, pp. 15–16, 80–1; vol. 9–2, pp. 37, 104.
41. Gassendi, *Disquisitio metaphysica*, pp. 398–9 (*Opera omnia*, vol. 3, p. 359).

tinued, besides its mistaken appeal to final causes, is precisely the fact that it is based on observation. "Nothing prevents all men from perceiving that they have the same ideas, unless they are too occupied by the perception of images of corporeal realities."[42]

Gassendi reaffirmed the role of final causes in physics, dismissing Descartes' objections out of hand. Most importantly, Gassendi argued, we need to consider final causes in order to recognize God as creator and governor of the universe, for without admitting final causes, we might be tempted to regard the universe as the product of chance.[43] We can indeed know many of God's ends, but God does not always reveal how second causes operate in a given instance.[44] For example, many of the facts of anatomy and physiology clearly reveal the uses to which the parts of the body were intended.[45] Sometimes the final causes are so evident that they are easier to know than efficient causes. Thus, it is clear that the purpose of the valves in the heart is to ensure the one-way flow of blood as it circulates through the heart, but it is not at all so easy to know how the valves are formed.[46] Note that Gassendi tacitly altered the meaning of "final cause." The traditional understanding of the term, stemming from Aristotle, referred to a teleology inherent in all natural processes. Closely tied to the metaphysics of matter and form, the concept of final cause embodied the idea that natural processes occur in order to actualize the potential form of the substance undergoing the process. Accordingly, acorns, which contain the form of oak *in potentia*, become actualized into oaks, and heavy objects seek their natural place at the center. Gassendi, who explicitly rejected the Aristotelian metaphysics of matter and form and replaced it with a mechanical reinterpretation of natural processes, considered finality to be imposed supernaturally by a providential God. The purpose evident in natural processes is divine purpose, imposed from without.[47] Descartes' criticism – that what Gassendi called final causes are really just efficient causes – was therefore on the mark.

42.   Ibid.
43.   Ibid., pp. 406–7 (pp. 360–1).
44.   Ibid., pp. 412–13 (p. 362).
45.   Ibid., pp. 408–11 (p. 361).
46.   Ibid., pp. 410–11 (p. 362).
47.   On various meanings of "teleology," see Ernst Mayr, "The Idea of Teleology," *Journal of the History of Ideas*, 53 (1992): 117–35. See also F. F. Centore, "Mechanism, Teleology, and Seventeenth Century English Science," *International Philosophical Quarterly*, 12 (1972): 553–71. Boyle adopted a view on final causes very similar to Gassendi's. See Boyle, *A Disquisition about the Final Causes of Natural Things: Wherein it is Inquired, Whether, and (if at all) with what Cautions, a Naturalist should*

The two philosophers disagreed about the metaphysical and epistemological status of universals and mathematical truths, and here their argument circles back to their underlying theological differences. Descartes believed that eternal and immutable natures exist, a view he affirmed in the Fifth Meditation:

> When . . . I imagine a triangle, even if perhaps no such figure exists, or has ever existed, anywhere outside my thought, there is still a determinate nature, or essence, or form of the triangle which is immutable and eternal, and not invented by me or dependent on my mind. This is clear from the fact that various properties can be demonstrated of the triangle, for example that its three angles equal two right angles, that is, the greatest side subtends its greatest angle, and the like; and since these properties are ones which I now clearly recognize whether I want to or not, even if I never thought of them at all when I previously imagined the triangle, it follows that they cannot have been invented by me.[48]

His ability to demonstrate various properties of the triangle led him to conclude that the triangle has an "immutable and eternal essence," independent of his own mind. The essence of the triangle thus has the status of an eternal truth, created freely by God, but created to be necessarily true. The necessity embedded in the nature of the triangle provided grounds for Descartes' claim to have demonstrative, a priori knowledge of its properties.[49]

Gassendi, the voluntarist, found "it hard to agree that there exists some *immutable and eternal nature* other than [that] of omnipotent God."[50] Descartes replied that God in fact created them that way. "I do not think that the essences of things and mathematical truths can be known to be independent of God, but I also think that because God wished it thus, and

admit them?, in *Works*, vol. 5, p. 410.

Newton wrote about final causes in terms very similar to Gassendi's: "Indeed, if God did reduce to order the System of the Sun and Planets, final causes will have a place in natural philosophy, and it will be legitimate to inquire to what end the world was founded, to what ends the limbs of animals were formed, and by what wisdom they have so elegant an arrangement." Isaac Newton, "Scholium Generale," MS. Add. 3965, fols. 357–65, translated in *Unpublished Scientific Papers of Isaac Newton*, edited by A. Rupert Hall and Marie Boas Hall (Cambridge University Press, 1962), p. 360.

48. Descartes, *Meditations*, in *PWD*, vol. 2, pp. 44–5 (AT, vol. 7, p. 64).
49. The equivocation here between the necessary relation between the essence and properties of a triangle, on the one hand, and the necessity of propositions about the triangle, on the other, is Descartes'.
50. Gassendi, *Disquisitio metaphysica*, pp. 468–9 (*Opera omnia*, vol. 3, p. 374).

has thus disposed them, they are immutable and eternal.[51] He created the essences of mathematical objects freely by his absolute power, but he created them to be immutable and eternal. That is to say, he cannot change the mathematical order once he has created it. His ability to intervene in the creation was limited by his creative act. His ordained power does not coincide with his absolute power. Here, Descartes returned to his theory of the created eternal truths, a theory he had repeated more than once since his letters to Mersenne in April and May 1630.

Gassendi replied to Descartes' theory of the created eternal truths by focusing on God's role as creator, arguing that Descartes had produced an unresolvable dilemma:

> When you say that "God wished that these natural realities be immutable and eternal and that he disposed them thus," either these natures in question existed by themselves and were not created, and God willed them and disposed them simply as a worker disposes to his will in building with the stones he finds already made and not created; or they were created by God, and God was their author in the sense that he wished and disposed them by a creative act.[52]

One horn of the dilemma places unacceptable limitations on God's absolute power by giving these eternal and immutable natures an existence independent of him: "In the first case, these natures exist by themselves, independently of God; and it is uselessly that God wished them thus, since they were such by themselves."[53] The other horn of the dilemma allows for divine creation of these essences but thereby implies that they are not eternal. "In the second case, they would not be eternal since in order to be created they must not have been and have thus in a given moment commenced to be, which is for an eternal thing, contradictory."[54] Gassendi's analysis of the problem thus focused on the question of God's absolute power and its relationship to the world.

Given Gassendi's voluntarism, it is clear that he had to opt for the second horn of the dilemma. Arguing against the view that God created some eternal, immutable things, like the triangle, Gassendi underscored his view that God is not bound by anything that he created. His absolute power enables him to destroy, if he wants to, anything he has created, even the nature of the triangle:

> If it is true that the thrice powerful God is not, as Jupiter of the fable is to the fates, subjected to things created by him, but can in virtue of his absolute power, destroy all that he has established, then I pray you,

51. Ibid., pp. 472–3 (p. 375).
52. Ibid., pp. 480–1 (p. 377).
53. Ibid.
54. Ibid.

how do you conceive that God could have disposed the Triangle in such a way (*Quaeso-te, qua ratione capias posse Deum disposuisse de Triangulo*) that if he wished it now absolutely, he could, destroying its nature that he created at the beginning, know that the Triangle is not formed from three sides and three angles?[55]

This passage resonates with the language of the absolute and ordained power of God. By his absolute power, God can do anything. By this power he created the triangle, but he was not compelled to create it according to some preexisting plan. By that same absolute power, God can destroy what he ordained in the initial creation: He could even change the nature of the triangle, if he chose. He is not bound by the created order. He is free to exercise his absolute power on every aspect of the created order. Here Gassendi's empiricism merged with his voluntarist conception of the deity. The nature of the triangle is contingent on divine will, just like everything else in the creation. This contingency lay at the heart of his empiricist epistemology.

Gassendi correctly perceived that Descartes was an intellectualist, although in this case he portrayed Descartes as a more extreme intellectualist than he actually was. Descartes' theory of the creation of the eternal truths had given those truths and truths about mathematical objects a necessity relative to God's having created them by his absolute power. Gassendi interpreted Descartes' reasoning about the essence of the triangle to mean that essences have an existence prior to and independent of God's absolute power. This Platonic position is not the one that Descartes had enunciated in his other writings, although it is an easy conclusion to draw from his statement that the triangle's nature is eternal. In either case, it is questions about God's power and the presence of necessity in the world on which the two philosophers disagreed. The old problem about the relationship between God's absolute and ordained powers emerged at the heart of their controversy.

It is important to note that there is an equivocation in both Descartes' and Gassendi's use of the word "eternal" (*aeternus*) here. When applied to God, "eternal" means timelessness; when applied to some aspect of the creation – such as the triangle that God creates – it means "sempiternity" or "infinite duration." Boethius (480–524) coined the term "*sempiternitas*" as distinct from "*aeternitas*" in *De Trinitate*, a distinction that was to have a major impact on medieval philosophy and theology.[56] Gassendi

55. Ibid.
56. See Richard Sorabji, *Time, Creation and the Continuum: Theories in Antiquity and the Early Middle Ages* (Ithaca, N.Y.: Cornell University Press, 1983), pp. 115–17.

interpreted Descartes to be using "eternal" in its strict sense as coeval with God. Given his theory of the creation of the eternal truths, Descartes was clearly using it in the sense of "sempiternal."

In objecting to Descartes' assertion that mathematical truths are eternal and immutable, Gassendi reiterated his view that mathematical concepts and, more generally, statements about universals are abstractions drawn by the human mind from the sensory experience of particulars:

> The triangle which you have in your mind is a kind of rule (*regula*) by means of which you examine if a thing merits the name of triangle, but it is not necessary to say for that this triangle is something with a real and true nature outside of the understanding: only that after having seen material triangles, it has formed this nature and rendered it common.[57]

Descartes had no patience with Gassendi's view. He replied that it is impossible to form ideas of mathematical figures from experience, because there are no truly straight lines or dimensionless points in the material world:

> But because the idea of a true triangle is nevertheless in us and it can be more easily conceived by our mind than the more complex figure of the drawn triangle, it follows that the complex figure, having been seen, it is not the same as the former, but rather the true triangle which we have grasped by thought. . . . It is certain that we could not recognize the geometrical triangle by the manner in which it is traced on paper unless our mind had received the idea in some other way.[58]

Since the material representation of the triangle is inadequate for teaching us its true nature, God must have implanted knowledge of its essence in our minds as one of the created eternal truths. Gassendi unequivocally rejected Descartes' metaphysics. "Besides the thrice-great God, we do not know a single thing that was not created by him . . . [to be] really singular."[59]

In the Fifth Meditation, Descartes appealed to the necessity of geometrical truths as supporting his argument for the existence of God. "It is quite evident that existence can no more be separated from the essence of God than the fact that its three angles equal two right angles can be separated from the essence of a triangle, or that the idea of a mountain can be

---

57. Gassendi, *Disquisitio metaphysica*, pp. 470–1 (*Opera omnia*, vol. 3, p. 375). Lewis and Short give the word "*regula*" the following definition as its first meaning: "a straight piece of wood, ruler, rule." See Charlton T. Lewis and Charles Short, *A Latin Dictionary* (Oxford University Press, 1879), p. 1553.
58. Gassendi, *Disquisitio metaphysica*, pp. 474–7 (*Opera omnia*, vol. 3, p. 376).
59. Gassendi, *Disquisitio metaphysica*, pp. 482–4 (*Opera omnia*, vol. 3, p. 377).

separated from the idea of a valley."[60] Invoking the clear and distinct rule to prove the necessity of geometrical truths, Descartes continued by saying that his knowledge of God's existence is even more self-evident than that of geometrical truths. Gassendi objected to Descartes' analogy between his proofs of the properties of geometrical figures and his proof for the existence of God, noting that there is no necessary relation between essence and existence in the case of the geometrical figures.[61] God's existence may follow necessarily from the fact that existence is part of his essence, but in the case of created things, like triangles or valleys, the existence in question is contingent. Such things exist only if God chooses to create them. They do not possess eternal and immutable essences. He perceived Descartes' position as dangerous to the true faith because it placed unacceptable limits on God's freedom and power. Quoting Giovanni Pico della Mirandola, Gassendi stated, "Nothing is more dangerous for a theologian than to know the *Elements* of Euclid."[62]

The issues that separated Gassendi and Descartes stemmed directly from their divergent understandings of God's relationship to the creation. In earlier chapters, I have demonstrated how Gassendi's voluntarism and Descartes' intellectualism were connected to their respective philosophical views about the nature of the world and the status of our knowledge. Not surprisingly, their confrontation following the publication of the *Meditations* found them at loggerheads on these very issues, and the confrontation becomes more fully intelligible when it is placed in the context of their theological differences. These differences had further implications when Gassendi and Descartes articulated their versions of the mechanical philosophy. Their respective accounts of the properties of matter and the status of mechanical explanations were closely related to their theories of knowledge and existence.

60. Descartes, *Meditations,* in *PWD,* vol. 2, p. 46 (AT, vol. 7, p. 66).
61. Descartes, "Fifth Set of Objections," in *PWD,* vol. 2, pp. 224–5 (AT, vol. 7, pp. 322–5).
62. "Nihil esse magis nocivum Theologo, quam nosse Elementa Euclidis." Gassendi, *Disquisitio metaphysica,* pp. 514–15 (*Opera omnia,* vol. 3, p. 384). Rochot seems correct in attributing this remark to Giovanni Pico della Mirandola, who wrote, "Nichil [*sic*] magis nociuum theologo, quam frequens et assidua in mathematici Euclidis exercitatio." Giovanni Pico della Mirandola, *Conclusiones, sive Theses DCCCC Romae anno 1486 publice disputandae sed non admissae,* introduction and critical annotation by Bohdahn Kiezkowski (Geneva: Librairie Droz, 1973), p. 74.

II

The mechanical philosophy and the formation of scientific styles

7

Introduction: Theories of matter and their epistemological roots

In Part I, I showed how the medieval discourse about divine power and its relationship to the created world was refracted in the prism of seventeenth-century natural philosophy, emerging as the two very different approaches to the epistemological status of knowledge about the natural world adopted by Gassendi and Descartes. In Part II, I turn to the mechanical philosophy itself, namely, the view that all natural phenomena can be explained solely in terms of matter and motion. Here I explore how the theories of matter these mechanical philosophers adopted reflected their respective epistemological inclinations and thereby engendered two distinct styles of science.

In its role as conceptual framework, the mechanical philosophy determined the ultimate terms of explanation in which the physical world was to be understood and the methods by which knowledge of that world could be acquired.[1] At the most abstract level, the mechanical philosophy replaced both the Aristotelian metaphysics of matter, form, and privation and the Neoplatonically inspired animistic philosophies of the Renaissance with a new metaphysics, which populated the natural world with only one kind of entity, matter. The motions, positions, and collisions of particles of matter were thought to be sufficient to explain everything in the physical world. In this mechanistic framework, matter was the only real existent, and causality was reinterpreted in terms of the impact of material particles. Action at a distance was deemed impossible.[2]

The concept of matter adopted by the mechanical philosophers differed in important ways from the older theories of matter. In designating matter

---

1. My understanding of "conceptual framework" owes a great deal to R. Harré, *Matter and Method* (London: Macmillan, 1964).
2. While Aristotelian natural philosophy had also ruled out action at a distance, its fourfold notion of causality allowed for more than mechanical action in the operation of the world. For a general account of the mechanical philosophy, see Richard S. Westfall, *The Construction of Modern Science: Mechanisms and Mechanics* (New York: Wiley, 1971), chap. 2.

as the fundamental component of the world, mechanical philosophers altered the concept of matter itself: Rather than being a correlative term, always associated with form (as it was on either the Aristotelian or the Platonic concept), matter came to be regarded as having a "free-standing" status.[3] This change brought with it a new set of philosophical problems and the reinterpretation of such traditional terms as "matter," "substance," "form," "quality," "cause," and "activity" in mechanical terms.[4]

In traditional Platonic and Aristotelian philosophies of nature, the term "matter" was always correlative to the term "form"; it did not signify an independent, self-subsistent actuality.[5] In the Aristotelian world, the ultimate existent was not prime matter per se, but individual corporeal substances, entities composed of prime matter and form. Some Aristotelians, like Thomas Aquinas, had thought that matter provides the principle of individuation to a substance. That is to say, where the form of a substance – say, the form of a cat – is the cat's essence, making it a cat rather than anything else, the particular bit of matter so informed determines that it is this individual cat rather than some other.[6] According to Aristotelian natural philosophy, prime matter does not exist in the world except in combination with form. It always exists as one component of a composite, corporeal substance.

Aristotelian forms endow corporeal substances with particular qualities. Some qualities were regarded as essential to the substance in question (say, four-footed, warm-blooded, and possessing the ability to purr, in the case of the cat), because they flow from the substance's es-

3. This account of the changes in the concept of matter owes a great deal to Ivor Leclerc, *The Nature of Physical Existence* (London: Allen & Unwin, New York: Humanities Press, 1972). See also Ivor Leclerc, "Atomism, Substance, and the Concept of Body in Seventeenth-Century Thought," *Filosofia, 18* (1967): 761–78. Thomas A. Spragens, Jr., makes a similar point. See Spragens, *The Poltics of Motion: The World of Thomas Hobbes* (Lexington: University of Kentucky Press, 1973), chap. 3.
4. See Norma E. Emerton, *The Scientific Reinterpretation of Form* (Ithaca, N.Y.: Cornell University Press, 1984).
5. Leclerc, *The Nature of Physical Existence*, p. 35. See also Daniel Garber, *Descartes' Metaphysical Physics* (Chicago: University of Chicago Press, 1992), pp. 68–9.
6. The question of the role of matter as the principle of individuation in Aristotelianism is complex. See Joseph Bobik, "Matter and Individuation," in *The Concept of Matter in Greek and Medieval Philosophy*, edited by Ernan McMullin (Notre Dame, Ind.: University of Notre Dame Press, 1963), pp. 281–98.

sence. Other qualities (say, the fact that this cat is gray or small or sleeping) were called accidental because they do not flow from the substance's essence.[7] Consequently, an important problem for Aristotelian matter theory was to discover, in any given case, which qualities are essential and which are only accidental. This conceptualization of matter was closely linked with the method of Aristotelian science: Many of the procedures outlined in the *Posterior Analytics,* Aristotle's treatise on scientific method, concern the problem of distinguishing between accidental and essential qualities. Once the essential qualities can be properly identified, according to Aristotle, the basis for a demonstrative science exists.[8]

Prime matter functioned as the substratum for change, where change was understood in terms of the actualizing of potential forms. It stood in contrast to life and mind, both of which were ascribed to the action of certain forms – the vegetative, sensitive, and rational souls. Matter played a role in explanation, one component of which was the material cause.[9] It was the ultimate subject of predication. In the medieval setting the Aristotelian theory of matter was also tied to important doctrinal matters. For example, it provided the basis for the Scholastic explanation of Christ's real presence in the Eucharist, transubstantiation. In all of these roles, however, matter was always understood as correlative to form; it never denoted a self-subsistent entity.[10]

As Aristotelian physics had developed in the Middle Ages, explanation in terms of matter and form was enhanced by the addition of substantial forms. The substantial form is what gives a substance its characteristic properties. As Aquinas said, "Whatever brings substantial existence into

7. See W. D. Ross, *Aristotle: A Complete Exposition of His Works and Thought,* 5th edition (New York: Meridian, 1959), pp. 66–9; and G. E. R. Lloyd, *Aristotle: The Growth and Structure of His Thought* (Cambridge University Press, 1968), pp. 129–32.
8. Aristotle, *Posterior Analytics,* bk. II, in *The Complete Works of Aristotle,* edited by Jonathan Barnes, 2 vols. (Princeton, N.J.: Princeton University Press, 1984), vol. 2, pp. 147–66.
9. For fuller discussion of these points, see Ernan McMullin, "The Concept of Matter in Transition," in *The Concept of Matter in Modern Philosophy,* edited by Ernan McMullin, revised edition (Notre Dame, Ind.: University of Notre Dame Press, 1978), pp. 1–55. See also Richard Sorabji, *Matter, Space, and Motion: Theories in Antiquity and Their Sequel* (Ithaca, N.Y.: Cornell University Press, 1988), chaps. 1–7.
10. On the Aristotelian theory of real qualities and substantial forms, see Garber, *Descartes' Metaphysical Physics,* pp. 94–111.

actuality is called substantial form."[11] For example, the heaviness of a stone, giving it the tendency to fall down toward the center of the world, cannot be explained by the stoniness of its constituent matter alone or by any of its more elementary properties:

> Since heavy and light things tend toward their natural places, though absent from that which produces them, they must necessarily have been given some means (*instrumentum*) that remains with them by virtue of which they are moved. But this can only be their substantial form and what follows from it, heaviness and lightness.[12]

Thus, the substantial form was thought to give a substance its its essential, as opposed to its accidental properties. This concept was one of the main targets that mechanical philosophers attacked.

The mechanical philosophers adopted a very different conception of matter. Although the vocabulary of matter remained much the same – words such as "matter," "form," "quality," "substance," "cause," and "activity" continued to appear frequently in the discourse of natural philosophy – the meanings of these words underwent a profound transformation. The writings of the mechanical philosophers can be understood, in part, as attempts to reinterpret these concepts in mechanical terms.[13] Boyle's *Origin of Forms and Qualities* (1666) is a particularly clear example of a mechanical reinterpretation of Aristotelian explanatory categories in mechanical terms. Having established "that there is one catholick or universal matter common to all bodies, by which I mean a substance extended, divisible, and impenetrable,"[14] Boyle proceeded to demonstrate that the qualities of bodies can be explained in terms of the configurations of their constituent corpuscles and that "form," "essence," "accident," and "substance" can all be redefined in mechanical terms, as can the Aristotelian processes of generation, corruption, and

---

11. Thomas Aquinas, *De principiis naturae*, translated by Mary T. Clark, in *An Aquinas Reader* (London: Anchor, 1974), pp. 164–8, as quoted by Emerton, *The Scientific Reinterpretation of Form*, p. 54.

12. *Commentarii in octo libros physicorum Aristotelis* (Conimbricae, 1592), I, 1, 2, 4, quoted by Étienne Gilson, *Index scolastico-cartésien*, 2d edition (Paris: J. Vrin, 1979), p. 226, translated by Garber, in *Descartes' Metaphysical Physics*, p. 96.

13. Boyle undertook precisely this kind of reinterpretation of Aristotelian terms in *The Origin of Forms and Qualities According to the Corpuscular Philosophy* (1666). See *Works of the Honourable Robert Boyle*, new edition, 6 vols. (London, 1772; reprinted, Hildesheim: George Olms, 1965), vol. 3, pp. 14–49. See also Emerton, *The Scientific Reinterpretation of Form*, chaps. 3 and 4.

14. Boyle, *The Origin of Forms and Qualities*, vol. 3, p. 15.

alteration.[15] He defined form as the spatial arrangement of the corpuscles composing material bodies and argued against the view that "these forms be true substantial entities, distinct from the other substantial principle of natural bodies, namely, matter."[16] The physical world was now thought to consist entirely of matter; and all qualities, bodies, and physical changes were to be explained in terms of the properties of that matter, its motions, and the collisions between different particles of matter. The word "matter" itself came to be understood as the ultimate stuff out of which the world is made. It became identified with the physical as such.[17] Whether matter was thought to consist of indivisible atoms, as Gassendi maintained, or to be infinitely divisible, as Descartes insisted, it was in all cases identified with substance. To be material was to be substantial. To exist in the physical world was to be material. Matter itself was the ultimate component of the world, created by God and in need of no further explanation. It was independent of any other created entity. Form was now reinterpreted in mechanical terms: The form of an object, a substance, or a living thing became the spatial arrangement of the particles of matter of which it was composed. "Substance" now referred to particular kinds of matter, each consisting of particles of certain shapes and sizes, arranged in characteristic ways.

With the new concept of matter came a new set of philosophical problems: the question of primary and secondary qualities, the problem of transdiction, problems about the activity of matter, and the status of the human soul. Each of these problems was related to problems that had arisen within Aristotelianism, but their particular seventeenth-century counterparts reflected matter's new ontological status.[18] One of the most important developments resulting from this new conceptualization of matter was the dramatic reduction in the number of qualities that matter was thought to possess. Most of the qualities the Aristotelians had thought to be really inherent in bodies – colors, tastes, and chemical properties, for example – were reduced to the few qualities still thought to belong to matter, usually a fairly short list including such properties as extension and impenetrability. All other qualities were relative to the

15. Ibid., pp. 1–49.
16. Ibid., p. 38.
17. Leclerc, *The Nature of Physical Existence*, p. 35.
18. For some of the more technical problems about matter arising within the mechanical philosophy, see Alan Gabbey, "The Mechanical Philosophy and Its Problems: Mechanical Explanation, Impenetrability, and Perpetual Motion," in *Change and Progress in Modern Science*, edited by Joseph C. Pitt (Dordrecht: Reidel, 1985), pp. 9–84.

interaction of these primary qualities either with other bodies or with our senses.[19]

If all phenomena in the natural world are to be explained in terms of the properties of matter alone, the question of which properties are essential to matter becomes extremely important: These essential properties become the most fundamental explanatory entities in natural philosophy. In seventeenth-century parlance, the essential properties came to be known as primary qualities, in terms of which all the secondary qualities could be explained.[20] The list of primary qualities varied from one natural philosopher to another. Consequently, a fundamental problem that each mechanical philosopher addressed was the formulation of a criterion by which to determine which qualities are in fact primary. The kinds of qualities designated as primary had close ties with epistemological convictions. Descartes' rationalism is reflected in the fact that matter, for him, possesses only one primary quality, geometrical extension, which can be known in a purely rational manner. Gassendi, the empiricist, in contrast, included in his list of primary qualities some that could only be known by utilizing empirical methods. It is precisely here that the connection between epistemology and matter theory is most evident.

Further questions arise: Is all matter substantially the same? That is to say, does all matter share the same primary qualities, or are there different kinds of matter possessing different primary qualities? Is matter infinitely divisible, or are there ultimate, discrete, indivisible atoms? Directly connected to the question of infinite divisibility and atomism was the question of the existence of an interstitial void: If one maintains that ultimately indivisible atoms exist, something must separate them from each other, and void space was the usual candidate called upon to serve that purpose. Historically, the question of the existence of the void was fraught with controversies – theological, metaphysical, and experimental.[21]

19. On this transformation in the "location" of qualities, see Keith Hutchison, "Individualism, Causal Location, and the Eclipse of Scholastic Philosophy," *Social Studies of Science, 21* (1991): 321–50.

20. The precise question of the relation between primary and secondary qualities is complex and has been addressed by many philosophers. For a taxonomical discussion of this problem, see John J. MacIntosh, "Primary and Secondary Qualities," *Studia Leibnitiana, 8* (1) (1976): 88–104.

21. See Edward Grant, *Much Ado About Nothing: Theories of Space and Vacuum from the Middle Ages to the Scientific Revolution* (Cambridge University Press, 1981). Grant's discussion deals comprehensively with the extracosmic void. There is relatively little discussion in Grant or other published sources of the problems associated with the interstitial void. In this connection, see Jane Elizabeth Jenkins, "Using Nothing: Vacuum, Matter,

Another problem generated by the mechanical philosophy was its appeal to microscopic mechanisms to explain macroscopic qualities and processes, a methodological problem that has been called transdiction.[22] Like induction, transdiction is a form of inference by which one reasons from the observed phenomena to some microscopic mechanism, as yet unobserved. While "induction" usually refers to an inference from one observable event or individual to another observable event or individual of the same kind, transdiction is a form of inference by which one reasons from the observed properties of macroscopic bodies to the unobservable properties of their microscopic components, a situation in which it is not known whether the inferred entities are of the same kind as the observed entities. Transdiction is a form of inductive reasoning that was intrinsic to the mechanical philosophy, which constantly sought to explain the phenomena of the macroscopic world in terms of microscopic mechanisms. Mechanical philosophers were not entirely unaware of the problematic character of this form of inference. The kinds of justification they offered for transdiction provide further insight into the links between matter theory and epistemology in the mechanical philosophy.

In presuming that all natural phenomena can be explained in terms of matter and motion alone and that there is no action at a distance, the mechanical philosophers departed from traditional philosophies of nature, which had endowed matter with various kinds of activity. For the Aristotelians, there exist natures, which endow bodies with tendencies to move in characteristic ways.[23] For example, heavy bodies, because of their nature, tend to move toward their natural place at the center of the world; or the nature of the oak tree, potentially contained in the acorn, causes the constituent matter to be formed into an oak tree, rather than a maple or a eucalyptus. Many Renaissance philosophers in the Neoplatonic, Hermetic, and Paracelsian traditions portrayed a highly animistic world, characterized by sympathies and antipathies, acting at a distance and endowing the material world with its own, innate activity. One reason why the mechanical philosophers were so averse to the ac-

and Spirit in the Seventeenth Century Mechanical Philosophy," M. A. thesis, University of Calgary, 1990. On the experimental aspects of the issue in the seventeenth century, see Steven Shapin and Simon Schaffer, *Leviathan and the Air-Pump: Hobbes, Boyle, and the Experimental Life* (Princeton, N.J.: Princeton University Press, 1985).

22. Maurice Mandelbaum, *Philosophy, Science, and Sense Perception: Historical and Critical Studies* (Baltimore: Johns Hopkins University Press, 1964).

23. See Keith Hutchison, "Dormitive Virtues, Scholastic Qualities, and the New Philosophies," *History of Science*, 29 (1991): 245–78.

tivity of matter is that active matter, insofar as it is self-moving, seemed capable of explaining the world without needing to appeal to God or the supernatural.[24] This danger could be avoided if matter were considered naturally inert and able to produce its effects only by mechanical impact; for then, the source of motion in the world must lie outside of the natural, material realm. God, as source of motion, seemed absolutely necessary to most seventeenth-century mechanical philosophers.[25]

In their determination to encompass all phenomena within the mechanical categories and to banish action at a distance from the natural world, the mechanical philosophers sought to explain the traditionally occult qualities in mechanical terms. Thus, we find in their works mechanical explanations of such phenomena as magnetism, static electricity, the notorious weapon salve, the purported fact that it is impossible to tune a string of sheep gut into perfect consonance with one of wolf gut, and a kind of sea creature that is supposedly able to stop a ship traveling at full speed.[26] Although the ontology of the mechanical philosophy was impoverished compared with that of the Renaissance naturalists,[27] its adherents believed that it was capable of explaining any possible phenomenon. The insistence on incorporating even the most bizarre and unlikely of the

24. And yet the mechanical philosophers were not entirely successful in expunging activity from matter. See John Henry, "Occult Qualities and the Experimental Philosophy: Active Principles in Pre-Newtonian Matter Theory," *History of Science*, 24 (1986): 335–81.

25. In this connection, see Keith Hutchison, "Supernaturalism and the Mechanical Philosophy," *History of Science*, 21 (1983): 297–333.

26. See, e.g., Pierre Gassendi, *Syntagma philosophicum*, in *Opera omnia*, 6 vols. (Lyon, 1658; facsimile reprint, Stuttgart-Bad Canstatt: Friedrich Fromann Verlag, 1964), vol. 1, pp. 449–547; Walter Charleton, *Physiologia-Epicuro-Charletoniana, or a Fabrick of Science Natural Built Upon the Hypothesis of Atoms* . . . (London, 1654; reprinted New York: Johnson Reprint, 1966), pp. 341–87; René Descartes, *Principia philosophiae*, in *Oeuvres de Descartes*, edited by Charles Adam and Paul Tannery, 11 vols. (Paris: J. Vrin, 1897–1983), vol. 8-1, pp. 283–314, and vol. 9-2, pp. 278–310. On the genesis of the canonical list of occult qualities, see Brian P. Copenhaver, "A Tale of Two Fishes: Magical Objects in Natural History from Antiquity Through the Scientific Revolution," *Journal of the History of Ideas*, 52 (1991): 373–98.

27. See Brian P. Copenhaver, "Natural Magic, Hermetism, and Occultism in Early Modern Science," in *Reappraisals of the Scientific Revolution*, edited by David C. Lindberg and Robert S. Westman (Cambridge University Press, 1990), pp. 270–5.

occult qualities within the mechanical framework highlights the importance – in the eyes of the mechanical philosophers – of denying the activity of matter.[28]

Removing all activity from matter and carefully separating matter from spirit – a move taken by all the mechanical philosophers – led to a consideration of the human soul, which most of them removed from the physical realm entirely. (One notable exception, of course, was Hobbes.) In regarding the soul as the form of the person, the Aristotelians and the Scholastics had been able to avoid any danger of slipping into materialism. Having eliminated the traditional concept of form and having declared matter to be inert and self-subsisting, the mechanical philosophers faced the problem of establishing the existence of an immaterial, immortal soul. Their discussions of the human soul established the limits of mechanization and provided a bulwark against the bugbear of materialism.[29]

28. On the role occult qualities played in the mechanical philosophy, see Keith Hutchison, "What Happened to Occult Qualities in the Scientific Revolution?" *Isis,* 73 (1982): 233–53. See also Ron Millen, "The Manifestation of Occult Qualities in the Scientific Revolution," in *Religion, Science, and Worldview: Essays in Honor of Richard S. Westfall,* edited by Margaret J. Osler and Paul Lawrence Farber (Cambridge University Press, 1985), pp. 185–216.

29. This problem continued to preoccupy mechanical philosophers throughout the seventeenth century. See, e.g., Thomas Harmon Jobe, "The Devil in Restoration Science: The Glanvill–Webster Debate," *Isis,* 72 (1981): 343–56; and Simon Schaffer, "Godly Men and Mechanical Philosophers: Souls and Spirits in Restoration Natural Philosophy," *Science in Context,* 1 (1987): 55–86.

8

Gassendi's atomism: An empirical theory of matter

> Physics can be defined as "the contemplative science of natural things," since through it we explore how complex a thing is, of what species it is, how much of it there is, of what principles it consists, by what cause it is produced, and what effect it brings about. . . . If these things are understood, then the nature of the thing . . . is understood.
>
> *Pierre Gassendi, Syntagma philosophicum*[1]

In earlier chapters, I discussed the reasons why Gassendi considered it necessary to modify Epicureanism in order to render it theologically acceptable. I then argued that a close relationship existed between his voluntarist theology and his empiricist theory of scientific knowledge. In this chapter, I argue that his theories of matter and causality were enmeshed in his epistemology and that they thus indirectly reflected the theological assumptions that informed his entire philosophical enterprise.

Gassendi was not the first advocate of atomism in the seventeenth century, but he was surely the most systematic. His coupling of atomism with the astronomy of Copernicus and the physics of Galileo, as well as his commitment to ridding atomism of the theologically objectionable doctrines traditionally associated with Epicureanism, set him apart from earlier advocates of the atomic theory and probably account for the ultimate success of his project. Like Gassendi, one early-seventeenth-century atomist, David van Goorle (d. 1612) was also concerned to refute Aristotle.[2] Sebastian Basso (fl. 1550–1600) advocated atomism as an alternative to Aristotelianism, which he criticized at length in *Philosophiae naturalis*

---

1. Pierre Gassendi, *Syntagma philosophicum*, in *Opera omnia*, 6 vols. (Lyon, 1658; facsimile reprint, Stuttgart-Bad Cannstatt: Friedrich Frommann Verlag, 1964), vol. 1, p. 125.
2. Tullio Gregory, "Studi sull'atomismo del seicento. II. David van Goorle e Daniel Sennert," *Giornale critico della filosofia italiana*, 20 (1966): 44–63.

adversus Aristotelem (1621).³ His atomism traced its paternity to the mathematical atomism of Plato's *Timaeus,* however, rather than to the ideas of Democritus and Epicurus.⁴ Moreover, he rejected the Earth's motion, thus dissociating himself from the community of natural philosophers promoting the new science.⁵ Around the turn of the century, the circle around Henry Percy, Ninth Earl of Northumberland, contained a number of advocates of atomism, including Walter Warner (c. 1570-c. 1642), Thomas Harriot (1560–1621), and possibly Nicholas Hill (c. 1570–1610).⁶ Although this group of natural philosophers supported Copernicanism, their conception of atomism was deeply influenced by the ideas of Giordano Bruno (1548–1600), whose works became known in England after his stay at Oxford in the mid-1580s.⁷ They made no effort to deal with the heterodox doctrines customarily associated with atomism.⁸ Hill's *Philosophia Epicurea* (1601) was the only work advocating atomism published by a member of this group. It does not develop a philosophy of nature systematically, and it shows clear signs of Bruno's influence in its claim that the universe is infinite and that there is a plurality of worlds, ideas that would have scared off more conservative natural philosophers.⁹ None of these earlier atomists published a systematic or

3. Lauge Olaf Nielsen, "A Seventeenth-Century Physician on God and Atoms: Sebastian Basso," in *Meaning and Inference in Medieval Philosophy: Studies in Memory of Jan Pinborg,* edited by Norman Kretzmann (Dordrecht: Kluwer, 1988), pp. 297–369; Lynn Thorndike, *A History of Magic and Experimental Science,* 8 vols. (New York: Columbia University Press, 1923–58), vol. 6, pp. 386–8; Tullio Gregory, "Studi sull'atomismo del seicento. I. Sebastiano Basson," *Giornale critico della filosofia italiana, 18* (1964): 38–65; and H. H. Kubbinga, "Les premières théories 'moléculaires': Isaac Beeckman (1620) et Sébastien Basson (1621): Le concept d' 'individu substantiel' et d' 'espèce substantielle,'" *Revue d'histoire des sciences, 37* (1984): 215–33.
4. Marie Boas, "The Establishment of the Mechanical Philosophy," *Osiris, 10* (1951): 427.
5. Thorndike, *Magic and Experimental Science,* vol. 7, p. 379.
6. Robert Hugh Kargon, *Atomism in England from Hariot to Newton* (Oxford University Press, 1966), chaps. 1–4; Grant McColley, "Nicholas Hill and the *Philosophia Epicurea,*" *Annals of Science, 4* (1939): 390–405; and Jean Jacquot, "Harriot, Hill, Warner, and the New Philosophy," in *Thomas Harriot: Renaissance Scientist,* edited by John W. Shirley (Oxford University Press, 1974), pp. 107–28. See also Thomas Franklin Mayo, *Epicurus in England (1650–1725)* (Dallas, Tex.: Southwest Press, 1934), chap. 2.
7. Kargon, *Atomism in England,* p. 9.
8. See Jean Jacquot, "Thomas Harriot's Reputation for Impiety," *Notes and Records of the Royal Society of London, 9* (1951–2): 164–87.
9. On Bruno's influence on Hill, see Daniel Massa, "Giordano Bruno's Ideas in

compelling defense of atomism, a fact that underscores the historical significance of Gassendi's project.

The "Physics," the second part of the *Syntagma philosophicum*, contains Gassendi's Christianized version of Epicurean atomism. The structure of the "Physics" reveals Gassendi's programmatic aims: to demonstrate how natural philosophy should look. To this end, he divided the "Physics" into three sections. The first section, entitled "On the Nature of Things Universally,"[10] describes the entities of which the world is composed. Here, Gassendi presented his accounts of space, time, matter, and cause and then argued that all the qualities of things can be explained in terms of these fundamental categories. The remaining sections of the "Physics" – "On Celestial Things" and "On Terrestrial Things," a section further divided into two parts, one "On Inanimate Things" and the other "On Living Things or Animals" – cover all kinds of phenomena found in the universe and attempt to explain them in the terms set out in the first section. Physics, Gassendi argued, must begin from a consideration of the ultimate components of the world. Only then can it embark upon a detailed analysis of various particular things. He understood that his enterprise was metaphysical as much as physical and that this part of his work was a direct reply to Aristotle's *Metaphysics*.[11] The particular mechanical explanations that he created (with an apparently boundlessly fertile imagination) were to serve as exemplars of how natural philosophy should look within the new conceptual framework.[12]

Atoms and the void

The natural world, according to Gassendi, consists of atoms and the void. The first section of the "Physics" begins with Gassendi's arguments for the existence of the void. The arguments he used to support its existence are

Seventeenth-Century England," *Journal of the History of Ideas, 38* (1977): 227–42. On Hill's biography, see Hugh Trevor-Roper, "Nicholas Hill: The English Atomist," in *Catholics, Anglicans, and Puritans: Seventeenth Century Essays*, edited by Hugh Trevor-Roper (London: Secker & Warburg, 1987), pp. 1–39.

10. Gassendi, *Syntagma philosophicum*, in *Opera omnia*, vol. 1, pp. 125–494.

11. Ibid., vol. 1, p. 133. See also Barry Brundell, *Pierre Gassendi: From Aristotelianism to a New Natural Philosophy* (Dordrecht: Reidel, 1987), esp. chap. 3.

12. On the difference between this aim of establishing the mechanical philosophy and the aims of later natural philosophers working within a mechanical framework, see Lynn Sumida Joy, "The Conflict of Mechanisms and Its Empiricist Outcome," *Monist, 71* (1988): 498–514.

largely empirical. Some of them are the arguments employed in antiquity by Epicurus, Lucretius, and Hero of Alexandria; most of them draw on the barometric experiments of Gassendi's contemporaries Torricelli, Pascal, and Auzout.

In accordance with traditional treatments of the subject, Gassendi classified the void into three traditional categories: the separate, extracosmic void; the interparticulate, interstitial, or disseminated void; and the *coacervatum*, or larger void, produced by collecting a number of interstitial voids, usually by means of a force created by some kind of mechanical device. He included his discussion of the void within the general context of space and time, the nature of which he subjected to rational, philosophical analysis. While this analysis was not, strictly speaking, empirical, it allowed Gassendi to set the parameters within which he could conduct empirical investigation. He argued that space and time are neither substances nor accidents, but altogether different categories of being, an idea that he borrowed from the Renaissance Platonist Francesco Patrizi (1529–97).[13] Patrizi had claimed that space is an incorporeal corporeal, thus avoiding Aristotle's confounding of dimensionality and corporeality.[14] Space continues to exist even when the matter in it moves away or ceases to exist. Patrizi thought of it as incorporeal extension.[15] Like Patrizi, whose work he knew,[16] Gassendi rejected the Aristotelian categories of substance and accident and included two additional kinds of things in the world, space and time.

The large extramundane void is the space in which God created the universe. It is boundless extension, and it is incorporeal. Gassendi demonstrated the possibility of an incorporeal nature by the following thought experiment: If God were to annihilate all the matter in the universe, space would nevertheless continue to exist. In this explicit rejection of Descartes' identification of spatial extension and matter, Gassendi defended the existence of the extramundane void.[17] This thought experiment had long roots extending back into medieval discussions of divine

13. Gassendi, *Syntagma philosophicum*, in *Opera omnia*, vol. 1, pp. 179–84. See Brundell, *Pierre Gassendi*, pp. 61–9. See also Edward Grant, *Much Ado About Nothing: Theories of Space and the Vacuum from the Middle Ages to the Scientific Revolution* (Cambridge University Press, 1981), p. 213; and John Henry, "Francesco Patrizi da Cherso's Concept of Space and Its Later Influence," *Annals of Science*, 36 (1979): 549–73.
14. Henry, "Patrizi da Cherso's Concept of Space," p. 560.
15. Ibid., p. 568.
16. Ibid., pp. 567–9.
17. Gassendi, *Syntagma philosophicum*, in *Opera omnia*, vol. 1, pp. 182–3.

power and the possibility of extraterrestrial void.[18] Hobbes appealed to the same idea, influenced either by Gassendi's ideas directly or by Walter Charleton's English rendition of them.[19] This thought experiment, which describes a situation that humans cannot possibly observe, is an example of Gassendi's use of conceptual analysis, a method involving the clarification of concepts rather than the observational testing of hypotheses. Gassendi himself said of this method, "It is frequently necessary to proceed this way [by assuming something impossible] in philosophy, as when they tell us to imagine matter without form in order to permit us to understand its nature."[20]

If Gassendi's discussion of the extracosmic void was largely theological and conceptual, his arguments for the existence of the interstitial void and the *coacervatum* void were generally empirical. The existence of interstitial or interparticulate void spaces disseminated among the particles of matter and separating them from each other is a central postulate of the atomic theory. Epicurus had argued that void is necessary in order for motion to occur. Since matter is impenetrable, a property he established on the basis of the observed tangibility of corporeal things,[21] a particle of matter could not move from one place to another unless there existed void space into which it could move. Motion is clearly observed to occur.[22] Therefore, void spaces must exist interspersed among the particles of

18. Imaginary annihilation was a common topos used by Scholastic writers to underscore the contingency of matter on God's absolute power. See Amos Funkenstein, *Theology and the Scientific Imagination from the Middle Ages to the Seventeenth Century* (Princeton, N.J.: Princeton University Press, 1986), p. 172. The seventeenth-century Aristotelian Bartholomeus Amicus (1562–1649) used a similar argument to prove that God could create a vacuum in the sublunar region. See Grant, *Much Ado About Nothing*, p. 166. Gassendi's thought experiment is suggestive of a similar one Newton employed in *De gravitatione et equipondio fluidorum*. See A. Rupert Hall and Marie Boas Hall, *Unpublished Scientific Papers of Isaac Newton* (Cambridge University Press, 1962), p. 139. On the possible historical connections between Gassendi's and Newton's conceptions of space, see Grant, *Much Ado About Nothing*, pp. 240–7.

19. Grant, *Much Ado About Nothing*, p. 390 n169.

20. Gasssendi, *Syntagma philosophicum*, in *Opera omnia*, vol. 1, p. 182; translated by Craig Brush in *The Selected Works of Pierre Gassendi* (New York: Johnson Reprint, 1972), p. 386.

21. Gassendi, *Syntagma philosophicum*, in *Opera omnia*, vol. 1, p. 231.

22. Epicurus, "Letter to Herodotus," 39–40, in A. A. Long and D. N. Sedley, *The Hellenistic Philosophers*, 2 vols. (Cambridge University Press, 1987), vol. 1, p. 27 (Greek text in vol. 2, p. 20).

matter.[23] Gassendi cited this argument in the section of the "Logic" devoted to the theory of signs, his theory of inference from visible to invisible things:

> The indicative sign pertains to things naturally hidden, not because it indicates a thing in such a way that the thing can ever be perceived and the sign can be visibly linked to the thing itself, so that it could be argued that where the sign is the thing is too, but on the contrary, because it is of such a nature that it could not exist unless the thing exists, and therefore whenever it exists, the thing also exists.[24]

"Such is . . . motion as it indicates the existence of the void (at least according to Epicurus)."[25] Gassendi regarded this Epicurean argument for the existence of void from motion as paradigmatic of reasoning from signs, an empirical method he repeatedly used to perform transdictive inferences.

Gassendi also invoked several arguments formulated by Hero of Alexandria (fl. 62 A.D.), who had defended the existence of void in antiquity.[26] Hero had drawn an analogy between the matter composing bodies and the properties of a heap of sand or wheat. Just as the individual grains of sand or wheat are separated from each other by air or water, so the particles composing bodies are separated by small void spaces.[27] The pneumatic cannon and aeolipile (a prototype of the steam engine), two of Hero's inventions, graphically illustrate the fact that air can be compressed and then rarefied. Condensation and rarefaction seem to require the existence of interparticulate void in order to be explained. Once again, Hero had drawn an analogy with the properties of a heap of wheat, this time the fact that it can be compressed into a smaller volume by shaking it down and thereby freeing some of the air separating its constituent particles.[28] Gassendi cited other phenomena to make the same point: the saturation of water with salt, the dissemination of dyes through water, the penetration of air and water by light, heat, and cold, all of which he

23. Gassendi, *Syntagma philosophicum*, in *Opera omnia*, vol. 1, pp. 192–3.
24. Ibid., p. 81; (translated by Brush in *The Selected Works of Pierre Gassendi*, p. 332).
25. Ibid.
26. Grant, *Much Ado About Nothing*, p. 97. For Hero's arguments, see Hero of Alexandria, *The Pneumatics of Hero of Alexandria*, translated by Joseph Gouge Greenwood, edited by Bennet Woodcroft, with an introduction by Marie Boas Hall (London: Taylor Walton & Maberly, 1851; facsimile reprint, London: MacDonald, 1971).
27. Gassendi, *Syntagma philosophicum*, in *Opera omnia*, vol. 1, p. 192.
28. Ibid., p. 194.

assumed to be corpuscular.[29] The interstitial void, Gassendi argued, is part of an atomic theory of matter.

The third kind of void, the *coacervatum* void, particularly interested Gassendi. Unlike the extramundane or the interstitial void, the *coacervatum* is not a product of nature, but is produced by the action of various instruments and machines.[30] Sometimes it is produced instantaneously, as when air is compressed in pneumatic cannon, pumps, siphons, bellows, and cupping glasses.[31] At other times a long-lasting void space is produced, as in the example to which Gassendi devoted by far the most attention, the space above the mercury in a barometer.[32] Rejecting the traditional explanation of these phenomena, which appealed to the paradigmatic occult quality, the *horror vacui*, Gassendi explained them in purely mechanical terms – the pressure and resistance of the air, both properties caused by the heaviness of the air.[33]

Gassendi was greatly impressed by the experiments of his contemporaries Torricelli, Galileo, Mersenne, Petit, Pascal, and Auzout. Their experiments with the mercury barometer culminated in Pascal's classic Puy de Dôme experiment, in which the height of a column of mercury was seen to vary inversely with the altitude at which it is measured. Gassendi repeated this experiment in Toulon in 1650.[34] It seemed to establish the fact that the mercury rises because of the heaviness of the air, not because of nature's fear of a vacuum. Barometric experiments raised the additional question of whether the space above the mercury is in fact void, an issue that Gassendi acknowledged was not entirely unproblematic. The physicist and mathematician Gilles Personne de Roberval (1602–75) accepted the void, but rejected the explanatory role of the column of air. Descartes accepted that the column of air explained the changing height of the mercury, but rejected the existence of the void. Pascal accepted both.[35]

29. Ibid., p. 196.
30. Ibid.
31. Ibid., pp. 197–201.
32. Ibid., pp. 203–16. Gassendi wrote this chapter, first published in the *Animadversiones in decimum librum Diogenis Laertii* (1649), sometime before May 1647. On the evolution of Gassendi's writings on the vacuum experiments, see Bernard Rochot, "Comment Gassendi interprétait l'expérience de Puy de Dôme," *Revue d'histoire des sciences*, 16 (1963): 53–76.
33. Gassendi, *Syntagma philosophicum*, in *Opera omnia*, vol. 1, pp. 198–9.
34. Rochot, "Comment Gassendi interprétait l'expérience de Dôme," p. 55.
35. For Pascal's account of the Puy de Dôme experiment, see Blaise Pascal, *The Physical Treatises of Pascal: The Equilibrium of Liquids and the Weight of the Mass of the Air*, translated by I. H. B. Spiers and A. G. H. Spiers, with an introduction and notes by Frederick Barry (New York: Columbia University

Although the height of the mercury was observed to be inversely proportional to the altitude at which it was measured – an observation supporting the view that the suspension of the mercury in the tube is caused by the pressure of the atmosphere – the question remained whether the space above the mercury is indeed void.[36] Gassendi argued that the fact that light, particles of heat and cold, magnetic particles, and the particles that flow from the earth to cause gravity all pass through the glass tube into the space above the mercury seems to negate the vacuity of that space.[37] But, he continued, even if there is some kind of rare, attenuated, subtle matter in the space above the mercury, it cannot be "rarer, more attenuated, and subtler [than ordinary air] without an intermixed void." He supported this contention by appealing to his earlier discussion of rarity and density in which he had shown that the only way to explain these qualities is by appealing to the interstitial void.[38] Furthermore, the matter of light, heat, cold, or magnetism could not penetrate the space above the mercury unless it contains some void. Otherwise, two bodies would occupy the same space simultaneously, a patent impossibility.[39] To those who claimed that there is actually air in the space above the mercury, Gassendi replied that fish, which somehow manage to extract air by breathing under water, continue to live in water exposed to the open air by opening the tube, whereas they will die if the tube is closed off. The presence of air in the water depends on the water's being exposed to air.[40] Drawing an analogy between the water containing the fish and the mercury in the barometer, Gassendi argued that the mercury does not contain air that it might release into the space above it in the barometer. He concluded that the

Press, 1937; reprinted, New York: Octagon, 1973), pp. 103–8. See also Richard S. Westfall, *The Construction of Modern Science: Mechanisms and Mechanics* (New York: Wiley, 1971), pp. 45–8; E. J. Dijksterhuis, *The Mechanization of the World Picture*, translated by C. Dikshoorn (Oxford University Press, 1961), pp. 444–5; and René Dugas, *Mechanics in the Seventeenth Century*, translated by Freda Jacquet (Neuchâtel: Éditions du Griffon, and New York: Central Book, 1958), pp. 229–33.

36. This point remained contentious through the remainder of the century. See Steven Shapin and Simon Schaffer, *Leviathan and the Air-Pump: Hobbes, Boyle, and the Experimental Life* (Princeton, N.J.: Princeton University Press, 1985). See also Jane Elizabeth Jenkins, "Using Nothing: Vacuum, Matter, and Spirit in the Seventeenth Century Mechanical Philosophy," M.A. thesis, University of Calgary, 1990.

37. Gassendi, *Syntagma philosophicum*, in *Opera omnia*, vol. 1, p. 205.

38. Ibid., p. 206.

39. Ibid.

40. Ibid.

barometric experiments can be interpreted without invoking the occult *horror vacui* and that they support the existence of the void, even if some particles of matter are interspersed within the void.

Gassendi had begun his account of the nature of the physical world in the *Syntagma philosophicum* with this discussion of the vacuum because he considered void space to be the container within which the physical world exists. He next considered the entities that constitute the physical world and the means by which they interact. In his terms, this task consisted of identifying a material principle and an efficient principle.[41] Arguing against various alternatives,[42] Gassendi adopted a modified version of Epicurean atomism as his theory of matter and his associated theory of causality.[43]

Gassendi began his discussion of matter by delineating the boundary between corporeal substance – matter – and incorporeal substances – God, the angels, and the rational soul. Corporeal substance differs from the incorporeal because it is endowed with mass, because it is tactile, and because it is capable of resistance. These properties render it tangible, its defining quality, for it is only by contact that matter can act.[44] Although he noted that matter is extended in three dimensions, he clearly distinguished between corporeal substance and purely mathematical extension. Corporeal bodies, unlike geometrical figures, cannot have length without breadth and cannot be without mass[45] and magnitude. They

41. Ibid., p. 229.
42. Ibid., pp. 229–56.
43. Ibid., pp. 230–1. The modifications all revolved around the need to ensure the theological orthodoxy of his physics. They included the addition of God, angels, and the rational soul – all incorporeal substances – to the list of entities comprising the world (ibid., p. 231); the denial of Epicurus' claim that atoms are eternal and the assertion that they, along with everything else in the universe, were created by God (ibid., pp. 234, 280); the rejection of the Epicurean swerve as a way of accounting for the impact of atoms (ibid., p. 277 – for Gassendi's discussion of the swerve in the context of ethics and free will, see Chapter 3); and the denial that atoms are infinite in number and shape and that they possess an innate activity (ibid., p. 280).
44. Ibid., p. 231. Citing the Gospel of John, Gassendi maintained that the resurrected Christ signaled this difference between body and spirit when he told Mary Magdalene not to touch him.
45. Gassendi used the Latin word *"moles,"* meaning "bulk." Although I have translated *"moles"* as "mass," it is incorrect to read any Newtonian meanings into his use of this term.

require these properties in order to act by touch.[46] Tangibility also distinguishes matter from the incorporeal void.[47] Gassendi's insistence on the tangibility of corporeal substance distinguished his theory of matter from Descartes'. It also injected an empiricist assumption into the very heart of his mechanical philosophy, since the tangibility of matter can be known only through the senses. To say that Gassendi's theory of matter presumed an empiricist epistemology is not to say that he actually performed experiments in order to determine the properties of matter.[48] The empiricist tenor of Gassendi's matter theory, however, stands in sharp contrast with Descartes', which presumed a rationalism underwriting a priori knowledge of the properties of matter.

Gassendi's atoms are perfectly full, solid, hard, indivisible particles.[49] Since they are so small that they fall below the threshold of sense, he had to argue for their existence and for their indivisibility indirectly by appealing to commonly observed phenomena from which he reasoned to the unobserved properties of atoms. Following Lucretius in this transdictive maneuver, he noted that various commonly observed phenomena lend support to the existence of such atoms.[50] Wind is evidence that invisible matter can produce visible, physical effects. So is the fact that paving stones and plowshares gradually wear away because of constant rubbing even though individual acts of rubbing produce no discernible change. Similarly, the passage of odors through the air can be explained in terms of tiny particles traveling from the original body to the nose.[51] How do we

46. Ibid., p. 232. On Gassendi's distinction between physical atoms and mathematical points, see Lynn Sumida Joy, *Gassendi the Atomist: Advocate of History in an Age of Science* (Cambridge University Press, 1987), chap. 5.
47. Gassendi, *Syntagma philosophicum*, in *Opera omnia*, vol. 1, p. 231.
48. On the uses of empirical arguments in support of corpuscularianism more generally in seventeenth-century natural philosophy, see Christoph Meinel, "Empirical Support for the Corpuscular Theory in the Seventeenth Century," in *Theory and Experiment: Recent Insights and New Perspectives on Their Relation*, edited by D. Batens and J. P. Van Bendegem (Dordrecht: Reidel, 1988); and Christoph Meinel, "Early Seventeenth-Century Atomism: Theory, Epistemology, and the Insufficiency of Experiment," *Isis*, 79 (1988): 68–103.
49. Gassendi, *Syntagma philosophicum*, in *Opera omnia*, vol. 1, pp. 257–8.
50. Lucretius presents his arguments for the existence of atoms in *De rerum natura*, bk. I, lines 265–328. See Titus Carus Lucretius, *De rerum natura*, translated by W. H. D. Rouse, revised by Martin Ferguson Smith, 2d edition (Cambridge. Mass.: Loeb Classical Library, 1982), pp. 23–31.
51. Gassendi, *Syntagma philosophicum*, in *Opera omnia*, vol. 1, p. 259.

know that the tiny particles that constitute the paving stones and odors are actually *indivisible* atoms? If such indivisible atoms did not exist, it would be necessary to postulate the infinite divisibility of matter. Infinite divisibility seemed to Gassendi, as it had to Epicurus, to lead to a host of absurdities, including Zeno's paradoxes. Therefore, there must be a finite limit to the process of dividing matter. Indivisible atoms must exist.[52] Atoms are perfectly full and contain no void; this is the reason for their indivisibility. Since they are indivisible, the interstitial void must exist to separate the atoms from each other.[53] The indivisibility of atoms does not present an obstacle to divine power and thus does not contradict Gassendi's voluntarism. Gassendi regarded Zeno's paradoxes as proving that the infinite divisibility of matter is self-contradictory. The necessary existence of indivisible atoms (given that God created matter at all) is thus an acceptable limitation of divine power since it follows from the principle of noncontradiction.

What are the properties of these atoms? Magnitude and figure, resistance (or solidity), and heaviness are the properties Gassendi ascribed to atoms.[54] Although they possess finite magnitude and are, for that reason, distinct from mathematical points, they are extremely small, falling far below the threshold of sense. For that reason, we must also be careful to distinguish between the minimum in atoms and the minimum of sense. In contrast to sensible bodies, which we can measure by the direct use of our senses, we measure the size of atoms by a complex process of reasoning.[55] Gassendi extrapolated from a number of observed phenomena to argue for the extreme smallness of atoms. The complex world revealed by the microscope, the new instrument that Gassendi called the "*Engyscopius*," lent credence to his claim that atoms are minute.[56] Particles of flour are seen to be complex, consisting of diversely shaped parts. The tiny mite is observed to possess a number of distinct organs. The dispersion of a small quantity of pigment in a large quantity of water leads to the conclusion that the pigment is comosed of minute particles. Gassendi drew a similar conclusion from the fact that a smoldering green log will continue to issue smoke for over eight hours. The microscope also reveals that small particles of things possess a great diversity of shapes. These

---

52. Ibid., pp. 258, 263–6. Joy discusses the issue of infinite divisibility at some length. See *Gassendi the Atomist*, chap. 7.
53. Gassendi, *Syntagma philosophicum*, in *Opera omnia*, vol. 1, p. 262.
54. Ibid., p. 266.
55. Ibid., pp. 267–8.
56. Ibid., p. 269. Gassendi's arguments here were typical of seventeenth-century arguments in defense of the corpuscular theory and atomism.

facts, coupled with the observation that no two grains or leaves or hands are precisely the same shape, led Gassendi to conclude that the atoms themselves are variously shaped.[57] Indeed, the fact that crystals of different substances are differently, but regularly shaped – the crystals of salt being cubical, those of pure alum being octahedral – and the fact that these crystals grow from smaller crystals of the same shape seem to justify the transdictive conclusion that the tiniest particles of salt or of pure alum have the same shape as the macroscopic crystals of these solutions.[58] Although atoms come in a variety of shapes, he thought that there were only a finite number of such shapes. The multitude of things in the world can be produced from a finite number of kinds of atoms in the same combinatorial way that the complexities of language can be produced from only twenty-two letters of the (Latin) alphabet.[59]

In addition to the geometrical properties of size and shape, atoms possess two properties necessary for them to be the *material* principle of things, resistance (or solidity) and weight. Solidity, or resistance, which is really the impenetrability of matter, is necessary in order for atoms to be material and to act by contact.[60] Their resistance, or solidity, along with their magnitude, distinguishes them from mathematical points. Atoms also possess an innate weight (*pondus*), or heaviness (*gravitas*). Gassendi considered this weight to be a natural or internal faculty or force by which the atoms can move. Unlike Epicurus, who had considered the propensity for motion to be innate to atoms, Gassendi believed that atoms are mobile and active because of the power of moving and acting that God instilled in them at their creation. If their mobility and activity were indeed innate, the dangers of materialism would be very real. Rather, he claimed, their mobility and activity function with divine assent, "for he compels all things just as he conserves all things."[61]

57. Ibid., p. 268.
58. Ibid., pp. 270–1.
59. Ibid., p. 271. The analogy between atoms and letters of the alphabet goes back at least to Lucretius. See *De rerum natura*, bk. I, lines 823–7, 912–14; bk. II, lines 688–94, 1013–18.
60. Gassendi, *Syntagma philosophicum*, in *Opera omnia*, vol. 1, p. 232.
61. Ibid., p. 280. The activity of Gassendi's matter is discussed by Marco Messeri, *Causa e spiegazione: La fisica di Pierre Gassendi* (Milan: Franco Angeli, 1985), pp. 74–93; and by Olivier René Bloch, *La philosophie de Gassendi: Nominalisme, matérialisme, et métaphysique* (The Hague: Martinus Nijhoff, 1971), pp. 210–29. I disagree with Bloch, who argues that Gassendi favored the activity of matter, finding in it the basis for materialism on which he superimposed a creationist theology (p. 214). Even Bloch allows that in the chapter "De Motu et Mutatione Rerum," Gassendi returned to a

Although motion is imposed on atoms by God, their motion persists perpetually. Blochrargues that "the principle of inertia proper to pure mechanics only works fully [in Gassendi's physics] for *res concretae*, where rest can exist the same as motion, but it is founded at the level of matter itself, i.e., atoms, on a dynamics in which mobility is primordial and movement perpetual."[62] In this way, Gassendi attempted to combine his atomism with Galileo's inertial mechanics, which he had endorsed in *De motu impresso*. Bloch uses this consistency in Gassendi's thought to call him a materialist, because it provides the basis for ascribing "all natural effects to the presence or action of matter and of matter alone."[63] What Bloch overlooks in his otherwise penetrating discussion of Gassendi's theory of motion is the importance of the theological framework that infuses his entire natural philosophy. The fact that God imparts motion to atoms in the first place and could, by an act of will, remove it, obviates the self-sufficiency of the material world that a genuine materialism demands.

Both properties – solidity, or resistance, and weight – are what I call empirical, because their existence can only be known by our experience of the behavior of material things. Although Gassendi's arguments for the existence of atoms and their innate properties are not always conclusive, they are all based on observed properties of matter. From these observations, Gassendi proceeded by analogy, speculation, and sometimes testable hypothesis to describe the properties of atoms.

Atoms, Gassendi concluded, must be accepted as the material principle of things. God created them at the beginning, and then he fashioned the first things he created from atoms. All subsequent generation and corruption and all change result from the motion, impact, and rearrangement of the original atoms.[64] One might well ask how Gassendi, the empiricist, justified all this discussion about the properties of atoms, which obviously fall below the threshold of sense. He used two lines of argument to justify these transdictive inferences. First, he often argued indirectly, by analogy with visible bodies and those rendered visible by use of the microscope. Second, on several occasions he said explicitly, citing Lucretius, that what we cannot see we know by the understanding, but the understanding is

Galilean, inertial concept of motion. I think that it is truer to the texts to interpret Gassendi as taking theological matters very seriously and modifying classical atomism to meet theological demands rather than to see him – as Bloch does – as a closet materialist.

62. Bloch, *La philosophie de Gassendi*, p. 227; my translation.
63. Ibid., p. 228.
64. Gassendi, *Syntagma philosophicum*, in *Opera omnia*, vol. 1, p. 280.

informed entirely by sense.[65] That is to say, on the basis of macroscopic observations, we make inferences about the microscopic mechanisms producing the phenomena. Although we use our reasoning ability to make these inferences, the content of the ideas and premises we use all come from experience.

Matter in the form of atoms is the material principle in Gassendi's world. The efficient principle explains how these atoms interact. Here Gassendi adopted the common seventeenth-century analysis of causation: The first cause is God, who created everything else including the second causes that operate in the established order of the world.[66] Rejecting Aristotle's fourfold analysis of cause into efficient, material, formal, and final, Gassendi maintained that the word "cause" is synonymous with the word "efficient" and that all of Aristotle's causes can be reduced to the efficient cause.[67] Although Gassendi apparently contradicted himself when he argued for the existence of final causes in his controversy with Descartes,[68] the final causes of which he wrote were not immanent, Aristotelian final causes. Rather, they were God's purposes, imposed on nature from without, and can therefore be understood as efficient causes. Hence, the natural world really contains only efficient causes, even if some of them express divine purpose.

Since atoms constitute the material principle of the natural world, questions about causality are transformed into questions about how atoms interact. And the answer, for Gassendi, is straightforward: The activity of atoms lies in their motion. Hence, the action of causes is simply the motion of atoms.[69] Although God, the first cause, does not act by any motion of his own, but rather by his mere command, by his command he instills motion and hence activity into the atoms.[70] The causal structure of the physical world is thus reduced to the motions of matter, to problems in mechanics. Gassendi adopted the Epicurean definition of "motion" as "migration from place to place, whether of the whole body or its parts."[71]

---

65. Ibid., p. 269. See Meinel, "Early Seventeenth-Century Atomism." See also Catherine Wilson, "Visual Surface and Visual Symbol: The Microscope and the Occult in Early Modern Science," *Journal of the History of Ideas*, 49 (1988): 85–108.

66. See Chapter 2.

67. Gassendi, *Syntagma philosophicum*, in *Opera omnia*, vol. 1, pp. 283–4.

68. See Chapter 6.

69. Gassendi, *Syntagma philosophicum*, in *Opera omnia*, vol. 1, pp. 336–8.

70. Ibid., p. 334.

71. Ibid., pp. 338–9. Since Gassendi understood "space" (*spatium*) in the Epi-

That motion exists is beyond question as we know it from the experiences of our senses. Gassendi dismissed Zeno's paradoxes out of hand.[72] Purely rational argument, he stated, cannot withstand the testimony of the senses.

Gassendi thus reduced all physical change to the local motion of atoms. Where Aristotle in the *Physics* had enumerated several kinds of change – growth, decay, generation, corruption, and qualitative change – Gassendi reduced them all to the motions of atoms.[73] Atoms communicate their motions to each other by contact and collision, thus making impact the primary agent of change in the physical world. In some cases, contact between the mover and the moved is not evident – for example, in the case of magnetic attraction or the transmission of heat from fire. Nevertheless, contact does occur, at the invisible, atomic level.[74] There is no action at a distance.

Through his extensive discussion of atoms and the void, motion and change, Gassendi had answered the fundamental ontological questions about the natural world: Of what kinds of entities does the world consist? By what means do these entities interact? In answering these questions, he had replaced traditional Aristotelianism with the mechanical conceptual framework within which he thought natural philosophy should be formulated.

Atoms and qualities: An empiricist mechanical philosophy

For the remainder of this chapter, I describe how Gassendi combined his empiricist account of knowledge with his atomistic philosophy to formulate a mechanical account of nature. Gassendi's emphasis on sensible experience gave questions about the qualities of things and how they affect us a central place in his natural philosophy since what the senses perceive of the physical world are the qualities of things. The subject matter of physics consists of our sensations of qualities, those things that lie open to sight, touch, and the other senses.[75] Given his mechanical

curean sense of "infinite void space stretching in all directions," he understood "place" (*locus*) to be a point in that three-dimensional space. See Brundell, *Pierre Gassendi*, pp. 61–9.

72. Gassendi, *Syntagma philosophicum*, in *Opera omnia*, vol. 1, pp. 340–1.
73. Ibid., pp. 362–4.
74. Ibid.
75. Ibid., p. 372.

account of the world, the doctrine of primary and secondary qualities became the centerpiece of his physics as well as of his theory of perception. The fundamental problems for Gassendist physics were to understand the ways in which the atomic structures of physical things produce their secondary qualities and to understand our perception of those qualities.

The mechanical (although not always atomistic) explanation of qualities held an equally prominent place in other systematic treatises on the mechanical philosophy. Kenelm Digby's *Two Treatises,* which is a transitional work aiming for mechanical explanations but still immersed in many Aristotelian assumptions, devotes several chapters to the explanation of qualities in mechanical terms.[76] Walter Charleton's *Physiologia Epicuro-Gassendo-Charltoniana* (1654), which is an English version of the physical part of Gassendi's *Animadversiones in decimum librum Diogenis Laertii* (1649), echoes Gassendi's treatment of qualities virtually verbatim.[77] Thomas Hobbes' treatise *De corpore* (1655) devotes most of Part IV, "Of Physics, or the Phenomena of Nature," to the mechanical explanation of qualities.[78] A similar emphasis is to be found in Descartes' *Principles of Philosophy,* as I will explain in Chapter 9.

For Gassendi, all bodies are alike insofar as they are composed of atoms and void. An individual body is a body of a certain kind, however, because of the particular arrangement of the atoms of which it is composed. The configuration of the atoms constituting a thing determines what that thing is like. This configuration is its individual nature and endows it with its particular, determinate qualities.[79] Gassendi's atomistic analysis of qualities thus replaced the Aristotelian account of qualities, which had employed the concepts of substance and accident, matter and form. Gassendi reinterpreted the concept of form or nature in atomistic terms: For him, the form is nothing but the texture of the organized corpuscles composing the object. All other qualities are to be explained in terms of

76. Kenelm Digby, *Two Treatises in the One of Which, The Nature of Bodies; in The Other, The Nature of Mans Soule; is looked into: In Way of Discovery, of the Immortality of Reasonable Soules* (Paris, 1644), chaps. XIV–XXII.
77. Walter Charleton, *Physiologia Epicuro-Gassendo-Charltoniana, or a Fabrick of Science Natural Upon the Hypothesis of Atoms* (London, 1654; reprinted, New York: Johnson Reprint, 1966), bk. 3, pp. 127–414.
78. Thomas Hobbes, *Elements of Philosophy. The First Section, Concerning Body,* in *The English Works of Thomas Hobbes of Malmesbury,* edited by Sir William Molesworth, 11 vols. (London, 1839–45; reprinted, Aalen: Scientia, 1962), vol. 1, pp. 445–532.
79. Gassendi, *Syntagma philosophicum,* in *Opera omnia,* vol. 1, p. 372.

the configuration of the atoms and void of which the object consists.[80] Qualities become manifest in two ways: by the effects they produce on our organs of sense and by the effects they produce on other material bodies and thus by the changes they produce in our sensations of these bodies. The perception of qualities results from the contact between atoms and our sense organs; therefore, qualities ultimately derive from the magnitude, figure, heaviness, and resultant mobility of atoms.[81]

All qualities – colors, sounds, odors, flavors – are caused by the motions of corpuscles, which are endowed with magnitude, figure, position, and certain motions.[82] When the corpuscles impinge on the sense organs, they cause motions in the nerves. The spirits in the nerves transmit these motions to the brain, where they produce sensations of the qualities in the observed body. Gassendi cited the example of pains in the phantom limb of an amputee to illustrate this account of the sensation of qualities. What really happens when the amputee "feels" a pain in the amputated limb is that the spirits in the nerves are caused to move in the same way as they would have if there were a real limb and corpuscles impinged on it in certain ways. These motions relay the same information to the brain, even though the limb is missing, thus causing the sensation of pain in the absent limb.[83] This example underscores the subjectivity of secondary qualities. Compared with the Aristotelian and chemical theories of matter, which ascribed various real qualities to matter, Gassendi's ontology was seriously impoverished. Although matter possesses a few real qualities, designated as primary, most qualities are secondary and are relegated to the realm of subjectivity. It may require a bit of mental acrobatics to understand that honey does not contain its own sweetness or milk its own whiteness, he said, but it is perfectly evident that wine does not contain its drunkenness and that a poisonous plant does not contain its own danger.[84] In general, matter does not possess any of the qualities that we perceive.[85]

---

80. Ibid., pp. 372–3. For how Gassendi and others reinterpreted this crucial Aristotelian concept in mechanical terms, see Norma E. Emerton, *The Scientific Reinterpretation of Form* (Ithaca, N.Y.: Cornell University Press, 1984).
81. Gassendi, *Syntagma philosophicum*, in *Opera omnia*, vol. 1, p. 394.
82. Ibid., vol. 2, p. 338.
83. Ibid., p. 339.
84. And, as Galileo said, the feather does not contain its own tickling. See Galileo Galilei, *The Assayer*, in *Discoveries and Opinions of Galileo*, translated by Stillman Drake (Garden City, N.Y.: Doubleday Anchor, 1957), p. 275.
85. Gassendi, *Syntagma philosophicum*, in *Opera omnia*, vol. 2, p. 340.

A particular act of sensation depends on two things: the corpuscles composing the object and the corpuscles composing the sense organs. The perceived quality, say the sweet taste, is not in the honey itself but is a consequence of the interaction between the corpuscular texture of the corpuscles composing the honey and the corpuscular texture of the organs of taste. This account of qualities enabled Gassendi to explain the fact that different kinds of animals, different people, and even the same person in different states of health or at different ages perceive the same object quite differently – the standard skeptical arguments as they had been set out by Sextus Empiricus. Since the atoms comprising the sense organs have different configurations in different people or animals or in the same person under different circumstances, these people and animals will all be affected differently by the same configuration of atoms coming from some external object.

Gassendi had begun his philosophical career with a skeptical critique of Aristotelianism, in the *Exercitationes paradoxicae adversus Aristoteleos,* using these arguments. Now, in the posthumous *Syntagma philosoph-icum,* his account of qualities provided him with an atomistic explanation of those very same arguments. He not only used the skeptical arguments as a way of establishing his probabilistic account of empirical knowledge; his empiricism coupled with his atomism led him to a physico-physiological explanation of those very skeptical arguments. In his mind, this explanation probably answered them quite adequately. Gassendi was thus able to provide not only an explanation of the phenomena of the natural world, but also physical grounds for his sensation-based theory of knowledge and scientific method.

Gassendi tried to explain all kinds of qualities mechanically: rarity and density, transparency and opacity; magnitude, figure, subtlety, smoothness, and roughness; mobility; gravity and levity; heat and cold; fluidity and hardness, moistness and dryness; softness, rigidity, flexibility, elasticity, and ductility; taste and odor; sound, light, and colors.[86] For example, he explained the phenomena of light by means of a corpuscular model. He established the corporeality of light by appealing to the fact that one of the defining characteristics of matter is impenetrability. Just as a bean rebounds from a wall because it cannot move through space already occupied by another body, so a ray of light is reflected from material bodies by impact; and this could not be the case unless the ray itself was corporeal.[87] The fact that some bodies are transparent and allow light to pass through them might be regarded as a counterexample to the corporeality of light. But Gassendi was able to explain this phenomenon in

86. Ibid., vol. 1, pp. 372–449.
87. Ibid., p. 427.

terms of his atomism. A body is transparent because there are many pores separating its constituent corpuscles, and the particles of light can thus pass between them easily. It is usually the case that not all of the impinging light is transmitted through transparent bodies, but some of it is reflected back toward the source. This phenomenon can be understood by the analogy of a sieve. Drop a handful of sand from some distance onto a sieve; those grains that fall on the holes go straight through; those that fall on the solid parts rebound. So too with a ray of light falling on some solid body: Those corpuscles that fall on the solid parts are reflected and those that fall on the pores are transmitted. Sometimes a pore does not penetrate all the way through the body but may twist around in its interior. Corpuscles of light that enter such pores are lost in the interior of the body, thus accounting for the absorption of light.[88] Using a similarly rich mix of observation, analogy, speculation, and fantasy, Gassendi effectively argued that every kind of known phenomenon could be explained in terms of his mechanical philosophy of nature – matter and motion.

Along with the so-called manifest qualities, Gassendi argued that the occult qualities could be given mechanical explanations.[89] The distinction between manifest and occult qualities was epistemological for Gassendi. In the former case, we know the mechanism by which they are produced; in the latter case, we do not. Our ignorance of the mechanism that produces them does not justify the conclusion that the normal processes of nature are violated. Examples of such occult qualities include all of the so-called sympathies and antipathies in the world, many of which Gassendi proceeded to explain in mechanical terms. Accordingly, the occult sympathies that have been invoked to account for otherwise inexplicable attractions can be explained in a mechanical way by means of "hooks, cords, goads, prods, and other such things, which although they are invisible, must not be called nothing."[90] For example:

> When you observe a chameleon seize a fly from half a palm away and draw it to its mouth, you see an organ of attraction, the vibration and retraction of the tongue by its great agility, the end of which is viscous and curves into itself. What would you otherwise judge to happen when, for example, amber, sealing wax, and other electrics, when you first rub them, seize, draw, and hold straw and other light things? Indeed, innumerable little rays like tongues seem to be emitted from electric bodies of this kind, which they fill, seize, and carry back and

88. Ibid.
89. Ibid., pp. 449–57.
90. Ibid., p. 450.

hold, by the insinuation of their ends into the little pores of those light things.[91]

All sorts of remarkable phenomena – the heliotrope's following the Sun, the aversion of sheep to wolves, the poisonous glance of the basilisk, the charming of snakes with music, the electric glow of the torpedo fish, the strange power of the remora, which can bring a ship to a dead stop,[92] the medicinal properties of various substances, and the long-distance healing power of the weapon salve – may incite wonder in us, but there is no reason to believe that they are brought about by any cause other than what produces the most familiar effects, namely, the motions and collisions of atoms in the void.[93]

Gassendi did not seem to be aware of the explanatory circularity of many of his mechanical explanations – for example, the use of hooked particles to explain attraction or the invention of special particles to explain certain phenomena, such as the frigorific and calorific particles he invoked to explain heat and cold.[94] But before simply condemning him of circularity, it is necessary to distinguish two issues here: first, his invention of a new, mechanical language for natural philosophy; and second, his attempts to create actual explanations of phenomena and the methodological flaws embedded in these attempts. I regard his discussion of qualities – both manifest and occult – in these sections of the "Physics" as part of the first enterprise, that is, the invention of a new set of explanatory terms for natural philosophy. His goal seems to have been less to provide the definitive explanation of the phenomena he discusses than to demonstrate the possibility of providing some kind of mechanical explanation of all of them.

With what epistemological status did Gassendi endow these mechanical explanations? Although he tried to base them on experience – for example, his appeal to macroscopic impact to explain the reflection of light – I think he regarded them as models for what mechanical explanations should be like rather than as well-established mechanisms. This interpretation fits with the programmatic quality of the *Syntagma philosophicum*.

91. Ibid.
92. On the history of accounts of the torpedo fish and the remora, see Brian P. Copenhaver, "A Tale of Two Fishes: Magical Objects in Natural History from Antiquity Through the Scientific Revolution," *Journal of the History of Ideas*, 52 (1991): 373–98.
93. Gassendi, *Syntagma philosophicum*, in *Opera omnia*, vol. 1, pp. 449–55. On the particular occult qualities that Gassendi considers, see Thorndike, *Magic and Experimental Science*, vol. 7, pp. 452–9.
94. Gassendi, *Syntagma philosophicum*, in *Opera omnia*, vol. 1, pp. 394–401.

Rather than being an exposition of a complete science of nature, it was a prescriptive work, instructing natural philosophers how to proceed to construct a mechanical science of nature.[95]

Gassendi's long discussion of matter, void, causality, and qualities in the first section of the "Physics" laid the foundations for his mechanical philosophy. It spelled out the ultimate constituents of the material world – atoms and void – and pointed to the kind of explanation that would find a place in the new natural philosophy established along mechanical lines. The theories of matter and causality that Gassendi articulated here had close ties with his empiricist epistemology: He employed empirical arguments to establish the existence of both the atoms and the void, then demonstrated that the atoms that constitute the material principle of the world possess primary properties, some of which can be known only empirically. And finally, the mechanisms in terms of which natural phenomena can be understood provided Gassendi with what he thought to be a physiological explanation for his epistemological beliefs. Gassendi's empiricism pervaded his mechanical philosophy and thus linked it to the voluntarist theology within which he had framed it.

95. Popkin agrees with this assessment of Gassendi's approach. See Richard H. Popkin, "Epicureanism and Scepticism in the Early 17th Century," in *Philomathes: Studies and Essays in the Humanities in Memory of Philip Merlan*, edited by Robert B. Palmer and Robert Hamerton-Kelly (The Hague: Martinus Nijhoff, 1971), pp. 346–57.

9

Mathematizing nature: Descartes' geometrical theory of matter

> I neither admit nor desire any principles in physics other than [those]
> in geometry or abstract mathematics; because all the phenomena of
> nature are thus explained, and certain demonstrations concerning
> them can be given.
>
> René Descartes, Principia philosophiae[1]

Like many other seventeenth-century natural philosophers, Descartes shared with Gassendi the goal of replacing Aristotelianism with a mechanical philosophy of nature. Descartes presumed that there was a close relationship between the metaphysical roots of this philosophy – which he worked out in the *Meditations* and Part I of *The Principles of Philosophy* – and the mechanical explanations that grew from those roots. Articulating what he considered the conceptual framework within which natural philosophy was to develop, Descartes specified what sorts of things exist in the world and the means by which they interact. Accordingly, he described a new theory of matter and redefined causality, replacing the old scheme root and branch.

From the time Descartes began composing *Le monde* in 1629 until the end of his life, he consistently held certain consistent views about natural philosophy.[2] The physical world, he thought, consists entirely of matter,

---

1. René Descartes, *Principia philosophiae*, in *Oeuvres de Descartes*, edited by Charles Adam and Paul Tannery, 11 vols. (Paris: J. Vrin, 1897–1983) (hereafter AT), vol. 8–1, pp. 78–9. All translations are my own unless otherwise noted.

2. To say that Descartes' outlook from *Le monde* to *The Principles of Philosophy* exhibits a certain consistency is not to say that he did not experience substantial intellectual development. For a detailed account of the steps leading to the formulation of his standpoint in *Le monde*, see John Andrew Schuster, "Descartes and the Scientific Revolution, 1618–1634: An Interpretation," Ph.D. dissertation, Princeton University, 1977. See also, John A. Schuster, "Descartes' *Mathesis Universalis*, 1619–1628," and Stephen

the one defining property of which is extension.[3] Matter is infinitely divisible and fills all space. There is no void. All physical causality is reduced to the motion and impact of particles of matter. From the beginning, Descartes sought a philosophy of nature that would reflect his claim that its first principles could be known a priori. This approach is evident in his theory of matter as well as in his methodological stance, which I have already delineated in Chapter 5.

Descartes was explicit about the programmatic nature of his project. Writing to Mersenne, who had asked him to explain the "Phainomene de Rome" – the "parhelia" or rainbow-like appearance of multiple suns observed by the Jesuit astronomer Christopher Scheiner[4] – Descartes laid bare, with his customary lack of modesty, his early vision of this project. "In place of explaining only one phenomenon, I have resolved to explain all the phenomena of nature, that is to say, all of physics."[5] Moreover, he believed that he had found incontrovertible foundations for this physical system. The aim of his project remained a constant of Descartes' intellec-

Gaukroger, "Descartes' Project for a Mathematical Physics," both in *Descartes: Philosophy, Mathematics, and Physics,* edited by Stephen Gaukroger (Sussex: Harvester Press, Totowa, N.J.: Barnes & Noble, 1980), pp. 41–96, 97–140.

3. For the development of Descartes' ways of describing matter, see Garber, *Descartes' Metaphysical Physics* (Chicago: University of Chicago Press, 1992), pp. 63–70.

4. Descartes to Mersenne, 8 October 1629, AT, vol. 1, p. 23. "The striking optical phenomenon, whose nature was not known at the time, occurs when the sun shines through a thin cloud composed of hexagonal ice crystals falling with their principal axis vertical. The refraction is through a 60° prism and results in the component colors of the solar spectrum being bent through a slightly different angle. The red end of the spectrum, being bent the least, appears on the inside, while the blue, when visible, appears on the outside. There is usually one circle, but in this case three were observed accompanied by four patches of shimmering light, the parhelia, which are refracted images of the sun." William R. Shea, *The Magic of Numbers and Motion: René Descartes' Scientific Career* (Canton, Mass: Science History Publications, 1991), p. 201. Gassendi also concerned himself with this puzzling phenomenon in *Phoenomenon rarum Romae observatum 20. Martii anno 1629. Subjuncta est causarum explicatio brevis clarissimi philosophi et mathematici D. Petri Gassendi ad illustrissimum Cardinalem Barbarini (sic)* (Amsterdam, 1629). A new edition of this work with a slightly different title was published in 1630 and then reprinted in Pierre Gassendi, *Opera omnia* (Lyon, 1658), vol. 3, pp. 651–62. See Bernard Rochot, *Les travaux de Gassendi sur Épicure et sur l'atomisme, 1619–1658* (Paris: J. Vrin, 1944), p. 99 n124.

5. Descartes to Mersenne, 13 November 1629, AT, vol. 1, p. 70.

tual life, continuing to guide him as he composed *The Principles of Philosophy*.[6] His confidence in his ultimate success was based in large part on the claim that he could identify and explain all the phenomena of nature in terms of his mechanical philosophy. "There are no phenomena of nature apart from those that are in this treatise."[7]

In pursuing this program, Descartes felt that he needed metaphysical foundations for physics, which he believed he had established in his *Meditations*. "These six Meditations contain all the foundations of my physics."[8] He explicitly viewed them as subverting the metaphysics of Aristotle: "But please do not tell people, for that might make it harder for supporters of Aristotle to approve them. I hope that readers will gradually get used to my principles, and recognize their truth, before they notice that they destroy the principles of Aristotle."[9] Viewed from this perspective, the *Meditations* and the very similar Part I of *The Principles of Philosophy* established the ultimate terms of explanation from which Descartes proceeded to construct his physics.[10] In these works, he attempted to prove that God, the soul, and matter are the ultimate constituents of the world. He established the fundamental properties of the soul and matter, and he enunciated his criterion for knowledge. "There, in sum, are all the Principles from which I deduce the truth of other things."[11]

6. Descartes to Mersenne, 11 November 1640, AT, vol. 3, pp. 231–3.
7. Descartes, *Principia philosophiae*, AT, vol. 8–1, p. 323. See also Peter Machamer, "Causality and Explanation in Descartes' Natural Philosophy," in *Motion and Time, Space and Matter,* edited by Peter K. Machamer and Robert G. Turnbull (Columbus: Ohio State University Press, 1976), pp. 168–99; and Daniel Garber, "'Semel in vita': The Scientific Background to Descartes' *Meditations,*" in *Essays on Descartes' Meditations,* edited by Amélie Oksenberg Rorty (Berkeley: University of California Press, 1986), pp. 81–116.
8. Descartes to Mersenne, 28 January 1641, AT, vol. 3, pp. 297–8; in *The Philosophical Writings of Descartes* (hereafter *PWD*), translated by John Cottingham, Robert Stoothoff, Dugald Murdoch, and Anthony Kenny, 3 vols. (Cambridge University Press, 1984, 1985, 1991), vol. 3, p. 173.
9. Ibid. Descartes' sanguine appraisal of his accomplishment was not fulfilled by the reception of his ideas by traditional Aristotelians, who failed to understand the revolutionary nature of his philosophical proposals. See Daniel Garber, "Descartes, the Aristotelians, and the Revolution That Did Not Happen in 1637," *Monist*, 71 (1988): 471–86.
10. Garber emphasizes this point in "'Semel in vita,'" p. 82.
11. Descartes, *Les principes de la philosophie*, "Letter to the Translator," in AT, vol. 9–2, p. 10.

Matter is the fundamental component of Descartes' mechanical philosophy. From his early work *Le monde* to the letters he wrote to Henry More in the final year of his life, Descartes was consistent in maintaining that matter has only one primary property, extension. It has three dimensions and possesses only geometrical properties.[12] Descartes justified this conclusion in *The Principles of Philosophy* on the basis of a mental exercise in which he considered various properties of matter to see which were absolutely necessary for conceiving it. "The nature of matter, or of body, considered universally, does not consist in the fact that it is hard, heavy, colored, or any other mode affecting the senses; but only in that it is a thing extended in length, breadth, and depth."[13] None of its empirically observed properties is essential because it is possible to imagine not observing these properties in some body and still recognizing it as material:

> For as far as hardness is concerned, our senses tell us nothing else about it than that the parts of hard bodies resist the motion of our hands when they meet them. If indeed every time our hands moved in a certain direction, all the bodies located there receded at the same speed at which our hands approach; we would never feel any hardness. Yet it cannot in any way be understood that the bodies which would thus recede would for that reason lose the nature of a body. Therefore, the nature of body does not consist in hardness. For the same reason, it can be shown that weight, color, and all the other properties of this kind which are observed in material substance, can be removed; leaving it intact. Whence it follows that its nature does not depend on any of these [properties].[14]

Descartes had used a similar approach in the Second Meditation. Ruminating on the properties of a piece of wax – in the course of arguing that his mind is better known to him than his body – he observed that although all of the properties of the piece of wax are subject to change, what remains constant throughout is the fact that it is a body. "The wax was not after all the sweetness of the honey, or the fragrance of the flowers, or the whiteness, or the shape, or the sound, but was rather a body which presented itself to me in these various forms a little while ago, but which now exhibits different ones."[15] Two points are especially sig-

12.   René Descartes, *The World*, translated by Michael Sean Mahoney (New York: Abaris Books, 1979), pp. 49–55 (AT, vol. 11, pp. 31–5) (hereafter cited as *Le monde*); *Principia philosophiae*, in AT, vol. 8-1, pp. 24–5, 42; Descartes to More, 5 February 1649, AT, vol. 5, p. 268.
13.   Descartes, *Principia philosophiae*, in AT, vol. 8-1, p. 42.
14.   Ibid.
15.   Descartes, *Meditations on First Philosophy*, in *PWD*, vol. 2, p. 20 (AT, vol. 7, p. 30).

nificant, one concerning the ontological status of matter and the second concerning Descartes' method of argument. These passages directly attack the Aristotelian understanding of substance as the product of matter and form. An Aristotelian would have conceptualized the piece of wax as a bit of prime matter variously informed, in this case with the forms producing its fragrance, color, and other observable qualities. Qualities, for the Aristotelian, are real and indicative of the nature or essence of the substance in question. Descartes fundamentally altered the ontological commitments embodied in the piece of wax. For him, it becomes simply "a body . . . in these various forms" presenting itself to him. The body, rather than its qualities, is the ultimate existent. It stands alone. No longer a union of matter and form, it is simply a body that affects our senses in various ways.

How did Descartes justify this radically new understanding of the nature of matter? Unlike Gassendi, Descartes appealed neither to ancient authorities nor to experiential data, but rather to the fact that he could form a clear and distinct idea of the nature of the wax:

> Here is the point, the perception I have of it is a case not of vision or touch or imagination – nor has it ever been, despite previous appearances – but of purely mental scrutiny; and this can be imperfect and confused, as it was before, or clear and distinct as it is now, depending on how carefully I concentrate on what the wax consists in.[16]

Having discovered a clear and distinct idea in his own mind, Descartes believed that he had the epistemological grounds for claiming that the world is as he understood it to be. He believed that he had discovered the nature of matter in the world by using purely a priori methods: "I now know that even bodies are not strictly perceived by the senses or the faculty of imagination but by the intellect alone, and that this perception derives not from their being touched or seen but from their being understood."[17]

Descartes determined the nature of matter by a priori methods, and he established its existence by invoking his clear and distinct rule as well. In this case, he applied the rule to a sensible perception rather than to a product of mental scrutiny alone. Starting with the assumption that anything we sense "depends on whatever is influencing our senses," Descartes continued, "Because we feel, or rather, because our senses lead us to clearly and distinctly perceive a certain matter which is extended in length,

16. Ibid., p. 21 (p. 31).
17. Ibid., p. 22 (p. 34).

breadth, and depth,"[18] we must conclude that matter – as extension – exists. It might be the case that no such matter exists and that God could cause these sensations in us directly, as voluntarists like Ockham and Gassendi claimed.[19] But, in that case, God would be a deceiver, something Descartes had already ruled out as "completely contrary to God's nature."[20] "It must therefore be concluded with certainty that there is a certain substance, extended in length, breadth, and depth, and possessing all those properties which we clearly conceive to be appropriate to extended things; and it is this extended substance which we call body or matter."[21] Extrapolating from the piece of wax, Descartes arrived at the conclusion that matter possesses only one defining property, spatial extension:[22]

> Extension in length, breadth, and depth constitutes the nature of corporeal substance. . . . For everything else which can be attributed to body presupposes extension, and is only a certain mode of an extended thing. . . . Thus, for example, figure cannot be understood except in an extended thing, nor can motion, except in an extended space. . . . But on the contrary, extension can be understood without figure or motion. [23]

Extension is a necessary condition for all the other qualities attributed to matter, but it presupposes none of them. Descartes' theory of matter is a priori in the sense that the primary qualities of matter – those qualities associated with extension – can be known by purely a priori methods. In this respect, Descartes' theory of matter differs from Gassendi's, which endowed matter with some properties – like tangibility – that could be known only empirically.

What implications did Descartes' a priori theory of matter have for his mechanical philosophy of nature? One immediate consequence of Descartes' having endowed matter with only geometrical properties – a consequence that has methodological as well ontological implications – is that physics can be reduced to geometry.[24] This conclusion is a conse-

---

18. Descartes, *Principia philosophiae*, in AT, vol. 8–1, pp. 40–1.
19. William of Ockham, *Quodlibeta*, vi, qu. vi, in Ockham, *Philosophical Writings*, translated and edited by Philotheus Boehner (Edinburgh: Nelson, 1957), pp. 28–9.
20. Descartes, *Meditations*, in *PWD*, p. 39 (AT, vol. 7–1, pp. 40–1).
21. Ibid.
22. Ibid., pp. 20–1 (pp. 30–1).
23. Descartes, *Principia philosophiae*, in AT, vol. 8–1, p. 25.
24. According to Descartes, space and matter share the same essential property – extension. The question of whether he thought that this shared essence implies a literal equivalence of space and matter is vexed. See Gregory

quence of the fact that "it is the same extension that is the nature of body and that constitutes the nature of space."[25] The only difference between space (*spatium*) and material substance is in our conception of them. Space itself has no independent existence. This idea of space is an abstraction from what is real – matter.[26] A stone, for example, possesses extension of a particular shape. If the stone is moved, it continues to possess the same extension. But,

> we judge that the extension of the place (*locus*) in which the stone was remains and is the same, although the stone's place may now be occupied by wood, water, air, or some other bodies, or even vacuum, if it exists. The reason for this is that it uses the same magnitude and figure and has the same location among external bodies that determine that space (*spatium*).[27]

The body can be understood solely in terms of its geometrical properties. "Indeed, the names 'place' (*loci*) or 'space' (*spatii*) do not signify something different from the body which is said to be in the place, but only designate its magnitude, figure, and location among other bodies."[28] Since size, shape, and location – all geometrical properties – are the primary qualities of the body, its nature and properties can be known mathematically.

For Descartes, who – like many seventeenth-century thinkers – esteemed geometrical methods for their truth-preserving qualities, this conclusion seemed to provide a direct route to the mathematization of natural philosophy. In this respect, however, his ambition outstripped his accomplishments. He was notoriously unsuccessful in mathematizing physics in the manner of Galileo, Huygens, or Newton because, in Westfall's phrase, "the mechanical philosophy's obsession with causation" inhibited him from using the method of idealization essential for that undertaking.[29] Descartes did not understand this limitation to his approach and believed himself to have provided suitable foundations for mathematical physics. Writing to Mersenne about his reaction to Galileo's *Discourses on Two New Sciences,* Descartes criticized Galileo for limiting himself to the analysis of particular problems without providing metaphysical foundations for physics:

Brown, "Mathematics, Physics, and Corporeal Substance," *Pacific Philosophical Quarterly,* 70 (1989): 281–302.

25. Descartes, *Principia philosophiae,* in AT, vol. 8–1, p. 46.
26. Garber, *Descartes' Metaphysical Physics,* p. 136.
27. Descartes, *Principia philosophiae,* in AT, vol. 8–1, pp. 46–7.
28. Ibid., p. 47.
29. Richard S. Westfall, *Force in Newton's Physics: The Science of Dynamics in the Seventeenth Century* (New York: American Elsevier, 1971), p. 89.

I found in general that he philosophizes much better than the vulgar, in that he leaves as much as he can the errors of the Schools and tries to examine physical matters by mathematical reasons. In this I am entirely in accord with him, and I hold that there is no other way of finding the truth. But it seems to me that he lacks much in that he continually makes digressions and does not stop to explain everything about a matter. This shows that he does not examine them in order and that without having considered the first causes of nature, he has only found the reasons for several particular effects; and thus he builds without foundation.[30]

It is precisely this foundation for a mathematical physics that Descartes believed he had discovered in the equivalence of matter and geometrical extension. "For I fully acknowledge that I know of no other matter of corporeal things than that which can be entirely divided, shaped, and moved, what Geometers call quantity and take as the object of their demonstrations."[31] Moreover, all the properties of *res extensa*, Descartes claimed, can be derived by mathematical methods. "And because all Natural Phenomena can thus be explained, as will appear in what follows; I think that no other principles of Physics should be accepted, or even desired."[32]

Descartes believed that identifying extension as the essence of matter endows physics with the demonstrative certainty of geometry. Since he considered the capacity for geometrical reasoning innate to the human mind, this idenitity provided him with an explanation of how it is that mathematics is applicable to physics.[33] It also enabled him to prove certain far-reaching conclusions about the properties of matter. He appealed to the equivalence of matter and extension to prove that matter fills all space:

That vacuum taken philosophically – that is, a space in which there is entirely no substance – cannot exist is manifest from the fact that the extension of space, or of internal place, does not differ from the extension of body. For since from this alone, that a body is extended

30. Descartes to Mersenne, 11 October 1638, AT, vol. 2, p. 380. Gaukroger compares Galileo's and Descartes' approaches to mathematical physics at some length. See Gaukroger, "Descartes' Project for a Mathematical Physics," pp. 97–8.

31. Descartes, *Principia philosophiae*, in AT, vol. 8-1, pp. 78–9.

32. Ibid. See also Descartes' extensive discussion of the equivalence of matter and extension in his letter to Henry More, 5 February 1649, AT, vol. 5, pp. 267–79.

33. See Nancy L. Maull, "Cartesian Optics and the Geometrization of Nature," in *Descartes: Philosophy, Mathematics, and Physics*, edited by Gaukroger, pp. 23–40.

in length, breadth, and depth; we rightly conclude that it is a substance, because it is wholly contradictory for that which is nothing to
be extension; and the same must also be concluded about space,
which is supposed to be empty: that, certainly since it has extension in
it, there also must necessarily be substance in it.[34]

Using the same kind of a priori reasoning, he proved that matter can be
divided without limit:

We also easily understand that there cannot be any atoms, or parts of
matter indivisible by their own nature. The reason is that if there were
any such things, they would necessarily have to be extended, no matter how small they are imagined to be. We can, therefore, still divide
them into two or more smaller ones in thought and thus know that
they are infinitely divisible.[35]

He justified this claim of infinite divisibility by appealing to his clear and
distinct criterion. Since we can conceive, clearly and distinctly, of these
small particles being further divided, they can, in fact, be so divided:

Even if we imagine that God wished to create a particle of matter
which could not be divided into smaller ones, nevertheless that particle could not be properly called indivisible. Even if he made it such
that it cannot be divided by any created thing, he could not have
removed his own faculty of dividing it, because he clearly cannot
diminish his own power. . . . Strictly speaking, this particle will remain divisible, since it is so by virtue of its own nature.[36]

Descartes continued, responding to two common objections to his plen-
ist account of matter. If all matter is qualitatively the same – an assumption Descartes made when he equated matter with extension – and if,
consequently, all space is equally full of matter, how is one to account for
the different densities of different bodies? Descartes attempted to explain
variations in density entirely in terms of the geometrical properties of
matter:

As for rarefaction and condensation, whoever attends to his thoughts
and wishes to admit nothing except what he perceives clearly, will not
believe that anything other than change of shape occurs. Thus, in
order that bodies be rarefied, there are many intervals between their
parts which are filled by other bodies. And they become denser because their parts, by approaching one another, either diminish or
completely eliminate these intervals; if the latter ever occurs, then the
body becomes so dense that it cannot become denser. However, it is
not then less extended than when, having its parts scattered, it filled a
greater space. For whatever extension is contained in the pores or

34. Descartes, *Principia philosophiae*, in AT, vol. 8-1, p. 49.
35. Ibid., pp. 51-2.
36. Ibid.

intervals remaining between its parts can in no way be attributed to it, but to whatever other bodies fill those spaces. Thus, when we see a sponge full of water or another liquid, we do not think that, in terms of its own individual parts, it has more extension than when it is compressed and dry; but only that its pores have greater openings, and that its parts are therefore diffused over more space.[37]

But how can the matter filling the sponge have more or less extension when it must fully occupy a given quantity of space? Descartes begged the question at this point: A given body's being more or less compressed requires precisely the same explanation that he was seeking for rarefaction and compression in the first place. In a similarly self-serving argument, he noted that the classical observations that others used to support the existence of the void could easily be explained away. For example, he explained the movement of air filling a bellows as resulting from the fact that when air is pushed out of the bellows, it has nowhere to go in the plenum unless some other air moves into the now emptied bellows.[38]

Descartes believed that all the matter in the universe is qualitatively the same inasmuch as it is equivalent to geometrical extension: "The matter of the heaven does not differ from that of the earth; and even if there were countless worlds in all, it would be impossible for them not to all be of one and the same [kind] of matter. . . . Nor can we discover in ourselves, the idea of any other [kind] of matter."[39] In the face of this uniformity of nature, he developed a theory of elements in order to explain some of the qualitative differences he observed in the world. He distinguished matter into three elements, which differ from each other according to the sizes of their constituent particles:[40]

We have two very different kinds of matter, which can be called the first two elements of this visible world. The first of these is that which has so much force of agitation that, by colliding with other bodies, it is divided into particles of indefinite smallness and adapts its figures to fill all the narrow parts of the little angles left by the others. The other is that which is divided into spherical particles, very small indeed if they are compared with those bodies that our eyes can discern; but nevertheless of a certain and determined quantity and divisble into others much smaller. We discover the third, consisting of parts that are either much thicker or have figures less suited to movement.

37. Ibid., p. 43.
38. Descartes to Mersenne, 2 February 1642, AT, vol. 3, p. 612.
39. Descartes, *Principia philosophiae*, in AT, vol. 8-1, p. 52.
40. On the development of Descartes' theory of the elements, see John W. Lynes, "Descartes' Theory of Elements: From *Le Monde* to *The Principles*," *Journal of the History of Ideas*, 43 (1982): 55–72.

And we shall show that all the bodies of this visible world are composed of these three elements: the Sun and the fixed Stars of the first, the Heavens of the second, and the Earth, the Planets, and the Comets of the third.[41]

Some matter consists of spherical particles, shaped by the rubbing of pieces of matter that were broken apart at the time the world was formed. This matter is the matter of which the sun and the stars are composed.[42] Descartes called this subtle matter. The particles of this subtle matter move very rapidly. Light is produced by the centrifugal pressure in this subtle matter, which forms the vortices filling space.[43] In a plenum, however, not all the particles of matter can be spherical since there can be no empty spaces between them. Consequently, there exist very small particles, the shavings and rubbings remaining from the formation of the spherical particles, that fill all the interstitial spaces. Finally, there is a third kind of matter, consisting of much larger pieces. This third matter is what forms the planets and the Earth.[44]

Descartes' theory of the elements reveals him to be in a state of transition between an Aristotelian conception of matter, which preserved the ordinary experience of qualitative differences between different kinds of matter in a philosophical account, which included real qualities as constituents of the world, and a purely mechanical conception of matter, which demanded a reeducation of common sense by reducing all qualitative differences to differences of the quantity and configuration of particles of extended matter. To the extent that he found it necessary to distinguish several different kinds of matter out of which all other things are composed, he retained some impulse toward the common-sense assumption that there are qualitatively different kinds of matter in the world. The remnants of Aristotelianism in his thinking are also apparent in the fact that he thought that each of his three elements has a natural place in the universe:

> And yet one need not think therefore that the elements have no places in the world that are particularly destined for them and where they can be perpetually conserved in their natural purity. On the contrary, each part of matter always tends to be reduced to one of their forms and, once having been reduced, tends never to leave that form.[45]

41. Descartes, *Principia philosophiae*, in AT, vol. 8-1, p. 110.
42. Ibid., pp. 103–4.
43. Ibid., pp. 116–17.
44. Ibid., pp. 151–2. Descartes gave a similar account of the three elements in *Le monde*, pp. 35–9, 43–5, 81–7 (AT, vol. 11, pp. 23–5, 43–5, 49–54).
45. *Le monde*, pp. 43–5 (AT, vol. 11, pp. 28–9). There are similar passages in the *Principia philosophiae*. See, e.g., AT, vol. 8-1, p. 105.

Although the idea that the elements have proper places in the world resembles Aristotle's notion of natural place, Descartes no longer retained any of the associated ideas about natural motion.

Since all the matter – understood as extension – is the same everywhere in the universe, qualitative differences and change must have some source other than matter itself:

> All the properties which we clearly perceive in it [matter] are reducible to this one, that it is divisible and its parts movable; and that it is therefore capable of all the affections that we perceive can follow from the motion of its parts. . . . all the variation of matter, or the diversity of its forms, depends on motion.[46]

Like Gassendi, Descartes considered God to be the first cause of motion in the world:

> And as for what pertains to the [general] cause, it seems manifest to me that this is none other than God Himself, who in the beginning created matter along with both motion and rest; and now, by his ordinary concourse (*concursum ordinarium*), conserves in the sum total of matter, the same quantity of motion and rest as he placed in it at that time.[47]

God is thus the efficient cause of the world and also its sustaining cause, causing not only its existence, but also all the motion in it.[48] All other causes in the physical world can be reduced to the motion and impact of particles of matter.

Descartes explicitly ruled final causes out of natural philosophy because he thought that God's intentions remain inscrutable. He thus attempted to reduce the four causes of Aristotelian physics and metaphysics to the efficient cause, which he now redefined in terms of motion and impact. Machamer has argued that Descartes was not entirely successful in eliminating final causes from natural philosophy. His "work makes extensive, though unacknowledged, use of final causes and teleological factors as traditionally conceived."[49] The fact that God created the world to "fit"

46. Descartes, *Principia philosophiae*, in AT, vol. 8-1, pp. 52–3.
47. Ibid., p. 61. "As for the general cause of all the motions which are in the world, I do not conceive of it as any other than God, who from the first instant that he created matter, began to move all its parts diversely, and now by the same action as he conserves this matter, he also conserves in it as much motion as he put into it." Descartes to the Marquis of Newcastle, October 1645, AT, vol. 4, p. 328.
48. See Daniel Garber, "How God Causes Motion: Descartes, Divine Sustenance, and Occasionalism," *Journal of Philosophy*, 84 (1987): 567–80.
49. Machamer, "Causality and Explanation in Descartes' Natural Philosophy," p. 195.

our modes of knowing is Machamer's prime example of final causes play-
ing a role in Cartesian philosophy. On some occasions, an almost Aristotelian kind of teleology did slip into Descartes' discourse. For example,
writing to Clerselier about the rules of impact, he said:

> Nor will you have further difficulty with these rules when you take
> account of the fact that they depend uniquely on the principle that
> *when two bodies collide, and they contain incompatible modes, then
> there must occur some change in these modes in order to make them
> compatible; but this change is always the least that may occur.* In
> other words, *if these modes can become compatible when a certain
> quantity of them is changed, then no larger quantity will change.*[50]

This principle states that there is an end state of compatibility that the
modes will seek, in much the way that bodies seek their natural place in
the Aristotelian world. Such remarks are evidence of the fact that
Descartes' thinking contained some unconscious carryover from traditional natural philosophy. They express ideas that he rejected when he
confronted them directly. "We shall not select any reasons about natural
things concerning the end that God or nature intended for himself in
creating these things: because we ought not be so arrogant as to think that
we are participants in his plans."[51] Despite this fragmentary evidence that
Descartes held some residual teleological views, he explicitly stated that
final causes have no role in physics, and they played no role in his mechanical explanations of various phenomena in the way that they had in traditional Aristotelian natural philosophy.[52]

Having created motion, God maintains it in the universe according to
the Laws of Nature which follow from his immutability. These Laws – the
conservation of motion, the principle of inertia, and Descartes' notable
but flawed formulation of the laws of impact[53] – ostensibly provided the

50. Descartes to Clerselier, 17 February 1645, AT, vol. 4, p. 185 (*PWD*, vol. 3, p.
     247).
51. Descartes, *Principia philosophiae*, in AT, vol. 8–1, p. 15. Descartes discussed
     his rejection of final causes in physics extensively in his controversy with
     Gassendi. See Chapter 6.
52. This point is underscored by Garber, *Descartes' Metaphysical Physics*, pp.
     119, 273–4.
53. On Descartes' treatment of impact and the problems he encountered, see
     Westfall, *Force in Newton's Physics*, chap. 2; and Garber, *Descartes' Meta-
     physical Physics*, chap. 8. See also Alan Gabbey, "Force and Inertia in the
     Seventeenth Century: Descartes and Newton," in *Descartes: Philosophy,
     Mathematics, and Physics*, edited by Stephen Gaukroger, pp. 230–320;
     Richard J. Blackwell, "Descartes' Laws of Motion," *Isis*, 57 (1966): 220–34;
     and Desmond M. Clarke, "The Impact Rules of Descartes' Physics," *Isis, 68*

foundations for the science of motion that Descartes thought could in principle be used to explain all the phenomena in the world, the task that he addressed in Books III and IV of *The Principles of Philosophy*. However, even he acknowledged that this optimistic claim was not literally true, since he deferred discussing living things and humans in *The Principles of Philosophy*, claiming insufficient knowledge and lack of time.[54] As we have seen in Chapter 5, he regarded these laws as necessary truths, which, he believed, can be demonstrated with certainty from first principles. Descartes believed that he had reached this stage of his project using only a priori methods, although empirical methods are needed to find out about the details of this world.[55]

Descartes was fully aware that his understanding of matter as geometrical extension was a radical break from the Aristotelian concept of matter as correlative to form.[56] He proceeded to show how the explanatory concepts of the traditional Scholastic framework were to be rejected and reinterpreted in mechanical terms. From the outset, he rejected the Scholastic view that material objects possess real qualities. The opening sentence of *Le monde* established the goal to which much of his philosophizing was directed:

> (1977): 55–66. For criticism of Westfall's and Gabbey's interpretations of the concept of force implicit in Descartes' mechanics, see Gary C. Hatfield, "Force (God) in Descartes' Physics," *Studies in History and Philosophy of Science, 10* (1979) 113–40. Hatfield argues that Descartes' mechanics does not contain a hidden concept of force, but that his insistence on totally passive matter requires God's activity in every impact and every conservation of motion. The problem to which this disagreement points is the same one discussed by John Henry, "Occult Qualities and the Experimental Philosophy: Active Principles in Pre-Newtonian Matter Theory," *History of Science, 24* (1986): 335–81. Henry notes that the same problems that drove Newton to introduce active principles and forces into his theory of matter plagued all earlier versions of the mechanical philosophy, which attempted to eliminate all innate activity from matter.

54. Descartes, *Principia philosophiae*, in AT, vol. 8-1, p. 188.
55. Steven M. Nadler, "Deduction, Confirmation, and the Laws of Nature in Descartes' *Principia philosophiae*," *Journal of the History of Philosophy, 28* (1990): 379–80.
56. At the most abstract level, Descartes' matter theory can be construed as similar to Aristotle's. On this reading, one might argue that geometrical extension is the form that Cartesian matter-as-extension assumes and that matter itself is an unknowable, unexplained, and undiscussed substratum – indeterminate extension. In any concrete body, it always comes in a determinate size and shape (form). I am grateful to Paul Teller and an anonymous reader for making this point.

The first thing I want to make clear to you is that there can be a difference between our sensation of light (i.e., the idea that is formed in our imagination through the intermediary of our eyes) and what is in the objects that produces that sensation in us (i.e., what is in the flame or in the sun that is called by the name of 'light').[57]

In other words, he began by rejecting the Aristotelian contention that sensation is a reliable source of knowledge about the natures of material objects. In a passage reminiscent of Galileo's *Il Saggiatore* (1623), Descartes graphically illustrated the subjectivity of secondary qualities:

Of all our senses, touch is the one thought least misleading and most certain, so that if I show you that even touch causes us to conceive many ideas that in no way resemble the objects that produce them, I do not think you will find it strange if I say that sight can do the same. Now, there is no one who does not know that the ideas of tickling and of pain, which are formed in our thoughts when bodies from without touch us, bear no resemblance whatever to those bodies. One passes a feather lightly over the lips of a child who is falling asleep, and he perceives that someone is tickling him. Do you think the idea of tickling that he conceives resembles anything in this feather?[58]

Descartes sought to eliminate Aristotelian real qualities and substantial forms (with the significant exception of the human soul) and replace them with matter and motion as the only explanatory terms for natural phenomena. Writing to Mersenne in the last decade of his life, he stated his view explicitly:

I do not believe there are in nature any real qualities attached to substances and separable from them by divine power like so many little souls in their bodies. Motion, and all the other modifications of substance which are called qualities, have no greater reality, on my view, than is commonly attributed to philosophers to shape, which they call only a mode and not a real quality.[59]

---

57. *Le monde*, p. 1 (AT, vol. 11, p. 3).
58. Ibid., p. 5 (pp. 5–6). Note the striking resemblance to Galileo: "A piece of paper or a feather drawn lightly over any part of our bodies . . . by touching the eye, the nose, or the upper lip, . . . excites in us an almost intolerable titillation, even though elsewhere it is scarcely felt. This titillation belongs entirely to us and not to the feather; if the live and sensitive body were removed it would remain no more than a mere word. I believe that no more solid an existence belongs to many qualities which we have come to attribute to physical bodies – tastes, odors, colors, and many more." Galileo Galilei, *The Assayer*, in *Discoveries and Opinions of Galileo*, translated by Stillman Drake (Garden City, N.Y.: Doubleday Anchor, 1957), p. 275.
59. Descartes to Mersenne, 26 April 1643, AT, vol. 3, pp. 648–9 (*PWD*, vol. 3, p. 216).

Descartes' main argument against real qualities was that the concept is not intelligible:

> My principal reason for rejecting these real qualities is that I do not see that the human mind has any notion, or particular idea, to conceive them by; so that when we talk about them and assert their existence we are asserting something we do not conceive, and doing something we do not understand.[60]

He thought, moreover, that his mechanical explanation of qualities gave a better account of the phenomena of nature.[61] The existence of real qualities had provided the Aristotelians with a warrant for claiming the reliability of sensory experience. If a body perceived to be hot really contains the quality of hotness, then it is possible to know something about its inner nature empirically. While Descartes did not reject Aristotle's essentialism, he did reject Aristotle's opinion that essences can be known empirically, claiming instead that they can be known only by a priori methods.[62]

In Part IV of *The Principles of Philosophy,* he undertook to provide mechanical explanations for all the qualities of corporeal things. In this respect, *The Principles of Philosophy,* like Gassendi's *Syntagma philosophicum,* is similar to other seventeenth-century treatises on the mechanical philosophy. They each began with a discussion of the nature of matter and motion and then proceeded to detailed explanations of the specific qualities found in the world, taking pains to show that the so-called occult, as well as the manifest, qualities can be explained mechanically.[63]

60. Ibid.
61. Ibid. Descartes expressed similar views in his letter to Regius of 24 January 1642, AT, vol. 3, pp. 500–10 (*PWD,* vol. 3, pp. 207–9).
62. In a similar way, Galileo's new conception of motion made it impossible to reach conclusions about the essences of bodies from knowledge of their motions. In Aristotelian physics, such inferences were possible: Natural upward or downward motion was a sign that a body was essentially light or heavy. See Margaret J. Osler, "Galileo, Motion, and Essences," *Isis, 64* (1973): 504–9. For a summary of the Aristotelian theory of perception and the role real qualities play it in, see W. D. Ross, *Aristotle: A Complete Exposition of His Works and Thought* (New York: Meridian, 1959), pp. 134–8, 159–61.
63. For the significance of these explanations of occult qualities, see Keith Hutchison, "What Happened to Occult Qualities in the Scientific Revolution?" *Isis, 73* (1982): 233–53; and Ron Millen, "The Manifestation of Occult Qualities in the Scientific Revolution," in *Religion, Science, and Worldview: Essays in Honor of Richard S. Westfall,* edited by Margaret J. Osler and Paul Lawrence Farber (Cambridge University Press, 1985), pp. 185–216.

Descartes enumerated all the qualities of bodies – manifest and occult – for which he provided mechanical explanations. The list of qualities is familiar from Aristotelian natural philosophy[64] and similar to what is found in other seventeenth-century treatises on the mechanical philosophy: gravity, light, heat, the rarefaction and compression of air, the freezing and evaporation of water, the tides, the particular properties of mercury and other chemical substances and metals, the nature of fire and of gunpowder, the properties of glass and the magnet, the amber effect, and finally sensation.[65] Descartes pointed especially to his putative success in explaining the phenomena of the magnet, which had traditionally been regarded as the paradigm case of an occult power.[66] He claimed that he had thus explained all the phenomena of nature:

> And thus, by simple enumeration, it is concluded that no phenomena of nature have been omitted by me in this treatise. For nothing is to be counted among the phenomena of nature, except what is perceived by sense. However, except for size, figure, and motion which I have explained as they are in each body, nothing located outside us is observed except light, color, odor, taste, sound, and tactile qualities; which I have now demonstrated are nothing in the objects other than, or at least are perceived by us as nothing other than, certain dispositions of size, figure, and motion.[67]

He was fully aware that explaining the macroscopic qualities of things involved unobservable, microscopic mechanisms. "I assign determinate figures, and sizes, and motions to the imperceptible particles of bodies, as if I had seen them, and yet I acknowledge that they are insensible."[68] Descartes' justification of transdiction rested on his a priori knowledge of the first principles of natural philosophy combined with an assumption about the uniformity of nature. Knowing how bodies interact, we know that they *must* interact that way even if they fall below the threshold of sense.[69] This conclusion follows from Descartes' claim that there is only

---

64. Alan Gabbey, "The Mechanical Philosophy and Its Problems: Mechanical Explanations, Impenetrability, and Perpetual Motion," in *Change and Progress in Modern Science,* edited by Joseph C. Pitt (Dordrecht: Reidel, 1985), p. 14.
65. Descartes, *Principia philosophiae,* in AT, vol. 8–1, pp. 216–323.
66. See Richard S. Westfall, *The Construction of Modern Science: Mechanisms and Mechanics* (New York: Wiley, 1971), pp. 37–8. See also J. L. Heilbron, *Elements of Early Modern Physics* (Berkeley: University of California Press, 1982), pp. 24–6.
67. Descartes, *Principia philosophiae,* in AT, vol. 8–1, p. 323.
68. Ibid., pp. 325–6.
69. Ibid.

one kind of matter in the world that behaves according to the same laws at both the macroscopic and microscopic levels.[70]

Descartes acknowledged that transdictive inferences might not be demonstrative. Indeed, even given the certainty of our knowledge of first principles, we may not be able to determine which of the several mechanisms compatible with first principles God could have created to produce the phenomena he actually did create. Here the inscrutability of God's absolute power renders us incapable of determining the actual mechanisms by reasoning alone. Consequently, our inferences to specific mechanisms have a conjectural quality and thus lack metaphysical certainty. Sounding much like Gassendi, Descartes compared the method by which we form conjectures about these mechanisms with the way we would approach the problem of deciphering a Latin message written in code. Our conclusions may only reach the level of moral certainty, but they suffice for the needs of everyday life.[71] In contrast to Gassendi, who was willing to settle for conjectural knowledge, however, Descartes thought that these conjectures could eventually be established as true and then incorporated into the demonstrative structure of his mechanical philosophy.

Despite the conjectural nature of knowledge of the actual mechanisms God produced in the world, Descartes considered the truth of first principles to be evident, not only because he had apparently deduced them from the *cogito,* but also because of the large number of phenomena he was able to explain by using them:

> Besides, there are, even among natural things, some that we judge to be absolutely and more than morally certain, having rested on the metaphysical foundation that God is the most good and the least deceitful, and that the faculty that he gave us for distinguishing the true from the false cannot err when we use it rightly and perceive something distinctly with its help. Such are mathematical demonstrations; such is the knowledge that material things exist; and such are all evident reasonings that concern them.[72]

He regarded these principles as sufficiently powerful to provide demonstrative explanations of all phenomena, once they are known to us. "These reasonings of ours will perhaps be included among the these absolutely certain things by those who consider how they have been deduced in a continuous series from the first and simplest principles of human knowledge."[73] Since his theory of perception was based on the motions of small

70. Garber, " 'Semel in Vita,' " p. 88.
71. Descartes, *Principia philosophiae,* in AT, vol. 8–1, pp. 327–8.
72. Ibid., pp. 328–9.
73. Ibid., p. 328.

particles of matter, ultimately even our empirical knowledge of the phenomena could be reduced to the laws of motion, which follow demonstratively from first principles. In this way, the entire physical world would be embraced by Descartes' mechanical philosophy. "for having accepted these, all the rest, at least the more general things which I have written about the World and the Earth, seem to be scarcely intelligible otherwise than as I have explained them."[74] Apart from a final statement acknowledging the authority of the Catholic church,[75] Descartes concluded his account of the universe with this optimistic assessment of his own achievement in The Principles of Philosophy.

Despite the fact that Descartes thoroughly mechanized the physical world at both the macroscopic and microscopic levels, he was no materialist. Like Gassendi, he insisted that the human mind or soul is not material. The distinction between soul and body, for Descartes, marked the limits of mechanization in his world, just as it had for Gassendi and the other mechanical philosophers. What explicitly pertains to the present discussion is the fact that Descartes drew his distinction between mind and body in a purely a priori manner. He did this by considering the essential attributes of these two substances. Just as he had proven to his own satisfaction that matter exists and is equivalent to geometrical extension, he used a similarly a priori argument – his famous "cogito ergo sum" – to prove that the mind exists and that its essence is thinking.[76] He considered extension and thinking to be completely different concepts; consequently matter and mind are entirely distinct from each other. "And thus we can easily have two clear and distinct notions or ideas, one of created thinking substance, the other of corporeal substance, if indeed we accurately distinguish all attributes of thought from the attributes of extension."[77] From clear and distinct ideas of this sort, Descartes concluded that a real distinction must exist. Accordingly, he believed that he had established the distinction between mind and matter – the limits of mechanization – with certainty and in a purely a priori way.

In order to explain sensation and motion in animals, he developed a theory of animal spirits, consisting of material particles elaborated from the blood by vital heat. The particles composing these spirits are "more agitated than those which compose air, but less than those that compose

74. Ibid., p. 329.
75. Ibid.
76. Descartes, *Meditations*, in *PWD*, vol. 2, pp. 16–23 (AT, vol. 7, pp. 23–34); and *Principia philosophiae*, in AT, vol. 8–1, pp. 6–7.
77. Descartes, *Principia philosophiae*, in AT, vol. 8–1, pp. 25–6. See also Descartes to Elizabeth, 21 May 1643, AT, vol. 3, pp. 663–8.

flame."[78] Comparable in both structure and function to Gassendi's *anima* or animal soul, Descartes' animal spirits flow through the nerves producing motion and sensation.[79] Also like Gassendi, Descartes argued that only humans have minds and that animals lack the self-consciousness that is the essence of *res cogitans*.[80] For both mechanical philosophers, the boundary between mind and matter fell along the same line.

Descartes was also concerned to establish the immortality of the soul, even though his determination to do so seems less passionate than Gassendi's. He mentioned his intentions in an early letter to Mersenne:

> I shall find out in the *Dioptrics* whether I am capable of explaining my conceptions and of persuading others of a truth after I have persuaded myself of it. I do not really think I can. But, if I should find by experience that such is the case, I do not say that someday I shan't complete a small treatise on metaphysics that I began when in Frisia and of which the principal points are to prove the existence of God and that of our souls, when they are separated from the body, whence follows their immortality.[81]

His only published argument for the immortality of the soul appeared in "Reply II":

> Our natural knowledge tells us that the mind is distinct from the body, and that it is a substance. But in the case of the human body, the difference between it and other bodies consists merely in the arrangement of the limbs and other accidents of that sort; and the final death of the body depends solely on a division or change of shape. Now we have no convincing evidence or precedent to suggest that the death or annihilation of a substance like the mind must result from such a trivial cause as a change in shape, for this is simply a mode, and what is more not a mode of the mind, but a mode of the body which is really distinct from the mind. Indeed, we do not even have convincing evidence or precedent to suggest that any substance can perish. And

78. Descartes to Vorstius, 19 June 1643, AT, vol. 3, p. 687. He also discussed animal spirits in *Traité de l'homme*. See René Descartes, *Treatise of Man*, translated by Thomas Steele Hall (Cambridge, Mass: Harvard University Press, 1972).
79. Descartes, *Treatise of Man*, pp. 19–25. See also Descartes, *Principia philosophiae*, AT, vol. 8–1, pp. 317–18.
80. Descartes' account of animals is more complex and less consistent than its frequent characterization as a *bête-machine* doctrine. See John Cottingham, "'A Brute to the Brutes?': Descartes' Treatment of Animals," *Philosophy*, 53 (1978): 551–9.
81. Descartes to Mersenne, 25 November 1630, AT, vol. 1, p. 182, translated by Michael Sean Mahoney, in the introduction to *Le monde*, p. xi.

this entitles us to conclude that the mind, in so far as it can be known
by natural philosophy, is immortal.[82]
Like his argument for the distinction between matter and mind, his argument for the immortality of the human soul rests entirely on definitions
and a priori methods.

Descartes' mechanical philosophy was closely tied to his rationalist
epistemology, which, I have argued in Chapter 5, had deep roots in his
underlying theological assumptions about God's relationship to the creation.[83] In this respect, my analysis of the relationship of his theology, his
epistemology, and his theory of matter parallels my analysis of the same
elements in Gassendi's thought. While both Descartes and Gassendi advocated a mechanical philosophy of nature to replace traditional Aristotelianism, they differed about many of the details of the new philosophy. The
differences between their theories of knowledge and the details of their
versions of the mechanical philosophy were clearly related to their
differences in their theological presuppositions.

82. Descartes, "Second Set of Replies," in *PWD*, vol. 2, pp. 108–9 (AT, vol. 7,
    pp. 153–4).
83. In this connection, see Gary Hatfield, "Reason, Nature, and God in
    Descartes," *Science in Context*, 3 (1989): 175–202.

10

Conclusion: Theology transformed – the emergence of styles of science

I am of your opinion that Gassendes and De Cartes are of different dispositions.

Charles Cavendish to John Pell, December 1644[1]

In this study, I have examined the transformation of medieval ideas about God's relationship to the creation into seventeenth-century ideas about matter and method as embodied in early articulations of the mechanical philosophy. Medieval thinkers were primarily concerned with the theological problem of God's relationship to the world he created. They discussed questions about necessity and contingency as related to divine power. By the seventeenth century, the focus had shifted to natural philosophy and the extent and certitude of human knowledge. Underlying theological assumptions continued to be reflected in the epistemological and metaphysical orientations incorporated into different versions of the mechanical philosophy.

I have argued that the differences between Gassendi's and Descartes' versions of the mechanical philosophy directly reflected the differences in their theological presuppositions. Gassendi described a world utterly contingent on divine will. This contingency expressed itself in his conviction that empirical methods are the only way to acquire knowledge about the natural world and that the matter of which all physical things are composed possesses some properties that can be known only empirically. Descartes, on the contrary, described a world in which God had embedded necessary relations, some of which enable us to have a priori knowledge of substantial parts of the natural world. The capacity for a priori knowledge extends to the nature of matter, which, Descartes claimed to demonstrate, possesses only geometrical properties. Gassendi's views can be traced

1. J. O. Halliwell, ed., *A Collection of Letters Illustrative of the Progress of Science in England from the Reign of Queen Elizabeth to that of Charles the Second* (London: Dawson's, 1965), p. 86.

back to the ideas of the fourteenth-century nominalists, while Descartes' can be linked to the Thomist tradition he imbibed at La Flèche. Refracted through the prism of the mechanical philosophy, these theological conceptualizations of contingency and necessity in the world were mirrored in different styles of science that emerged in the second half of the seventeenth century. By "style" of science I mean specific manifestations of the general epistemological and metaphysical assumptions that govern a articular scientific practice.[2] Do scientists presume that abstract reasoning and mathematics alone can determine scientific truth? Or do they emphasize observational and experimental methods? Are the results of this enterprise thought to be certain? Or are they thought to achieve at best some degree of probability? What role does chance play in the world described by the scientist? Can all events be completely explained scientifically, or do some lie beyond the reach of scientific theory?[3] The style of science is the practical expression of the epistemological and metaphysical assumptions embedded within the conceptual framework. Even though scientists may not be fully conscious of the philosophical assumptions lying at the root of their practice, their choice of problems, method of proof, and criteria for choosing explanations all reflect a presumed metaphysics and epistemology. As these assumptions have been translated or transplanted from one field of discourse to another, they have led to – or,

2. The concept of style as applied to the history of science is fraught with difficulties. See, e.g., Michael Otte, "Style as a Historical Category," *Science in Context,* 4 (1991): 233–64; and Anna Wessely, "Transposing 'Style' from the History of Art to the History of Science," *Science in Context,* 4 (1991): 265–78. Hacking dismisses the concept of style as useless: "Doubtless the very word 'style' is suspect. It is cribbed from art critics and historians, who have not evolved a uniform connotation for the word. Nor would all their remarks about style tidily transfer to modes of reasoning." Ian Hacking, "Language, Truth, and Reason," in *Rationality and Relativism,* edited by Martin Hollis and Steven Lukes (Oxford: Blackwell, 1982), p. 51. Hacking discusses the concept of style at greater length in "'Style' for Historians and Philosophers," *Studies in History and Philosophy of Science,* 23 (1992): 1–20. In this article, Hacking uses the term "style" to refer to what I mean by "conceptual framework." I use the concept in the restricted sense defined in my text. For a recent attempt to apply the art historians' concept of style to the history of science, see W. L. Wisan, "Galileo and the Emergence of A New Scientific Style," in *Pisa Conference Proceedings,* edited by J. Hintikka, D. Gruender, and E. Agazzi (Dordrecht: Reidel, 1980), vol. 1., pp. 311–39.
3. A modern example of the themes of contingency and necessity can be found in Paolo Palladino, "Stereochemistry and the Nature of Life: Mechanist, Vitalist, and Evolutionary Perspectives," *Isis, 81* (1990): 44–67.

perhaps at a psychological level, have reflected[4] – different styles of thinking about the world.

In the generation of natural philosophers that followed Gassendi and Descartes – a generation that included such notable figures as Thomas Hobbes, Henry More, Robert Boyle, Christiaan Huygens, and Isaac Newton – at least two distinct styles of science developed. These styles reflected the influence of the two traditions, the development of which I have traced in this study. Although these natural philosophers started by assuming that the mechanical philosophy was the proper framework within which to conduct investigations of the natural world,[5] their ap-

---

4. On different psychological styles of thinking, see Carol Gilligan, *In a Different Voice: Psychological Theory and Women's Development* (Cambridge, Mass.: Harvard University Press, 1982); and Mary Field Belenky, Blythe McVicker Clinchy, Nancy Rule Goldberger, and Jill Mattuck Tarule, *Women's Ways of Knowing: The Development of Self, Voice, and Mind,* (New York: Basic, 1986).

5. Having rejected Aristotelianism in his student days, Hobbes, whose long lifetime spanned the generation of Gassendi and Descartes and that of their followers, adopted a mechanistic account of sensation during his Parisian sojourn in the 1630s. See R. S. Peters, "Hobbes, Thomas," in *The Encyclopedia of Philosophy*, edited by Paul Edwards, 8 vols. (New York: Macmillan and Free Press, 1967), vol. 4, pp. 30–1; and Thomas A. Spragens, Jr., *The Politics of Motion: The World of Thomas Hobbes* (Lexington: University Press of Kentucky, 1973), pp. 32–3. Boyle began writing within a mechanical framework at a young age. For example, see his treatise *Some Considerations Touching the Usefulness of Experimental Natural Philosophy* (1663), in *The Works of the Honourable Robert Boyle,* edited by Thomas Birch, new edition, 6 vols. (London, 1772; reprinted, Hildesheim: Georg Olms, 1965), vol. 2, pp. 1–48. Although not published until 1663, parts of this exposition of the mechanical philosophy were written as early as 1647–9. See Marie Boas Hall, *Robert Boyle on Natural Philosophy* (Bloomington: Indiana University Press, 1965), p. 402. Henry More, the Cambridge Platonist, was an early disciple of Descartes and debated with Boyle about the interpretation of the air pump experiments. See A. Rupert Hall, *Henry More: Magic, Religion and Experiment* (Oxford: Blackwell, 1990); and various articles in *Henry More (1614– 1687): Tercentenary Studies,* edited by Sarah Hutton (Dordrecht: Kluwer, 1990). Huygens' scientific interests focused on problems in mechanics central to the Cartesian philosophy of nature. See articles by Robert S. Westman and Abn Gabbey in *Studies on Christiaan Huygens: Invited Papers from the Symposium on the Life and Work of Christiaan Huygens, Amsterdam, 22–25 August 1979,* edited by H. J. M. Bos, M. J. S. Rudwick, H. A. M. Snelders, and R. P. W. Visser (Lisse: Swets & Zeitlinger, 1980). See also Aant Elzinga, *On a Research Program in Early Modern Physics* (New York: Humanities

proaches differed in ways that continued to reflect the theological, metaphysical, and epistemological positions that had permeated Gassendi's and Descartes' respective versions of the new philosophy of nature. In the face of increasing specialization and the growing autonomy of scientific discourse, explicit discussion of theological issues often – but not always – vanished. Nevertheless, differences in scientific practice, approaches to scientific controversy, and the occasional methodological comment reveal their ancestry in the earlier formulations of the mechanical philosophy.

Clear examples of these different styles can be found in Boyle's controversies with Henry More and Thomas Hobbes concerning the interpretation of the air pump experiments.[6] More rejected Boyle's conclusion that the evacuated chamber in the air pump is void. Deeply influenced by Platonism and Neoplatonism, More was attracted to the philosophy of Descartes, which seemed to combine an interest in natural philosophy with proofs for the existence of God and an immortal soul and was thus preferable to Hobbes' mechanism.[7] Like Descartes, More was an intellectualist, although an intellectualist of a more extreme, Platonic kind. He believed that God was bound to create the world according to certain absolute and preexisting standards.[8] "But though the Divine Goodness

Press, 1972); Joella G. Yoder, *Unrolling Time: Christiaan Huygens and the Mathematization of Nature* (Cambridge University Press, 1988); and Richard S. Westfall, *Force in Newton's Physics* (New York: American Elsevier, 1971), chap. 4. On the young Newton's endorsement of the mechanical philosophy, see J. E. McGuire and Martin Tamny, *Certain Philosophical Questions: Newton's Trinity Notebook* (Cambridge University Press, 1983); Richard S. Westfall, "The Foundations of Newton's Philosophy of Nature," *British Journal for the History of Science*, 1 (1962): 171–82; Richard S. Westfall, *Never at Rest: A Biography of Isaac Newton* (Cambridge University Press, 1980), pp. 89–99; and B. J. T. Dobbs, "Stoic and Epicurean Doctrines in Newton's System of the World," in *Atoms, Pneuma, and Tranquillity: Epicurean and Stoic Themes in European Thought*, edited by Margaret J. Osler (Cambridge University Press, 1991), pp. 225–7.

6. I rely heavily on Shapin and Schaffer's account of these controversies. See Steven Shapin and Simon Schaffer, *Leviathan and the Air-Pump: Hobbes, Boyle, and the Experimental Life* (Princeton, N.J.: Princeton University Press, 1985).

7. William H. Austin, "More, Henry," in *Dictionary of Scientific Biography*, edited by Charles Coulton Gillispie, 16 vols. (New York: Scribner, 1972), vol. 9, p. 509; and Michael Boylan, "Henry More's Space and the Spirit of Nature," *Journal of the History of Philosophy*, 18 (1980): 395–405.

8. John Henry, "A Cambridge Platonist's Materialism: Henry More and the Concept of the Soul," *Journal of the Warburg and Courtauld Institutes*, 49 (1986): 186–7.

acts necessarily yet it does not blindly, but according to the Laws of Decorum and Justice."[9] It was this necessity that made it possible to discover the nature of things a priori. More never fully accepted the Cartesian mechanical philosophy, however, because he thought many phenomena – including the tides, the loadstone, and gravity, not to mention more occult phenomena such as women under the influence of the imagination giving birth to devils, apes, and crabs – were impossible to explain in purely mechanical terms.[10] In order to explain such phenomena, More introduced another causal agent into the world, the Spirit of Nature:

> The Spirit of Nature . . . is A substance incorporeal, but without Sense and Animadversion, pervading the whole Matter of the Universe, and exercising a Plastical Power therein according to the sundry predispositions and occasions in the parts it works upon, raising such *Phaenomena* in the World, by directing the parts of Matter and their Motion, as cannot be resolved into mere Mechanical powers.[11]

More interpreted Boyle's hydrostatical experiments as evidence for the existence of this Spirit of Nature, something he had proven to his own satisfaction on purely rational grounds.[12] Boyle rejected the Spirit of Nature on theological grounds: It would be an agent intermediate between God and the creation, which would interfere with God's ability to act directly on the world.[13] Boyle rejected such entities, adopting the

9. Henry More, *Annotations upon the Two foregoing Treatises, Lux Orientalis, or, An Enquiry into the Opinion of the Eastern Sages Concerning the Praeexistence of Souls; and the Discourse of Truth* (London, 1682), pp. 46–7.

10. Henry More, *The Immortality of the Soul, So farre forth as it is demonstrable from the Knowledge of Nature and the Light of Reason*, in Henry More, *A Collection of Several Philosophical Writings*, 2 vols. (London: James Fletcher, 1662; reprinted, New York: Garland, 1978), vol. 2, pp. 170–4; Alan Gabbey, "Henry More and the Limits of Mechanism," in *Henry More (1624–1687): Tercentenary Studies*, edited by Hutton, pp. 19–35; and Robert A. Greene, "Henry More and Robert Boyle on the Spirit of Nature," *Journal of the History of Ideas*, 23 (1962): 461–2.

11. More, *The Immortality of the Soul*, vol. 2, p. 193.

12. John Henry, "Henry More Versus Robert Boyle: The Spirit of Nature and the Nature of Providence," in *Henry More (1624–1687): Tercentenary Studies*, edited by Hutton, pp. 58–9; and Shapin and Schaffer, *Leviathan and the Air-Pump*, pp. 207–12. See also Jane Elizabeth Jenkins, "Using Nothing: Vacuum, Matter, and Spirit in the Seventeenth Century Mechanical Philosophy," M.A. thesis, University of Calgary, 1990, chap. 2.

13. Henry, "Henry More Versus Robert Boyle," pp. 63–4; Margaret J. Osler, "The Intellectual Sources of Boyle's Philosophy of Nature: Gassendi's Voluntarism and Boyle's Physico-Theological Project," in *Philosophy, Science, and*

voluntarist position that all activity in the world comes directly from God. "For aught I can clearly discern, whatsoever is performed in the merely material world, is really done by particular bodies, acting according to the laws of motion, rest, &c. that are settled and maintained by God among things corporeal."[14] More's intellectualism and Boyle's voluntarism found expression in their disagreement over the interpretation of the air pump experiments and their conceptions of the nature of matter.[15] Their two styles of reasoning – experimental and rational – and two theories of matter are expressions of underlying theological differences.[16]

The disagreement between Boyle and Hobbes hinged on their respective conceptions of scientific method and the kind of knowledge it generated. Boyle advocated the use of experiments, granting that the knowledge so gained is provisional, pending the outcome of further experiments:

> That then, that I wish for, as to our systems, is this, that men, in the first place, would forbear to establish any theory, till they have consulted with . . . a considerable number of experiments, in proportion to the comprehensiveness of the theory to be erected on them. And in the next place, I would have such kind of super-structure looked upon only as temporary ones; which though they may be preferred before any others, as being the best in their kind that we have, yet they are

Religion, 1640–1700, edited by Richard Ashcraft, Richard Kroll, and Perez Zagorin (Cambridge University Press, 1991), p. 186. For a thorough exposition of Boyle's voluntarism, see Edward Bradford Davis, Jr., "Creation, Contingency, and Early Modern Science: The Impact of Voluntaristic Theology on Seventeenth-Century Natural Philosophy," Ph.D. dissertation, Indiana University, 1984, chap. 4. See also Eugene M. Klaaren, *Religious Origins of Modern Science: Belief in Creation in Seventeenth-Century Thought* (Grand Rapids, Mich.: Eerdmans, 1977).

14. Robert Boyle, *The Origin of Forms and Qualities,* in *Works,* vol. 3, p. 15.

15. This is the point of Henry's eloquent article, "Henry More Versus Robert Boyle." I disagree with Henry's claim that Descartes was a voluntarist (p. 75 n57). On More's intellectualism, see Henry, "A Cambridge Platonist's Materialism," pp. 189–95.

16. John Henry is correct to emphasize that the theological differences between Boyle and More do not follow confessional lines. Both were Anglicans. See "Henry More Versus Robert Boyle," p. 56. The Protestant–Catholic, English–French dichotomy suggested by Peter Dear in "Miracles, Experiments, and the Ordinary Course of Nature," *Isis, 81* (1990): 663–83, breaks down in the face of Henry's discussion and also in the face of the differences between Gassendi and Descartes, who were both French Catholics. The salient issue is theological rather than confessional.

not entirely to be acquiesced in, as absolutely perfect, or uncapable of improving alterations.[17]

He claimed only probability for the outcome of experimental work.[18] He eschewed problems that he considered "metaphysical" – for example, the question of whether matter is infinitely divisible or whether indivisible atoms exist – and was comfortable using the word "vacuum" to describe the evacuated chamber of his air pump without being able to ascertain whether it was *really* devoid of all matter. Thus, he claimed to "write for the Corpuscularians in general, [rather] than any party of them," including both Gassendists and Cartesians under this rubric. Operating within a framework determined by the mechanical philosophy, he thought questions in natural philosophy could be solved experimentally.[19] Hobbes, on the contrary, did not believe that experimental methods could determine whether void exists. Modeling his approach to natural philosophy on geometry, he thought that philosophy must start from first principles, demonstrating effects from causes. Mere experimentation was not capable of determining the real causes of phenomena with geometrical certainty.[20] Hobbes had deduced from first principles that the world is a plenum; consequently, he argued, there is no way that Boyle's experiments could possibly establish the existence of a vacuum.[21]

One might think that since Boyle and Hobbes disagreed about the interpretation of particular experiments, clear criteria for resolving their differences should exist. That such criteria did not exist, despite the fact that both men were mechanical philosophers – that Boyle and Hobbes differed even about what might constitute such criteria – points to a difference at the level of conceptual framework. Although the differences in their assumptions about method were at the most general level similar to the differences between Gassendi and Descartes, the expression of these

17. Robert Boyle, *Certain Physiological Essays*, in *Works*, vol. 1, p. 303.
18. Robert Boyle, *The Excellency of Theology, Compared with Natural Philosophy (As both are Objects of Men's Study.)* (1674), in *Works*, vol. 4, pp. 41–2. See Shapin and Schaffer, *Leviathan and the Air-Pump*, chap. 2.
19. See Robert Boyle, *The Origin of Forms and Qualities According to the Corpuscular Hypothesis*, in *Works*, vol. 3, p. 7; Robert Boyle, *Some Considerations Touching the Usefulness of Experimental Natural Philosophy* (1663), in *Works*, vol. 2, pp. 5–15. On the difference between Boyle's approach to the mechanical philosophy and that of Gassendi and Descartes, see Lynn Joy, "The Conflict of Mechanisms and Its Empiricist Outcome," *Monist*, 71 (1988): 498–514. See also Antonio Clericuzio, "A Redefinition of Boyle's Chemistry and Corpuscular Philosophy," *Annals of Science*, 47 (1990): 561–89.
20. Shapin and Schaffer, *Leviathan and the Air-Pump*, chaps. 3 and 4.
21. Ibid., p. 99.

differences in the context of discussion of particular experimental results provides an opportunity for a more fine-grained analysis of their styles of approaching scientific problems.

Both Hobbes and Boyle wrote works in which they discussed the presuppositions underlying their philosophies of nature. Boyle accepted a voluntarist theology very similar to Gassendi's, by whom he may well have been influenced. His writings on theology emphasize the contingency of nature on divine will. Accordingly, he advocated empirical methods for natural philosophy:[22]

> And if the miracles vouchsafed either for the Christian, or for any other religion, be any of them granted to be true; (as almost all mankind agrees in believing in general, that there have been true miracles;) it cannot well be denied, but that physical propositions are but limited, and such as I called collected truths, being gathered from the settled phaenomena of nature, and are liable to this limitation or exception, that they are true, where the irresistable power of God, or some other supernatural agent, is not interposed to alter the course of nature.[23]

Although Hobbes did not write extensively on theological topics, he conceived of God as Aristotle's prime mover:[24]

> He that can from any effect he seeth come to pass, should reason to the next and intermediate cause thereof, and from thence to the cause of that cause, and plunge himself profoundly in the pursuit of causes; shall at last come to this, that there must be, as even the heathen philosophers confessed, one first mover; that is, a first, and an eternal cause of all things; which is that which men mean by the name of God.[25]

As first cause, God stands as the beginning of in an inexorable chain of causes and effects:

> And seeing a necessary cause is defined to be that, which being supposed, the effect cannot but follow; this also maybe collected, that whatsoever effect is produced at any time, the same is produced by a necessary cause. For whatsoever is produced, in as much as it is produced, had an entire cause, that is, had all those things which

22. See Osler, "The Intellectual Sources of Robert Boyle's Philosophy of Nature," pp. 178–98.
23. Robert Boyle, *Advices in Judging of Things Said to Transcend Reason* (1681), in *Works*, vol. 4, pp. 462–3.
24. Leopold Damrosch, Jr., "Hobbes as Reformation Theologian: Implications of the Free-Will Controversy," *Journal of the History of Ideas, 40*, (1979): 344.
25. Thomas Hobbes, *Leviathan* (New York: Meridian, 1963), p. 130.

being supposed, it cannot be understood but that the effect follows; that is, it had a necessary cause.[26] The natural necessity embedded in these causes is precisely what voluntarists ruled out of their world. Whether or not it is possible to find an articulated theology in Hobbes' thought, the kind of determinism he accepted was characteristic of the intellectualist tradition. It also provided him with the grounds for reasoning with necessity from the nature of things in general to the nonexistence of the void.

The relative prominence of mathematics in scientific discourse is frequently cited as a mark of different styles of science. Kuhn proposed such a characterization, tracing its origins to the seventeenth century, where he finds a "Baconian," experimental tradition coming into conflict with a mathematical, theoretical tradition.[27] This characterization by itself does not provide the key to the stylistic difference that I am tracing in this study. Although no mathematician, Boyle respected mathematics and mustered a series of arguments to demonstrate its usefulness for the experimental philosopher.[28] Nevertheless, he did not use mathematical methods in his work, and he argued against using them on theological, methodological, and social grounds.[29] The same contingency in nature that caused him to adopt empirical methods led him to avoid using mathematical methods in natural philosophy. He thought that mathematics would produce inflated expectations about the degree of certainty attainable by natural philosophy, a certainty rendered impossible by the utter contingency of the creation on divine will.

Hobbes became acquainted with mathematics accidentally, but his epiphany made a great impression on his thinking about the proper method for natural philosophy:

26. Thomas Hobbes, *Elements of Philosophy. The First Section, Concerning Body,* in *The English Works of Thomas Hobbes of Malmesbury,* edited by Sir William Molesworth, 11 vols. (London, 1839–45; reprinted, Aalen: Scientia, 1962), vol. 1, p. 123.

27. Thomas S. Kuhn, "Mathematical Versus Experimental Traditions in the Development of Physical Science," in *The Essential Tension: Selected Studies in Scientific Tradition and Change,* edited by Thomas S. Kuhn (Chicago: University of Chicago Press, 1977), pp. 41–50.

28. For example, see the section of Boyle's *The Usefulness of Experimental Philosophy* entitled "Of the Usefulness of Mathematicks to Natural Philosophy," in *Works,* vol. 3, pp. 425–34. See also Steven Shapin, "Robert Boyle and Mathematics: Reality, Representation, and Experimental Practice," *Science in Context,* 2 (1988): 28–33.

29. Ibid., pp. 33–52.

He was 40 yeares old before he looked on Geometry; which happened accidentally. Being in a Gentleman's Library, Euclid's Elements lay open, and 'twas the 47 *El. libri* I. He read the Proposition. *By G –*, sayd he (he would now and then sweare an emphaticall Oath by way of emphasis) *this is impossible!* So he reads the Demonstration of it, which referred him back to another, which he also read. *Et sic deinceps* . . . that at last he was demonstratively convinced of that trueth. This made him in love with Geometry.[30]

Hobbes' love affair with geometry led him to endorse a geometrical method for natural philosophy. Because all of physics "cannot be understood except as we know first what motions are in the smallest parts of bodies" and because these motions can be fully described geometrially, "they that study natural philosophy, study in vain except they begin at geometry; and such writers or disputers thereof, as are ignorant of geometry, do but make their readers and hearers lose their time."[31] Clearly, the deductive certainty of geometry appealed to Hobbes because he thought it was a powerful way of expressing the natural necessity inherent in nature. Nevertheless, his philosophy is no more mathematical, in actual fact, than Boyle's.

Another example of differences in scientific style can be found in the work of two of the most mathematical natural philosophers of the seventeenth century, Christiaan Huygens and Isaac Newton. While both men used mathematics to great effect in physics, their interpretations of mathematical physics differed considerably. Profoundly influenced by Descartes – adopting both his version of the mechanical philosophy and his methodological prescriptions for physics – , Huygens undertook a mathematical treatment of important problems in mechanics, achieving notable success with his analysis of impact.[32] Improving on Descartes' notoriously problematic treatment of impact, Huygens retained the kinematic approach, which characterized Cartesian mechanics.[33] He also retained Descartes' realist interpretation of the metaphysical significance of mathematical physics. He believed that his mathematical models described reality.[34] In fact, he believed that physical systems could be reduced to the mathematics that we use to describe them, and we can

30. Oliver Lawson Dick, ed., *Aubrey's Brief Lives* (Ann Arbor: University of Michigan Press, 1962; first published, London: Secker & Warburg, 1949), p. 150.
31. Hobbes, *Elements of Philosophy*, p. 73.
32. Elzinga, *On A Research Program in Early Modern Physics*, pp. 73–80.
33. Westfall, *Force in Newton's Physics*, chap. 4.
34. Elzinga, *On a Research Program in Early Modern Physics*, p. 80.

know that mathematics with certainty.[35] "Whatever you will have supposed not impossible either concerning gravity or motion or any other matter, if thence you prove something concerning the magnitude of a line, surface, or body, it will be true."[36] Huygens thus retained Descartes' view that it is possible to acquire certain knowledge of physics by a priori methods. In Huygens' writings on physics, which focus on particular problems rather than the philosophical foundations of science,[37] these views were completely divorced from theological or even metaphysical considerations; but they continued to embody the style of thinking – if not all the details[38] – that characterized Cartesian natural philosophy.

Contrary to Huygens, Newton made no such claims for the certainty of his physics or the identity of his mathematical concepts with physical reality. Unlike Huygens, he did not divorce his physics from his theological concerns. Hence, it is possible to draw a direct connection in Newton's thought between the style of his science and the theological issues whose influence I have been tracking in this study. Although committed to the mechanical philosophy from his student days, Newton was always troubled by the theological implications of Cartesian mechanism.[39] This concern dominates the "General Scholium." In this resounding final word of his *Principia*, Newton fulminated against Descartes' system by arguing that it cannot adequately describe the physical world and that it does not incorporate a proper understanding of the deity. As a voluntarist, Newton was particularly concerned to ensure divine activity in the world.[40] His voluntarism found its most articulate expression in the famous Leibniz–Clarke correspondence, in which Samuel Clarke, Newton's spokesman, argued for Newtonian voluntarism against Leibnizian intellectualism.[41]

35. Yoder, *Unrolling Time*, pp. 172–3.
36. Christiaan Huygens, draft of *De Vi Centrifuga*, in *Oeuvres complètes de Christiaan Huygens*, 22 vols. (The Hague: Martinus Nijhoff, 1888–1950), vol. 17, p. 286, translated by Yoder, in *Unrolling Time*, p. 173.
37. Alan Gabbey, "Huygens and Mechanics," in *Studies on Christiaan Huygens*, edited by Bos et al., p. 885; and Yoder, *Unrolling Time*, p. 172.
38. For example, Huygens' theory of matter differed from Descartes' in including hardness among the primary qualities and accepting the existence of indivisible atoms. See H. A. M. Snelders, "Christiaan Huygens and the Concept of Matter," in *Studies on Christiaan Huygens*, edited by Bos et al., pp. 104–25.
39. B. J. T. Dobbs, *The Janus Faces of Genius: The Role of Alchemy in Newton's Thought* (Cambridge University Press, 1991), pp. 33–5.
40. Dobbs traces the complex threads of Newton's thought on divine activity in ibid., chaps. 4–7.
41. H. G. Alexander, ed., *The Leibniz–Clarke Correspondence* (Manchester: University of Manchester Press, 1956).

Newton's voluntarist theology was reflected in his scientific style. Methodologically, he was an empiricist, recognizing explicitly that all conclusions are subject to revision in light of further empirical evidence. He highlighted this view in the *Principia*:

> In experimental philosophy we are to look upon propositions inferred by general induction from phenomena as accurately or so very nearly true, notwithstanding any contrary hypotheses that may be imagined, till such time as other phenomena occur, by which they may either be made more accurate, or liable to exceptions.[42]

Mathematics played a central role in Newton's natural philosophy, but mathematical description was not identical to physical truth. Writing to Oldenburg about his experiments on colors, he said:

> I should take notice of a casuall expression wch intimates a greater certainty in these things than I ever promised, viz: The certainty of *Mathematical Demonstrations*. I said indeed that the *Science of Colours was Mathematicall and as certain as any other part of Optiques;* but who knows not that Optiques and many other Mathematicall Sciences depend as well on Physicall Principles as on Mathematicall Demonstrations: And the absolute certainty of a Science cannot exceed the certainty of its Principles.[43]

Since matter has physical properties other than those that are purely geometrical, the science of the material world cannot be known by mathematical demonstrations alone.

Newton regarded even the results of the mathematical parts of his natural philosophy as subject to revision in light of further evidence:

> As in Mathematicks, so in Natural Philosophy, the Investigation of difficult Things by the Method of Analysis, ought ever to precede the Method of Composition. This Analysis consists in making Experiments and Observations, and in drawing general Conclusions from them by Induction, and admitting of no Objections against the Conclusions, but such as are taken from Experiments, or other Certain Truths. . . . And although the arguing from Experiments and Observations by Induction be no Demonstration of general Conclusions; yet it is the best way of arguing which the Nature of Things admits of, and may be looked upon as so much the stronger, by how much the Induction is more general. And if no Exception occur from Pha-

42. Isaac Newton, *Mathematical Principles of Natural Philosophy and His System of the World*, translated by Andrew Motte in 1729 and revised by Florian Cajori, 2 vols. (Berkeley and Los Angeles: University of California Press, 1962), vol. 2, p. 400.

43. Newton to Oldenburg, 11 June 1672, in *The Correspondence of Isaac Newton*, edited by H. W. Turnbull, J. P. Scott, A. R. Hall, and Laura Tilling, 7 vols. (Cambridge: Published for the Royal Society at the University Press, 1959–77), vol. 1, p. 187.

enomena, the Conclusion may be pronounced generally. But if at any time afterwards any Exception shall occur from Experiments, it may then begin to be pronounced with such Exceptions as occur. By this way of Analysis we may proceed from Compounds to Ingredients, and from Motions to the Forces producing them, and in general, from Effects to their Causes, and from particular Causes to more general ones, till the Argument end in the most general. This is the Method of Analysis: And the Synthesis consists in assuming the Causes discover'd, and establish'd as Principles, and by them explaining the Phaenomena proceeding from them, and proving the Explanations.[44]

Throughout the seventeenth century, assumptions about contingency and necessity in the world were translated from theological conceptions about God's relationship to the creation into different styles of approaching scientific investigation. In the centuries after Newton, scientists have generally abandoned the theological concerns that had dominated seventeenth-century thought. The styles took on lives of their own, losing connection with the theological roots from which they had grown.

The resounding success of Newtonian gravitational theory fostered an optimism about the power of physical theory to describe the world as a deterministic system, leading physicists to interpret the laws of physics as true descriptions of a law-bound, deterministic universe. By the end of the eighteenth century, Laplace (1749–1827) – in his massive *Traité de la méchanique céleste* (1798–1827) – was able to demonstrate that the solar system is a gravitationally stable Newtonian system, leading him to assert the paradigmatic statement of classical determinism:

> Given for one instant an intelligence which could comprehend all the forces by which nature is animated and the respective situation of all the beings who compose it – an intelligence sufficiently vast to submit these data to analysis – it would embrace in the same formula the movements of the greatest bodies of the universe and those of the lightest atoms; for it, nothing would be uncertain and the future, as the past, would be present to its eyes.[45]

Laplace's confidence in this deterministic approach was unambiguous. "Being assured that nothing will interfere between these causes and their effects, we venture to extend our views into futurity, and contemplate the

44. Isaac Newton, *Opticks, or a Treatise of the Reflections, Refractions, Inflections & Colours of Light*, based on 4th London edition of 1730 (New York: Dover, 1952), pp. 404–5.
45. Pierre Simon, Marquis de Laplace, *A Philosophical Essay on Probabilities*, translated from the 6th French edition by F. W. Truscott and F. L. Emory (New York: Dover, 1951), p. 4.

series of events which time alone can develop."[46] In Laplace's discourse, physical necessity and mathematical analysis replaced theological considerations. Indeed, he is reputed to have replied to Napoleon's query about the role of God in his system with the quip "I have no need of that hypothesis."[47] Nevertheless, the physical necessity embedded in Laplace's determinism continued to reflect the themes of an abandoned intellectualist theology. Despite twentieth-century challenges to classical determinism and the relativizing of space and time, to the extent that modern physicists continue to find purely mathematical reasons for accepting the truth of physical theories, their style of science resonates with an intellectualist past.

A style of science emphasizing the contingency of natural processes and empirical methods also persists. Darwin's theory of evolution by natural selection, with its insistence on the random appearance of variations and the impact of unpredictable environmental factors on the development of species, is a case in point.[48] The world of the evolutionists is not so different from the world as understood by Gassendi and Boyle, at least as far as contingency is concerned:

> Darwin recognized that the primary evidence for evolution must be sought in quirks, oddities, and imperfections that lay bare the pathways of history. Whales, with their vestigial pelvic bones, must have descended from terrestrial ancestors with functional legs. Pandas, to eat bamboo, must build an imperfect "thumb" from a nubbin of wrist bone, because carnivorous ancestors lost the requisite mobility of their first digit. . . . If whales retained no trace of their terrestrial heritage, if pandas bore perfect thumbs, if life on the Galápagos neatly matched the curious local environments – then history would not inhere in the productions of nature. But contingencies of "just history" do shape our world, and evolution lies exposed in the panoply of

46. Pierre Simon de Laplace, *The System of the World*, translated by J. Pond (London, 1809; first published, Paris, 1798), vol. 1, p. 206, as quoted by R. Harré, "Laplace, Pierre Simon de," in *The Encyclopedia of Philosophy*, edited by Paul Edwards, 8 vols. (New York: Macmillan and Free Press, 1967), vol. 4, p. 393.
47. See Roger Hahn, "Laplace and the Vanishing Role of God," in *The Analytic Spirit: Essays in the History of Science in Honor of Henry Guerlac* (Ithaca, N.Y.: Cornell University Press, 1981); and Roger Hahn, "Laplace and the Mechanistic Universe," in *God and Nature: Historical Essays on the Encounter Between Christianity and Science*, edited by David C. Lindberg and Ronald L. Numbers (Berkeley: University of California Press, 1986), p. 256.
48. See Stephen Jay Gould, *Wonderful Life: The Burgess Shale and the Nature of History* (New York: Norton, 1989).

structures that have no other explanation than the shadow of their past.[49]

The unpredictable actions of an omnipotent God have been replaced by unpredictable variations, which respond to selective pressures in ways that only empirical study can determine. Contingency rather than reason is the basis for explanation in this world:

> And so, if you wish to ask the question of the ages – why do humans exist? – a major part of the answer, touching those aspects of the issue that science can treat at all, must be: because *Pikaia* survived the Burgess decimation. This response does not cite a single law of nature; it embodies no statement about predictable evolutionary pathways, no calculation of probabilities based on general rules of anatomy or ecology. The survival of *Pikaia* was a contingency of "just history."[50]

Evolution does not have a predictable and determinate course of a kind that can be known by a priori and deductive methods. The metaphysical assumptions underlying this style of evolutionary science can be traced back to a voluntarist interpretation of the biblical worldview.[51]

Although theological language has dropped out of scientific discourse, contemporary styles of science are historically linked to the dialectic of the absolute and ordained powers of God. The interplay between necessity and contingency in the world is now constructed entirely in naturalistic terms, but it grew from roots embedded in an earlier, theological understanding.

49. Ibid., pp. 300–1.
50. Ibid., p. 323.
51. See James R. Moore, *The Post-Darwinian Controversies: A Study of the Protestant Struggle to Come to Terms with Darwin in Great Britain and America, 1870–1900* (Cambridge University Press, 1979), pp. 326–40.

Bibliography

PRIMARY SOURCES

1 Gassendi

a Original Sources

Gassendi, Petrus, *Opera omnia*, 6 vols. (Lyon, 1658; facsimile reprint, Stuttgart-Bad Cannstatt: Friedrich Frommann Verlag, 1964).

b Translations

Adelmann, Howard B., *Marcello Malpighi and the Evolution of Embryology*, 5 vols. (Ithaca, N.Y.: Cornell University Press, 1966), vol. 2, pp. 798–817 (translation of Gassendi, *Syntagma philosophicum*, sections from "De plantis" and "De generatione animalium").

Bernier, François, *Three Discourses of Happiness, Virtue, and Liberty, collected from the works of the Learn'd Gassendi* (London: Awnshawm & John Churchil, 1699).

Gassendi, Pierre, *Pierre Gassendi: Lettres familières à François Lullier pendant l'hiver 1632–1633*, edited by Bernard Rochot (Paris: J. Vrin, 1944).

Dissertations en forme de paradoxes contre les Aristotéliciens (Exercitationes paradoxicae adversus Aristoteleos), Books I and II, translated into French by Bernard Rochot (Paris: J. Vrin, 1959).

Disquisitio metaphysica seu dubitationes et instantiae adversus Renati Cartesii metaphysicam et responsa, edited and translated into French by Bernard Rochot (Paris: J. Vrin, 1962).

The Selected Works of Pierre Gassendi, translated by Craig Brush (New York: Johnson Reprint, 1972).

Peiresc, 1580–1637: Vie de l'illustre Nicolas-Claude Fabri de Peiresc, Conseiller au Parlement d'Aix, translated by Roger Lasalle with the collaboration of Agnès Bresson (Paris: Belin, 1992).

Gassendus, Petrus, *The Mirrour of True Nobility and Gentility. Being the Life of the Renowned Nicolaus Lord of Peiresck, Senator of the Parliament of Aix*, translated by W. Rand (London, 1657).

The Vanity of Judiciary Astrology. Or Divination by the Stars. Lately written in Latin by that Great Schollar and Mathematician, the Illustrious PETRUS GASSENDUS: Mathematical Professor to the King of FRANCE, translated into English by a Person of Quality (London: Humphrey Moseley, 1659).

Jones, Howard, *Pierre Gassendi's* Institutio Logica, *1658* (Assen: Van Gorcum, 1981).

Moxon, Joseph, *A Tutor to Astronomy and Geography. Or, An Easie and Speedy Way to Know the Use of Both the Globes, Coelestial and Terrestrial* (London, 1699).

2 Descartes

a Original Sources

Descartes, René, *Oeuvres de Descartes,* edited by Charles Adam and Paul Tannery, 11 vols. (Paris: J. Vrin, 1897–1983).

b Translations

Descartes, René, *Discourse on Method, Optics, Geometry, and Meteorology,* translated by Paul J. Olscamp (New York: Library of Liberal Arts, 1965).
Treatise of Man, translated by Thomas Steele Hall (Cambridge, Mass.: Harvard University Press, 1972).
The World, translated by Michael Sean Mahoney (New York: Abaris, 1979).
Principles of Philosophy, translated by Valentine Rodger Miller and Reese P. Miller (Dordrecht: Reidel, 1983).
The Philosophical Writings of Descartes, translated by John Cottingham, Robert Stoothoff, Dugald Murdoch, and Anthony Kenny, 3 vols. (Cambridge University Press, 1984, 1985, 1991).

Descartes, René, and Martin Schoock, *La querelle d'Utrecht,* translated and edited by Theo Verbeek (Paris: Les Impressions Nouvelles, 1988).

3 Other Primary Sources

Alexander, H. G., editor, *The Leibniz–Clarke Correspondence* (Manchester: University of Manchester Press, 1956).

Ames, William, *The Marrow of Theology,* translated by John D. Eudsden (Boston: United Church Press, 1968).

Aquinas, St. Thomas, *Summa contra Gentiles,* translated by Charles J. O'Neil, 4 vols. (Notre Dame, Ind.: University of Notre Dame Press, 1957).
Commentary on Aristotle's Physics, translated by Richard J. Blackwell, Richard J. Spath, and W. Edmund Thirlkel (London: Routledge & Kegan Paul, 1963).
Summa theologiae, Latin text and English translation, 61 vols. (Blackfriars, in conjunction with London: Eyre & Spottiswoode, and New York: McGraw-Hill, 1964).

Summa theologica, translated by the Fathers of the English Dominican Province, 3 vols. (New York: Benziger, Inc., 1947).

Aristotle, *The Complete Works of Aristotle,* edited by Jonathan Barnes, 2 vols. (Princeton, N.J.: Princeton University Press, 1984).

Augustine of Hippo, Saint, *The City of God,* translated by Marcus Dods (New York: Random House, 1950).

Bacon, Francis, *The Works of Francis Bacon, Baron of Verulam, Viscount St. Alban, and Lord High Chancellor of England,* collected and edited by James Spedding, Robert Leslie Ellis, and Douglas Denon Heath, 14 vols. (London: Longman, 1857–74).

Bernier, François, *Abrégé de la philosophie de M. Gassendi,* 8 vols. (Lyon, 1678). *Abrégé de la philosophie de M. Gassendi,* edited by Sylvia Murr and Geneviève Stefani, 2d edition, 7 vols. (Lyon, 1684; reprinted, Paris: Fayard, 1992).

Boyle, Robert, *The Works of the Honourable Robert Boyle,* edited by Thomas Birch, new edition, 6 vols. (London, 1772; reprinted, Hildesheim: Georg Olms, 1965).

Calvin, John, *Institutes of the Christian Religion,* edited by John T. McNeill and translated by Ford Lews Battles, 2 vols. (Philadelphia: Westminster Press, 1960).

Charleton, Walter, *The Darknes of Atheism dispelled by the Light of Nature: A physico-theologicall treatise* (London: 1652).
Physiologia Epicuro-Gassendo-Charltoniana, or a Fabrick of Science Natural Upon the Hypothesis of Atoms, Founded by Epicurus, Repaired by Petrus Gassendus, Augmented by Walter Charleton (London, 1654; reprinted, New York: Johnson Reprint, 1966).
The Immortality of the Humane Soul (London, 1657).
The Ephesian Matron, edited by Achsah Guibbory (Los Angeles: William Andrews Clark Memorial Library, University of California, 1975; first published, London, 1659).

Cicero, Marcus Tullius, *De fato,* translated by H. Rackham (Cambridge, Mass.: Loeb Classical Library, 1942).
De divinatione, translated by William Armistead Falconer (Cambridge, Mass.: Loeb Classical Library, 1953).
De natura deorum, translated by H. Rackham (Cambridge, Mass.: Loeb Classical Library, 1956).

Cohen, I. Bernard, editor, *Isaac Newton's Papers and Letters on Natural Philosophy* (Cambridge, Mass.: Harvard University Press, 1958).

Cornford, Francis MacDonald, *Plato's Cosmology: The Timaeus of Plato* (Indianapolis, Ind.: Bobbs-Merrill, n.d.).

Dante Alighieri, *The Divine Comedy,* translated by Charles S. Singleton, 3 vols. (Princeton, N.J.: Princeton University Press, 1970).

Dick, Oliver Lawson, editor, *Aubrey's Brief Lives* (Ann Arbor: University of Michigan Press, 1962; first published, London: Secker & Warburg, 1949).

Digby, Kenelm, *Observations upon Religio medici* (London, 1644; facsimile reprint, Menston: Scolar Press, 1973).

Two Treatises in One of which The Nature of Bodies; in the other The Nature of Mans Soule; is looked into: in Way of Discovering, of the Immortality of Reasonable Soules (Paris: Gilles Blaizot, 1644; facsimile reprint, New York: Garland, 1978).

Drake, Stillman, translator and editor, *Discoveries and Opinions of Galileo* (Garden City, N.Y.: Doubleday Anchor, 1957).

Galilei, Galileo, *Dialogue Concerning the Two Chief World Systems – Ptolemaic and Copernican*, translated by Stillman Drake (Berkeley: University of California Press, 1953).

Grant, Edward, *A Sourcebook in Medieval Science* (Cambridge, Mass.: Harvard University Press, 1974).

Hall, A. Rupert, and Marie Boas Hall, editors, *Unpublished Scientific Papers of Isaac Newton* (Cambridge University Press, 1962).

Halliwell, J. O., editor, *A Collection of Letters Illustrative of the Progress of Science in England from the Reign of Queen Elizabeth to that of Charles the Second* (London: Dawson's, 1965).

Hero of Alexandria, *The Pneumatics of Hero of Alexandria*, translated by Joseph Gouge Greenwood, edited by Bennet Woodcroft, introduction by Marie Boas Hall (London: Taylor Walton & Maberly, 1851; facsimile reprint, London: MacDonald, 1971).

Hobbes, Thomas, *The English Works of Thomas Hobbes of Malmesbury*, edited by Sir William Molesworth, 11 vols. (London, 1839–45; reprinted, Aalen: Scientia, 1962).

   Computatio sive logica. Logic (Part I of *De corpore*), translated by Aloysius Martinich (New York: Abaris, 1981).

   Leviathan (New York: Meridian, 1965).

Isidore of Seville, *Etymologies*, edited by W. M. Lindsay (Oxford University Press, 1911).

Kepler, Johannes, *Harmonies of the World*, translated by Charles Glenn Wallis, in *Ptolemy, Copernicus, Kepler* (Great Books of the Western World, vol. 16) (Chicago: Encyclopedia Britannica, 1952).

Kristeller, Paul Oskar, *Eight Philosophers of the Renaissance* (Stanford, Calif.: Stanford University Press, 1964).

Laplace, Pierre Simon, Marquis de, *A Philosophical Essay on Probabilities*, translated from the 6th French edition by F. W. Truscott and F. L. Emory (New York: Dover, 1951).

Lipsius, Justus, *Physiologiae Stoicorum, libri tres* (Antwerp, 1604).

Long, A. A., and D. N. Sedley, *The Hellenistic Philosophers*, 2 vols. (Cambridge University Press, 1987).

Loux, Michael J., editor and translator, *Ockham's Theory of Terms. Part I of the Summa logicae* (Notre Dame, Ind.: University of Notre Dame Press, 1974).

Lucretius, Titus Carus, *De rerum natura*, translated by Cyril Bailey, 3 vols. (Oxford University Press, 1947).

   De rerum natura, translated by W. H. D. Rouse, revised by Martin Ferguson Smith, 2d edition (Cambridge, Mass.: Loeb Classical Library, 1982).

Mcguire, J. E., and Martin Tamny, *Certain Philosophical Questions: Newton's Trinity Notebook* (Cambridge University Press, 1983).

Mersenne, Marin, *Correspondance du P. Marin Mersenne,* edited by Cornelius de Waard and Armand Beaulieu, 15 vols. (Paris: CNRS, 1933–83).

La vérité des sciences contre les sceptiques ou pyrrhoniens (Paris, 1625; facsimile edition, Stuttgart-Bad Cannstatt: Friedrich Frommann Verlag, 1969).

Mirandola, Giovanni Pico della, *Conclusiones, sive theses DCCCC, Romae anno 1486 publice disputandae, sed non admissae,* with an introduction and critical annotations by Bohdan Kieszkowski (Geneva: Librairie Droz, 1973).

Mitchell, Stephen, *The Book of Job* (San Francisco: North Point, 1987).

Molina, Luis de, *On Divine Foreknowledge (Part IV of the Concordia),* translated, with an introduction and notes by Alfred J. Freddoso (Ithaca, N.Y.: Cornell University Press, 1988).

Montaigne, Michel de, *The Complete Essays of Montaigne,* translated by Donald M. Frame (Stanford, Calif.: Stanford University Press, 1948).

More, Henry, *A Collection of Several Philosophical Writings,* 2 vols. (London: James Fletcher, 1662; reprinted, New York: Garland Publishing, 1978).

Annotations upon the Two foregoing Treatises, Lux Orientalis, or, An Enquiry into the Opinion of the Eastern Sages Concerning the Prae-existence of Souls; and the Discourse of Truth (London, 1682).

Newton, Isaac, *Mathematical Principles of Natural Philosophy and His System of the World,* translated by Andrew Motte in 1729 and revised by Florian Cajori, 2 vols. (Berkeley and Los Angeles: University of California Press, 1962).

Philosophiae naturalis principia mathematica, 3d edition (1726) with variant readings, edited by Alexandre Koyré and I. Bernard Cohen, with the assistance of Anne Whitman, 2 vols. (Cambridge, Mass.: Harvard University Press, 1972).

Opticks, or A Treatise of the Reflections, Refractions, Inflections & Colours of Light, with a forward by Albert Einstein, introduction by Sir Edmund Whittaker, and preface by I. Bernard Cohen. Based on the 4th edition, London, 1730 (New York: Dover, 1952).

The Correspondence of Isaac Newton, edited by H. W. Turnbull, J. P. Scott, A. R. Hall, and Laura Tilling, 7 vols. (Cambridge: Published for the Royal Society at the University Press, 1959–77).

Ockham, William, *Philosophical Writings,* translated and edited by Philotheus Boehner (Edinburgh: Nelson, 1957).

Predestination, God's Foreknowledge, and Future Contingents, translated with an introduction, notes, and appendices by Marilyn McCord Adams and Norman Kretzmann (New York: Appleton-Century-Crofts, 1969).

Quodlibetal Questions, translated by Alfred J. Freddoso and Francis E. Kelley, 2 vols. (New Haven, Conn.: Yale University Press, 1991).

O'Connor, Eugene, editor and translator, *The Essential Epicurus: Letters, Principal Doctrines, Vatican Sayings, and Fragments* (Buffalo, N.Y.: Prometheus, 1993).

Pascal, Blaise, *The Physical Treatises of Pascal: The Equilibrium of Liquids and The Weight of the Mass of the Air*, translated by I. H. B. Spiers and A. G. H. Spiers, with an introduction and notes by Frederick Barry (New York: Columbia University Press, 1937; reprinted, New York: Octagon, 1973).

Peiresc, N. Cl. Fabri de, *Lettres de Peiresc*, edited by Philippe Tamizey de Larroque, in *Documents inédits sur l'histoire de France*, 7 vols. (Paris: Imprimerie Nationale, 1888–98).

Radbertus, Paschasius, *Paschasii Radberti Abbatis Corbeiensis opera omnia*, in *Patrologiae Latinae tomus 120*, edited by J.-P. Migne (Turnholti: Typographi Brepols Editores Pontificii, 1844–1902).

Roberts, Alexander, and James Donaldson, *The Ante-Nicene Fathers. Translations of the Fathers down to A.D. 325*, revised by A. Cleveland Coxe, 10 vols. (Grand Rapids, Mich.: Eerdmans, 1986–87).

Rupp, E. Gordon, and Philip S. Watson, editors and translators, *Luther and Erasmus: Free Will and Salvation* (Philadelphia: Westminster Press, 1969).

Sextus Empiricus, *Outlines of Pyrrhonism*, translated by R. G. Bury, 4 vols. (Cambridge, Mass.: Loeb Classical Library, 1976).

Stanley, Thomas, *The History of Philosophy: Containing The Lives, Opinions, Actions and Discourses of the Philosophers of Every Sect*, 2d edition, 3 vols. (London: Thomas Bassett, 1687).

Voltaire, *The Complete Works of Voltaire*, edited by Theodore Besterman et al., 135 vols. (Geneva: Institut et Musée Voltaire, and Toronto: University of Toronto Press, 1969).

SECONDARY SOURCES

Adams, Marilyn McCord, "Intuitive Cognition, Certainty, and Scepticism in William of Ockham," *Traditio*, 1970, 26: 389–98.

William Ockham, 2 vols. (Notre Dame, Ind.: University of Notre Dame Press, 1987).

"William Ockham: Voluntarist or Naturalist?" in *Studies in Medieval Philosophy*, edited by John F. Wippel (Washington, D.C.: Catholic University of America Press, 1987).

Alberti, A., "La canonica di Epicuro nell' interpretazione di Gassendi," *Annali dell' Instituto di filosofia dell' Università di Firenze*, 1980, 2: 151–94.

"Atomo e 'materia prima' nell'epicureismo di Gassendi," *Studi filosofici: Annali del Seminario di Studi dell'Occidente Medioevale e Moderno dell'Instituto Universitario Orientale di Napoli*, 1981, 4: 95–126.

Gassendi et l'atomismo epicureo (Firenze: Instituto Universitario Europeo, 1981).

Alexander, Peter, *Ideas, Qualities, and Corpuscles: Locke and Boyle on the External World* (Cambridge University Press, 1985).

Ariew, Roger, "Descartes and Scholasticism: The Intellectual Background to Descartes' Thought," in *The Cambridge Companion to Descartes*, edited by John Cottingham (Cambridge University Press, 1992).

Armogathe, J.-R., *Theologia cartesiana: L'explication physique de l'eucharistie chez Descartes et Dom Desgabets* (The Hague: Martinus Nijhoff, 1977). "Les sources scolastiques du temps cartésien: Éléments d'un débat," *Revue internationale de philosophie,* 1983, 37: 326–36.

Armstrong, A. H., editor, *The Cambridge History of Later Greek and Early Medieval Philosophy,* corrected edition (Cambridge University Press, 1967).

Armstrong, D. M., *Nominalism and Realism: Universals and Scientific Realism,* 2 vols. (Cambridge University Press, 1978).

Arthur, Richard T. W., "Continuous Creation, Continuous Time: A Refutation of the Alleged Discontinuity of Cartesian Time," *Journal of the History of Philosophy,* 1988, 26: 349–75.

Ashley, Kathleen, "Divine Power in the Chester Cycle and Late Medieval Thought," *Journal of the History of Ideas,* 1978, 39: 387–404.

Asmis, Elizabeth, *Epicurus' Scientific Method* (Ithaca, N.Y.: Cornell University Press, 1984).

Austin, William H., "More, Henry," in *Dictionary of Scientific Biography,* edited by Charles Coulton Gillispie, 16 vols. (New York: Scribner, 1972), vol. 9, pp. 509–10.

Bailey, Cyril, *The Greek Atomists and Epicureans* (New York: Russell & Russell, 1964; first published 1928).

Baillet, Adrien, *La vie de Monsieur Des-Cartes,* 2 vols. (Paris: Daniel Horthemels, 1691; reprinted, Geneva: Slatkine, 1970).

Barker, Peter, "Jean Pena and Stoic Physics in the Sixteenth Century," in *Recovering the Stoics,* edited by Ronald H. Epp, *Southern Journal of Philosophy,* 1985, 23, supplement: 93–108.
"Stoic Contributions to Early Modern Science," in *Atoms, Pneuma, and Tranquillity: Epicurean and Stoic Themes in European Thought,* edited by Margaret J. Osler (Cambridge University Press, 1991).

Barker, Peter, and Bernard R. Goldstein, "Is Seventeenth-Century Physics Indebted to the Stoics?" *Centaurus,* 1984, 27: 148–64.

Bechler, Zev, "The Essence and Soul of the Seventeenth-Century Scientific Revolution," *Science in Context,* 1987, 1: 87–104.

Beck, Daniel A., "Miracle and the Mechanical Philosophy: The Theology of Robert Boyle in Its Historical Context," Ph.D. dissertation, University of Notre Dame, 1986.

Belenky, Mary Field, Blythe McVicker Clinchy, Nancy Rule Goldberger, and Jill Mattuck Tarule, *Women's Ways of Knowing: The Development of Self, Voice, and Mind* (New York: Basic, 1986).

Berr, Henri, *Du scepticisme de Gassendi,* translated by Bernard Rochot (Paris: Albin Michel, 1960; first published in Latin in 1898).

Beyssade, Jean-Marie, "Création des vérités éternelles et doute métaphysique," *Studia Cartesiana,* 2 (Amsterdam: Quadratures, 1981), pp. 86–104.

Blackwell, Richard J., "Descartes' Laws of Motion," *Isis,* 1966, 57: 220–34.

Blake, R. M., "The Role of Experience in Descartes' Theory of Method," in *Theories of Scientific Method,* edited by E. H. Madden (Seattle: University of Washington Press, 1960).

244   *Bibliography*

Bloch, Olivier René, "Gassendi critique de Descartes," *Revue philosophique de la France et de l'étranger*, 1966, 91: 217–36.
   La philosophie de Gassendi: Nominalisme, matérialisme, et métaphysique (The Hague: Martinus Nijhoff, 1971).
   "Gassendi and the Transition from the Middle Ages to the Classical Era," *Yale French Studies*, 1973, 49: 43–55.
Blom, John J., *Descartes: His Moral Philosophy Psychology* (New York: New York University Press, 1978).
Boas, Marie, "The Establishment of the Mechanical Philosophy," *Osiris*, 1952, 10: 412–541.
Bobik, Joseph, "Matter and Individuation," in *The Concept of Matter in Greek and Medieval Philosophy*, edited by Ernan McMullin (Notre Dame, Ind.: University of Notre Dame Press, 1963).
Boehner, Philotheus, *Collected Articles on Ockham*, edited by Eligius M. Buytaert (St. Bonaventure, N.Y.: Franciscan Institute, 1958).
Bordo, Susan R., *The Flight to Objectivity: Essays on Cartesianism and Culture* (Albany: State University of New York Press, 1987).
Bos, H. J. M., "Huygens and Mathematics," in *Studies on Christiaan Huygens: Invited Papers from the Symposium on the Life and Work of Christiaan Huygens, Amsterdam, 22–25 August 1979*, edited by H. J. M. Bos, M. J. S. Rudwick, H. A. M. Snelders, and R. P. W. Visser (Lisse: Swets & Zeitlinger, 1980).
Bougerel, Joseph, *Vie de Pierre Gassendi, Prévôt de l'Église de Digne et Professeur de Mathématiques au Collège Royal* (Paris, 1737; reprinted, Geneva: Slatkine, 1970).
Boutroux, Émile, *Des vérités éternelles chez Descartes*, translated by M. Canguilhem (Paris: Félix Alcan, 1927; reprinted, J. Vrin, 1989).
Boylan, Michael, "Henry More's Space and the Spirit of Nature," *Journal of the History of Philosophy*, 1980, 18: 395–405.
Brandt, Frithiof, *Thomas Hobbes' Mechanical Conception of Nature*, translated by Vaughan Maxwell and Annie I. Fausbøll (Copenhagen: Levin & Munksgaard and London: Librairie Hachette, 1928).
Braun, L., *Histoire de l'histoire de la philosophie* (Paris: Editions Ophrys, 1973).
Bréhier, Émile, "La création des vérités éternelles," *Revue philosophique de la France et de l'étranger*, 1937, 113: 15–29; republished as "The Creation of the Eternal Truths in Descartes' System," in *Descartes: A Collection of Critical Essays*, edited by Willis Doney (Notre Dame, Ind.: University of Notre Dame Press, 1968), pp. 192–208.
Brett, G. S., *The Philosophy of Gassendi* (London: Macmillan, 1908).
Brockliss, L. W. B., "Aristotle, Descartes, and the New Science: Natural Philosophy at the University of Paris, 1600–1740," *Annals of Science* 1981, 38: 33–77.
   "Philosophy Teaching in France, 1600–1740," *History of Universities*, 1981, 1: 131–68.
   French Higher Education in the Seventeenth and Eighteenth Centuries: A Cultural History (Oxford University Press, 1987).

Brooke, John Hedley, *Science and Religion: Some Historical Perspectives* (Cambridge University Press, 1991).

Brown, Gregory, "Mathematics, Physics, and Corporeal Substance," *Pacific Philosophical Quarterly*, 1989, 70: 281–302.

Brown, Theodore, "Descartes, René du Perron," in *Dictionary of Scientific Biography*, edited by Charles Coulton Gillispie, 16 vols. (New York: Scribner, 1972), vol. 4, pp. 51–65.

Brundell, Barry, *Pierre Gassendi: From Aristotelianism to a New Natural Philosophy* (Dordrecht: Reidel, 1987).

Buckley, Michael J., S.J., *At the Origins of Modern Atheism* (New Haven, Conn.: Yale University Press, 1987).

Burns, Norman T., *Christian Mortalism from Tyndale to Milton* (Cambridge, Mass.: Harvard University Press, 1972).

Burtt, Edwin A., *The Metaphysical Foundations of Modern Science*, 2d edition (Garden City, N.Y.: Doubleday, 1954; first published 1932).

Busson, Henri, *La pensée religieuse française de Charron à Pascal* (Paris: J. Vrin, 1933).

Cariou, Marie, *L'atomisme: Gassendi, Leibniz, Bergson, et Lucrèce* (Paris: Aubier Montaigne, 1978).

Carré, Meyrick H., "Pierre Gassendi and the New Philosophy," *Philosophy*, 1958, 33: 112–20.

Cassirer, Ernst, Paul Oskar Kristeller, and John Herman Randall, Jr., editors, *The Renaissance Philosophy of Man* (Chicago: University of Chicago Press, 1948).

Céard, Jean, "Matérialisme et théorie de l'âme dans la pensée padouane, le *Traité de l'immortalité de l'âme* de Pomponazzi," *Revue philosophique de France et l'étranger*, 1981, 171: 25–48.

Centore, F. F., "Mechanism, Teleology, and Seventeenth Century English Science," *International Philosophical Quarterly*, 1972, 12: 553–71.

Centre International de Synthèse, *Pierre Gassendi* (Paris, 1955).

Chappell, Vere, and Willis Doney, editors, *Twenty-five Years of Descartes Scholarship, 1960–1984: A Bibliography* (New York: Garland, 1987).

Chartier, Roger, Dominique Julia, and Marie-Madeleine Compère, *L'éducation en France du XVIe au XVIIIe siècle* (Paris: Société d'Édition d'Enseignement Supérieure, 1976).

Cioffari, Vincenzo, "Fate, Fortune, and Chance," in *Dictionary of the History of Ideas*, edited by Philip P. Wiener, 4 vols. (New York: Scribner, 1973).

Clark, David J., "Ockham on Human and Divine Freedom," *Franciscan Studies*, 1978, 38: 122–60.

Clark, J. T., "Pierre Gassendi and the Physics of Galileo," *Isis*, 1963, 54: 351–70.

Clark, Stuart, "The Scientific Status of Demonology," in *Occult and Scientific Mentalities in the Renaissance*, edited by Brian Vickers (Cambridge University Press, 1988), pp. 351–74.

Clarke, Desmond M., "The Impact Rules of Descartes' Physics," *Isis*, 1977, 68: 55–66.

Bibliography

"Physics and Metaphysics in Descartes' Principles," *Studies in History and Philosophy of Science,* 1979, 10: 89–110.

Descartes' Philosophy of Science (Manchester: Manchester University Press, 1982).

Occult Powers and Hypotheses: Cartesian Natural Philosophy Under Louis XIV (Oxford University Press, 1989).

"Descartes' Philosophy of Science and the Scientific Revolution," in *The Cambridge Companion to Descartes,* edited by John Cottingham (Cambridge University Press, 1992).

Clericuzio, Antonio, "A Redefinition of Boyle's Chemistry and Corpuscular Philosophy," *Annals of Science,* 1990, 47: 561–89.

Cochois, Paul, *Bérulle et l'École française* (Paris: Éditions du Seuil, 1963).

Cohen, I. Bernard, *The Newtonian Revolution: With Illustrations of the Transformation of Scientific Ideas* (Cambridge University Press, 1980).

Coirault, Gaston, "Gassendi et non Locke créateur de la doctrine sensualiste moderne sur la génération des idées," in *Actes du Congrès Tricentenaire de Pierre Gassendi (1655–1955),* edited by Comité du Tricentaire de Gassendi (Digne: CNRS, 1955), pp. 69–94.

Colish, Marcia L., *The Stoic Tradition from Antiquity to the Early Middle Ages. I. Stoicism in Classical Latin Literature; II. Stoicism in Christian Latin Thought Through the Sixth Century,* 2 vols. (Leiden: E. J. Brill, 1985).

Collier, Raymond, "Gassendi et le spiritualisme ou Gassendi était-il un libertin?" in *Actes du Congrès du Tricentenaire de Pierre Gassendi (1655–1955),* edited by Comité du Tricentenaire de Gassendi (Digne: CNRS, 1955), pp. 97–135.

Collingwood, R. G., *The Idea of Nature* (New York: Oxford University Press, 1960).

Collins, James, "Descartes' Philosophy of Nature," *American Philosophical Quarterly,* Monograph Series, no. 5 (Oxford, 1971).

Copenhaver, Brian P., "Natural Magic, Hermetism, and Occultism in Early Modern Science," in *Reappraisals of the Scientific Revolution,* edited by David C. Lindberg and Robert S. Westman (Cambridge University Press, 1990).

"A Tale of Two Fishes: Magical Objects in Natural History from Antiquity Through the Scientific Revolution," *Journal of the History of Ideas,* 1991, 52: 373–98.

Copleston, Frederick C., *Aquinas* (Harmondsworth: Penguin Books, 1955).

A History of Philosophy, 9 vols. (Garden City, N.Y.: Doubleday, 1962; first published 1950).

Cottingham, John, "'A Brute to the Brutes?': Descartes' Treatment of Animals," *Philosophy,* 1978, 53: 551–9.

"Cartesian Dualism: Theology, Metaphysics, and Science," in *The Cambridge Companion to Descartes,* edited by John Cottingham (Cambridge University Press, 1992).

A Descartes Dictionary (Oxford: Blackwell, 1993).

"A New Start? Cartesian Metaphysics and the Emergence of Modern Philosophy," in *The Rise of Modern Philosophy: The Tension Between the New and*

Traditional Philosophies from Machiavelli to Leibniz, edited by Tom Sorell (Oxford University Press, 1993).

Courtenay, William J., "Nominalism and Late Medieval Thought: A Bibliographical Essay," *Theological Studies*, 1972, *33*: 716–34.

"Nominalism in Late Medieval Religion," in *The Pursuit of Holiness in Late Medieval and Renaissance Religion*, edited by Charles Trinkaus and Heiko A. Oberman (Leiden: E. J. Brill, 1974).

"Late Medieval Nominalism Revisited: 1972–1982," *Journal of the History of Ideas*, 1983, *44*: 159–64.

"The Dialectic of Omnipotence in the High and Late Middle Ages," in *Divine Omniscience and Omnipotence in Medieval Philosophy*, edited by Tamar Rudavsky (Dordrecht: Reidel, 1985).

Capacity and Volition: A History of the Distinction of Absolute and Ordained Power (Bergamo: Pierluigi Lubrina, 1990).

Courtenay, William J., editor, *Covenant and Causality in Medieval Thought: Studies in Philosophy, Theology, and Economic Practice*, (London: Variorum Reprints, 1984).

Craig, William Lane, *The Problem of Divine Foreknowledge and Future Contingents from Aristotle to Suarez* (Leiden: E. J. Brill, 1988).

Cranefield, Paul F., "On the Origin of the Phrase '*Nihil est in intellectu quod non prius fuerit in sensu*,'" *Journal of the History of Medicine*, 1970, *25*: 77–80.

Crombie, Alistair C., "Marin Mersenne (1588–1648) and the Seventeenth-Century Problem of Scientific Acceptability," *Physis*, 1975, *17*: 186–204.

Cronin, Timothy J., "Eternal Truths in the Thought of Descartes and His Adversary," *Journal of the History of Ideas*, 1960, *21*: 553–9.

"Eternal Truths in the Thought of Suarez and Descartes," *Modern Schoolman*, May 1961, *38* (4): 269–88; November 1961, *39* (1): 23–38.

"Objective Reality of Ideas in Human Thought: Descartes and Suarez," in *Wisdom in Depth: Essays in Honor of Henri Renard, S. J.* (Milwaukee, Wis.: Bruce, 1966), pp. 68–79.

Cunningham, Andrew, "How the *Principia* Got Its Name: Or, Taking Natural Philosophy Seriously," *History of Science*, 1991, *29*: 377–92.

Curley, Edwin, "Recent Work on Seventeenth-Century Continental Philosophy," *American Philosophical Quarterly*, 1974, *11*: 235–55.

Descartes Against the Skeptics (Cambridge, Mass.: Harvard University Press, 1978).

"Descartes on the Creation of the Eternal Truths," *Philosophical Review*, 1984, *93*: 569–97.

Dalbiez, R., "Les sources scolastiques de la théorie cartésienne de l'être objectif: À propos du 'Descartes' de M. Gilson," *Revue d'histoire de la philosophie*, 1929, *3*: 464–72.

Damiron, Jean-Philibert, *Essai sur l'histoire de la philosophie en France au XVIIe siècle*, 2 vols. (Paris: Hachette, 1846; reprinted, Geneva: Slatkine, 1970).

Damrosch, Leopold, Jr., "Hobbes as Reformation Theologian: Implications of the Free-Will Controversy," *Journal of the History of Ideas*, 1979, *40*: 339–52.

Daston, Lorraine, *Classical Probability in the Enlightenment* (Princeton, N.J.: Princeton University Press, 1988).

Davis, Edward Bradford, Jr., "Creation, Contingency, and Early Modern Science: The Impact of Voluntaristic Theology on Seventeenth-Century Natural Philosophy," Ph.D. dissertation, Indiana University, 1984.

"God, Man and Nature: The Problem of Creation in Cartesian Thought," *Scottish Journal of Theology,* 1991, 44: 325–48.

Davis, Leo Donald, S.J., "Knowledge According to Gregory of Rimini," *New Scholasticism,* 1981, 55: 331–47.

Dear, Peter, "Marin Mersenne and the Probabilistic Roots of 'Mitigated Scepticism,' " *Journal of the History of Philosophy,* 1984, 22: 173–205.

Mersenne and the Learning of the Schools (Ithaca, N.Y.: Cornell University Press, 1988).

"Miracles, Experiments, and the Ordinary Course of Nature," *Isis,* 1990, 81: 663–83.

Deason, Gary B., "Reformation Theology and the Mechanistic Conception of Nature," in *God and Nature: Historical Essays on the Encounter Between Christianity and Science,* edited by David C. Lindberg and Ronald L. Numbers (Berkeley: University of California Press, 1986).

Debus, Allen G., "Pierre Gassendi and His 'Scientific Expedition' of 1640," *Archives internationales d'histoire des sciences,* 1963, 6: 129–42.

The Chemical Dream of the Renaissance (Cambridge: Heffer, 1968).

Science and Education in the Seventeenth Century: The Webster–Ward Debate (New York: Science History Publications, 1970).

The Chemical Philosophy: Paracelsian Science and Medicine in the Sixteenth and Seventeenth Centuries, 2 vols. (Chicago: University of Chicago Press, 1977).

Man and Nature in the Renaissance (Cambridge University Press, 1978).

The French Paracelsians: The Chemical Challenge to Medical and Scientific Tradition in Early Modern France (Cambridge University Press, 1991).

Defarrari, Roy J., and Sister M. Inviolata Barry, with Ignatius McGuiness, *A Lexicon of St. Thomas Aquinas, Based on the* Summa Theologica *and Selected Passages of His Other Works* (Washington, D.C.: Catholic University of America Press, 1948).

Degrood, David H., *Philosophies of Essence: An Examination of the Category of Essence,* 2d edition (Amsterdam: B. R. Grüner, 1976).

Desharnais, Richard Paul, "The History of the Distinction Between God's Absolute and Ordained Power and Its Influence on Martin Luther," Ph.D. Dissertation, Catholic University of America, 1966.

Dijksterhuis, E. J., *The Mechanization of the World Picture,* translated by C. Dikshoorn (Oxford University Press, 1961).

Dillenberger, John, *Protestant Thought and Natural Science: A Historical Study* (Nashville, Tenn.: Abingdon, 1960).

Dobbs, Betty Jo Teeter, "Studies in the Natural Philosophy of Sir Kenelm Digby," *Ambix,* 1971, 18: 1–25.

"Studies in the Natural Philosophy of Sir Kenelm Digby: Part II, Digby and Alchemy," *Ambix*, 1973, 20: 143–63.

"Studies in the Natural Philosophy of Sir Kenelm Digby: Part III, Digby's Experimental Alchemy – the Book of *Secrets*," *Ambix*, 1974, 21: 1–28.

The Foundations of Newton's Alchemy or 'The Hunting of the Greene Lyon' (Cambridge University Press, 1975).

"Newton and Stoicism," in *Recovering the Stoics*, edited by Ronald H. Epp, *Southern Journal of Philosophy*, 1985, 23, supplement: 109–24.

"Stoic and Epicurean Doctrines in Newton's System of the World," in *Atoms, Pneuma, and Tranquillity: Epicurean and Stoic Themes in European Thought*, edited by Margaret J. Osler (Cambridge University Press, 1991).

The Janus Faces of Genius: The Role of Alchemy in Newton's Thought (Cambridge University Press, 1991).

Doig, James C., "Suarez, Descartes, and the Objective Reality of Ideas," *New Scholasticism*, 1977, 51: 350–71.

Doyle, John P., "Suarez on the Reality of the Possibles," *Modern Schoolman*, 1967, 45: 29–48.

Drake, Stillman, *Galileo at Work: His Scientific Biography* (Chicago: University of Chicago Press, 1978).

Driscoll, Edward A., "The Influence of Gassendi on Locke's Hedonism," *International Philosophical Quarterly*, 1972, 12: 87–110.

Dugas, René, *Mechanics in the Seventeenth Century*, translated by Freda Jacquet (Neuchâtel: Éditions du Griffon, and New York: Central Book, 1958).

Duhem, Pierre, *To Save the Phenomena: An Essay on the Idea of Physical Theory from Plato to Galileo*, translated by Edmund Doland and Chaninah Maschler (Chicago: University of Chicago Press, 1969; first published 1908).

Medieval Cosmology: Theories of Infinity, Place, Time, Void, and the Plurality of Worlds, edited and translated by Roger Ariew (Chicago: University of Chicago Press, 1985).

Duncan, David, "An International and Interdisciplinary Bibliography of Marin Mersenne (with a forward by A. Beaulieu)," *Bollettino di Storia della filosofia dell' Università degli Studi di Lecce*, 1986–89, 9: 201–42.

Edwards, Paul, editor, *The Encyclopedia of Philosophy*, 8 vols. (New York: Macmillan and Free Press, 1969).

Egan, Howard T., *Gassendi's View of Knowledge: A Study of the Epistemological Basis of His Knowledge* (Lanham, Md.: University Press of America, 1984).

Elzinga, Aant, *On A Research Program in Early Modern Physics* (New York: Humanities Press, 1972).

Emerton, Norma E., *The Scientific Reinterpretation of Form* (Ithaca, N.Y.: Cornell University Press, 1984).

Feldhay, Rivka, "Knowledge and Salvation in Jesuit Culture," *Science in Context*, 1987, 1: 195–214.

Ferguson, Wallace K., *The Renaissance in Historical Thought* (Boston: Houghton Mifflin, 1948).

Ferrier, Francis, *Un Oratorien ami de Descartes: Guillaume Gibieuf et sa philosophie de la liberté* (Paris: J. Vrin, 1980).

Force, James E., "Hume and the Relation of Science to Religion Among Certain Members of the Royal Society," *Journal of the History of Ideas*, 1984, 45: 517–36.

Force, James E., and Richard H. Popkin, *Essays on the Context, Nature, and Influence of Isaac Newton's Theology* (Dordrecht: Kluwer, 1990).

Foster, M. B., "The Christian Doctrine of Creation and the Rise of Modern Science," *Mind*, 1934, 43: 446–68.

"Christian Theology and Modern Science of Nature," *Mind*, 1935, 44: 439–66, and 1936, 45: 1–28.

Foster, Michael, "Sir Kenelm Digby (1603–1665) as Man of Religion and Thinker – I. and II. Intellectual Formation," *Downside Review*, 1988, 106: 35–58, 101–25.

Frankfurt, Harry, "Descartes on the Creation of the Eternal Truths," *Philosophical Review*, 1977, 86: 36–57.

Franklin, James, "The Ancient Legal Sources of Seventeenth-Century Probability," in *The Uses of Antiquity: The Scientific Revolution and the Classical Tradition*, edited by Stephen Gaukroger (Dordrecht: Kluwer, 1991).

Freudenthal, Gideon, *Atom and Individual in the Age of Newton: On the Genesis of the Mechanistic World View* (Dordrecht: Reidel, 1986).

Frijhoff, Willem, and Dominique Julia, *École et société dans la France d'ancien régime* (Paris: Libraire Armand Colin, 1975).

Fumagelli, Mariateresa Beonio-Brocchieri, et al., editors, *Sopra la volta del mondo: Omnipotenza e potenza assoluta di Dio tra medioevo e eta moderna* (Bergamo: Pierluigi Lubrina, 1986).

Funkenstein, Amos, "Descartes, Eternal Truths, and the Divine Omnipotence," *Studies in History and Philosophy of Science*, 1975, 6: 185–99. Reprinted in Stephen Gaukroger, editor, *Descartes: Philosophy, Mathematics, and Physics* (Brighton: Harvester, 1980).

Theology and the Scientific Imagination from the Middle Ages to the Seventeenth Century (Princeton, N.J.: Princeton University Press, 1986).

"The Body of God in the Seventeenth Century Theology and Science," in *Millenarianism and Messianism in English Literature and Thought, 1650–1800*, edited by Richard H. Popkin (Leiden: E. J. Brill, 1988).

Furley, David J., *Two Studies in the Greek Atomists: I. Invisible Magnitude; II. Aristotle and Epicurus on Voluntary Action* (Princeton, N.J.: Princeton University Press, 1967).

"Knowledge of Atoms and Void in Epicureanism," in *Essays in Ancient Greek Philosophy*, edited by J. P. Anton and G. L. Kustas (Albany: State University of New York Press, 1971). Reprinted in David Furley, editor, *Cosmic Problems: Essays on Greek and Roman Philosophy of Nature* (Cambridge University Press, 1989).

The Greek Cosmologists: Volume 1. The Formation of the Atomic Theory and Its Earliest Critics (Cambridge University Press, 1987).

Furley, David J., editor, *Cosmic Problems: Essays on Greek and Roman Philosophy of Nature* (Cambridge University Press, 1989).

Gabbey, Alan, "Force and Inertia in the Seventeenth Century: Descartes and Newton," in *Descartes: Philosophy, Mathematics, and Physics*, edited by Stephen Gaukroger (Sussex: Harvester, 1980).

"Huygens and Mechanics," in *Studies on Christiaan Huygens: Invited Papers from the Symposium on the Life and Work of Christiaan Huygens, Amsterdam, 22–25 August 1979*, edited by H. J. M. Bos, M. J. S. Rudwick, H. A. M. Snelders, and R. P. W. Visser (Lisse: Swets & Zeitlinger, 1980).

"Philosophia Cartesiana Triumphata: Henry More (1646–1671)," in *Problems of Cartesianism*, edited by Thomas M. Lennon, John W. Nicholas, and John W. Davis (Kingston and Montreal: McGill-Queen's University Press, 1982).

"The Mechanical Philosophy and Its Problems: Mechanical Explanations, Impenetrability, and Perpetual Motion," in *Change and Progress in Modern Science*, edited by Joseph C. Pitt (Dordrecht: Reidel, 1985).

"Henry More and the Limits of Mechanism," in *Henry More (1614–1687): Tercentenary Studies*, edited by Sarah Hutton (Dordrecht: Kluwer, 1990).

"Descartes's Physics and Descartes's Mechanics: Chicken and Egg?" in *Essays on the Philosophy and Science of René Descartes*, edited by Stephen Voss (New York: Oxford University Press, 1993).

Garber, Daniel, "Science and Certainty in Descartes," in *Descartes: Critical and Interpretative Essays*, edited by Michael Hooker (Baltimore: Johns Hopkins University Press, 1978).

"*Semel in vita*': The Scientific Background to Descartes' *Meditations*," in *Essays on Descartes' Meditations*, edited by Amélie Oksenberg Rorty (Berkeley: University of California Press, 1986).

"How God Causes Motion: Descartes, Divine Sustenance, and Occasionalism," *The Journal of Philosophy*, 1987, *84*: 567–80.

"Descartes, the Aristotelians, and the Revolution That Did Not Happen in 1637," *Monist*, 1988, *71*: 471–87.

Descartes' Metaphysical Physics (Chicago: University of Chicago Press, 1992).

"Descartes' Physics," in *The Cambridge Companion to Descartes*, edited by John Cottingham (Cambridge University Press, 1992).

"Descartes and Experiment in the *Discourse* and *Essays*," in *Essays on the Philosophy and Science of René Descartes*, edited by Stephen Voss (New York: Oxford University Press, 1993).

Garin, Pierre, *Thèses cartésiennes et thèses thomistes* (Paris: Desclée de Brouwer, 1931).

Gaukroger, Stephen, *Explanatory Structures: Concepts of Explanation in Early Physics and Philosophy* (Atlantic Highlands, N.J.: Humanities Press, 1978).

"Descartes' Project for a Mathematical Physics," in *Descartes: Philosophy, Mathematics, and Physics*, edited by Stephen Gaukroger (Brighton: Harvester, 1980), pp. 97–140.

Gelbart, Nina Rattner, "The Intellectual Development of Walter Charleton," *Ambix*, 1971, *18*: 149–68.

Gilligan, Carol, *In a Different Voice: Psychological Theory and Women's Development* (Cambridge, Mass.: Harvard University Press, 1982).

Gilson, Étienne, *Index scholastico-cartésien*, 2d edition (Paris: J. Vrin, 1979; first published 1913).

La liberté chez Descartes et la théologie (Paris: Félix Alcan, 1913).

The Philosophy of St. Thomas Aquinas, edited by Rev. G. A. Elrington, translated by Edward Bullough (New York: Dorset, 1929).

Études sur le rôle de la pensée médiévale dans la formation du système cartésien, 2d edition (Paris: J. Vrin, 1951; first published 1930).

History of Christian Philosophy in the Middle Ages (London: Sheed & Ward, 1955).

"Autour de Pomponazzi: Problématique de l'immortalité de l'âme en Italie au début du XVIe siècle," *Archives d'histoire doctrinale et littéraire du Moyen Âge*, 1961, 28: 163–279.

Glidden, David K., "Hellenistic Background for Gassendi's Theory of Ideas," *Journal of the History of Ideas*, 1988, 49: 405–24.

Goddu, André, *The Physics of William of Ockham* (Leiden: E. J. Brill, 1984).

Golinski, Jan V., "Hélène Metzger and the Interpretation of Seventeenth-Century Chemistry," *History of Science*, 1987, 25: 85–97.

"Robert Boyle: Scepticism and Authority in Seventeenth-Century Chemical Discourse," in *The Figural and the Literal: Problems of Language in the History of Science and Philosophy, 1630–1800*, edited by Andrew E. Benjamin, Geoffrey N. Cantor, and John R. R. Christie (Manchester: Manchester University Press, 1987).

"Chemistry in the Scientific Revolution: Problems of Language and Communication," in *Reappraisals of the Scientific Revolution*, edited by David C. Lindberg and Robert S. Westman (Cambridge University Press, 1990).

Gouhier, Henri, "La crise de la théologie au temps de Descartes," *Revue de théologie et de philosophie*, 3d ser., 1954, 4: 19–54.

Les premières pensées de Descartes: Contribution à l'histoire de l'anti-Renaissance (Paris: J. Vrin, 1958).

La pensée religieuse de Descartes, 2d edition (Paris: J. Vrin, 1972; first published 1924).

Cartésianisme et Augustinisme au XVIIe siècle (Paris: J. Vrin, 1978).

Gould, Josiah B., "The Stoic Conception of Fate," *Journal of the History of Ideas*, 1974, 35: 17–32.

Gould, Stephen Jay, *Wonderful Life: The Burgess Shale and the Nature of History* (New York: Norton, 1989).

Grant, Edward, "Hypotheses in Late Medieval and Early Modern Science," *Daedalus*, 1962, 91: 599–616.

"Medieval and Seventeenth-Century Conceptions of an Infinite Void Space Beyond the Cosmos," *Isis*, 1969, 60: 39–60.

Physical Science in the Middle Ages (New York: Wiley, 1971).

"Aristotelianism and the Longevity of the Medieval World View," *History of Science*, 1978, 16: 93–106.

"The Condemnation of 1277, God's Absolute Power, and Physical Thought in the Late Middle Ages," *Viator: Medieval and Renaissance Studies*, 1979, 10: 211–44.

Much Ado About Nothing: Theories of Space and the Vacuum from the Middle Ages to the Scientific Revolution (Cambridge University Press, 1981).

"Celestial Perfection from the Middle Ages to the Late Seventeenth Century," in *Religion, Science, and Worldview: Essays in Honor of Richard S. Westfall,* edited by Margaret J. Osler and Paul Laurence Farber (Cambridge University Press, 1985), pp. 137–62.

"Ways to Interpret the Terms 'Aristotelian' and 'Aristotelianism' in Medieval and Renaissance Natural Philosophy," *History of Science,* 1987, *25:* 335–58.

Gray, Robert, "Hobbes' System and His Early Philosophical Views," *Journal of the History of Ideas,* 1978, *39:* 199–215.

Greene, Robert A., "Henry More and Robert Boyle on the Spirit of Nature," *Journal of the History of Ideas,* 1962, *23:* 451–74.

Gregory, Tullio, *Scetticismo ed empirismo: Studio su Gassendi* (Bari: Laterza, 1961).

"Studi sull'atomismo del seicento. I. Sebastiano Basson," *Giornale critico della filosofia italiana,* 1964, *18:* 38–65.

"Studi sull'atomismo del seicento. II. David van Goorle e Daniel Sennert," *Giornale critico della filosofia italiana,* 1966, *20:* 44–63.

Grene, Marjorie, *Descartes* (Minneapolis: University of Minnesota Press, 1985).

Grimaldi, Nicolas, *Six études sur la volonté et la liberté chez Descartes* (Paris: J. Vrin, 1988).

Gueroult, Martial, "The Metaphysics and Physics of Force in Descartes," in *Descartes: Philosophy, Mathematics, and Physics,* edited by Stephen Gaukroger (Brighton: Harvester, 1980), pp. 196–229.

Descartes' Philosophy Interpreted According to the Order of Reasons, translated by Roger Ariew, 2 vols. (Minneapolis: University of Minnesota Press, 1984–5).

Guibert, A. J., *Descartes: Bibliographie des oeuvres publiées au XVIIe siècle* (Paris: Éditions du Centre National de la Recherche Scientifique, 1976).

Hacking, Ian, *The Emergence of Probability: A Philosophical Study of Early Ideas About Probability, Induction and Statistical Inference* (Cambridge University Press, 1975).

"Proof and Eternal Truths: Descartes and Leibniz," in *Descartes: Philosophy, Mathematics, and Physics,* edited by Stephen Gaukroger (Brighton: Harvester, 1980), pp. 169–80.

"Language, Truth, and Reason," in *Rationality and Relativism,* edited by Martin Hollis and Steven Lukes (Oxford: Blackwell, 1982).

"'Style' for Historians and Philosophers," *Studies in History and Philosophy of Science,* 1992, *23:* 1–20.

Hahn, Roger, "Laplace and the Vanishing Role of God," in *The Analytic Spirit: Essays in the History of Science in Honor of Henry Guerlac* (Ithaca, N.Y.: Cornell University Press, 1981).

"Laplace and the Mechanistic Universe," in *God and Nature: Historical Essays on the Encounter Between Christianity and Science,* edited by David C. Lindberg and Ronald L. Numbers (Berkeley: University of California Press, 1986).

Halbronn, Jacques E., "The Revealing Process of Translation and Criticism," in *Astrology, Science, and Society: Historical Essays,* edited by Patrick Curry (Suffolk: Boydell, 1987).

Hall, A. Rupert, "Sir Isaac Newton's Notebook, 1661–1665," *Cambridge Historical Journal*, 1948, 9: 239–50.

Henry More: Magic, Religion, and Experiment (Oxford: Blackwell, 1990).

Hall, Marie Boas, "Matter in Seventeenth-Century Science," in *The Concept of Matter*, edited by Ernan McMullin (Notre Dame, Ind.: Notre Dame University Press, 1963).

Robert Boyle on Natural Philosophy (Bloomington: Indiana University Press, 1965).

"Digby, Kenelm," in *The Encyclopedia of Philosophy*, edited by Paul Edwards, 8 vols. (New York: Macmillan and Free Press, 1967).

Harnack, Adolph, *History of Dogma*, translated by James Millar, 7 vols. (London: Williams & Norgate, 1898).

Harré, R., *Matter and Method* (London: Macmillan, 1964).

"Laplace, Pierre Simon de," in *The Encyclopedia of Philosophy*, edited by Paul Edwards, 8 vols. (New York: Macmillan and Free Press, 1967).

Harrison, Charles, "Bacon, Hobbes, and the Ancient Atomists," *Harvard Studies and Notes in Philology and Literature*, 1933, *15*: 191–213.

"The Ancient Atomists and English Literature of the Seventeenth Century," *Harvard Studies in Classical Philology*, 1934, *45*: 1–79.

Hatfield, Gary C., "Force (God) in Descartes' Physics," *Studies in History and Philosophy of Science*, 1979, *10*: 113–40.

"The Senses and the Fleshless Eye: The *Meditations* as Cognitive Exercises," in *Essays on Descartes' Meditations*, edited by Amélie Oksenberg Rorty (Berkeley: University of California Press, 1986).

"Metaphysics and the New Science," in *Reappraisals of the Scientific Revolution*, edited by David C. Lindberg and Robert S. Westman (Cambridge University Press, 1990).

"Reason, Nature, and God in Descartes," *Science in Context*, 1989, *3*: 175–201. Revised version in *Essays on the Philosophy and Science of René Descartes*, edited by Stephen Voss (New York: Oxford University Press, 1993).

Heilbron, John L., *Elements of Early Modern Physics* (Berkeley: University of California Press, 1982).

Henry, John, "Francesco Patrizi da Cherso's Concept of Space and Its Later Influence," *Annals of Science*, 1979, *36*: 549–73.

"A Cambridge Platonist's Materialism: Henry More and the Concept of the Soul," *Journal of the Warburg and Courtauld Institutes*, 1986, 49: 172–95.

"Occult Qualities and the Experimental Philosophy: Active Principles in Pre-Newtonian Matter Theory," *History of Science*, 1986, 24: 335–81.

"Henry More Versus Robert Boyle: The Spirit of Nature and the Nature of Providence," in *Henry More (1624–1687): Tercentenary Studies*, edited by Sarah Hutton (Dordrecht: Kluwer, 1990).

Heyd, Michael, "From a Rationalist Theology to Cartesian Voluntarism: David Derodon and Jean-Robert Chouet," *Journal of the History of Ideas*, 1979, 40: 527–42.

Between Orthodoxy and the Enlightenment: Jean-Robert Chouet and the In-

troduction of Cartesian Science in the Academy of Geneva (The Hague: Martinus Nijhoff, 1982).

Hillerbrand, Hans J., *Men and Ideas in the Sixteenth Century* (Prospect Heights, Ill.: Waveland, 1969).

Hine, William L., "Marin Mersenne: Renaissance Naturalism and Renaissance Magic," in *Occult and Scientific Mentalities in the Renaissance,* edited by Brian Vickers (Cambridge University Press, 1984).

Hoenen, P. H. J., " Descartes' Mechanicism," in *Descartes: A Collection of Critical Essays,* edited by Willis Doney (Notre Dame, Ind.: Notre Dame University Press, 1967).

Holtrop, Philip C., "The Bolsec Controversy: The Origin of Theodore Beza's Doctrine of Predestination," Ph. D. dissertation, Harvard University, 1985.

Hooker, Michael, editor, *Descartes: Critical and Interpretative Essays* (Baltimore: Johns Hopkins University Press, 1978).

Hooykaas, R., "Science and Religion in the Seventeenth Century: Isaac Beeckman (1588–1637), *Free University Quarterly,* 1951, 1: 169–83.

"Science and Theology in the Middle Ages," *Free University Quarterly,* 1954, 3: 77–163.

"Beeckman, Isaac," in *Dictionary of Scientific Biography,* edited by Charles Coulton Gillispie, 16 vols. (New York: Scribner, 1972), vol. 1, pp. 566–8.

Religion and the Rise of Modern Science (Grand Rapids, Mich.: Eerdmans, 1972).

"The Rise of Modern Science: When and Why?" *British Journal for the History of Science,* 1987, 20: 453–74.

Hutchison, Keith, "What Happened to Occult Qualities in the Scientific Revolution?" *Isis,* 1982, 73: 233–53.

"Supernaturalism and the Mechanical Philosophy," *History of Science,* 1983, 21: 297–333.

"Reformation Politics and the New Philosophy," *Metascience,* 1984, 1-2: 4–14.

"Dormitive Virtues, Scholastic Qualities, and the New Philosophies," *History of Science,* 1991, 29: 245–78.

"Individualism, Causal Location, and the Eclipse of Scholastic Philosophy," *Social Studies of Science,* 1991, 21: 321–50.

Hutton, Sarah, editor, *Henry More (1614–1687): Tercentenary Studies* (Dordrecht: Kluwer, 1990).

Jacquot, Jean, "Thomas Harriot's Reputation for Impiety," *Notes and Records of the Royal Society of London,* 1951-2, 9: 164–87.

"Harriot, Hill, Warner, and the New Philosophy," in *Thomas Harriot: Renaissance Scientist,* edited by John W. Shirley (Oxford University Press, 1974).

Jeannel, Charles, *Gassendi spiritualiste* (Montpellier: Félix Seguin, 1859).

Jedin, Hubert, *A History of the Council of Trent,* translated by Dom Ernest Graf, 2 vols. (London: Thomas Nelson; first published in German, Freiburg im Breisgau: Verlag Herder, 1949; and St. Louis: B. Herder, 1961).

Jenkins, Jane Elizabeth, "Using Nothing: Vacuum, Matter, and Spirit in the Seventeenth Century Mechanical Philosophy," M.A. thesis, University of Calgary, 1990.

Jobe, Thomas Harmon, "The Devil in Restoration Science: The Glanvill-Webster Debate," *Isis*, 1981, 72: 343–56.

Jones, Howard, *Pierre Gassendi, 1592–1655: An Intellectual Biography* (Nieukoop: B. de Graaf, 1981).

The Epicurean Tradition (London: Routledge, 1989).

Joy, Lynn Sumida, *Gassendi the Atomist: Advocate of History in an Age of Science* (Cambridge University Press, 1987).

"The Conflict of Mechanisms and Its Empiricist Outcome," *Monist*, 1988, 71: 498–514.

"Epicureanism in Renaissance Philosophy," *Journal of the History of Ideas*, 1992, 53: 573–83.

"Humanism and the Problem of Traditions in Seventeenth-Century Natural Philosophy," in *Philosophical Imagination and Cultural Memory: Appropriating Historical Traditions*, edited by Patricia Cook (Durham, N.C.: Duke University Press, 1993).

Kargon, Robert, "Walter Charleton, Robert Boyle, and the Acceptance of Epicurean Atomism in England," *Isis*, 1964, 55: 184–92.

Atomism in England from Hariot to Newton (Oxford University Press, 1966).

Keeling, S. V., *Descartes*, 2d edition (Oxford University Press, 1968).

Kish, John, "The Influence of Pierre Gassendi on John Locke's Theory of the Material World," Ph.D. dissertation, Johns Hopkins University, 1984.

Klaaren, Eugene M., *Religious Origins of Modern Science: Belief in Creation in Seventeenth-Century Thought* (Grand Rapids, Mich.: Eerdmans, 1977).

Konstan, David, "Problems in Epicurean Physics," in *Essays in Ancient Greek Philosophy*, edited by John P. Anton and Anthony Preus, vol. 2 (Albany: State University of New York Press, 1983).

Koyré, Alexandre, "Galileo and Plato," in *Metaphysics and Measurement: Essays in the Scientific Revolution*, edited by Alexandre Koyré (London: Chapman & Hall, 1968). First published in *Journal of the History of Ideas*, 1943, 4: 400–28.

"Gassendi and Science in His Time," translated by R. E. W. Maddison, in *Metaphysics and Measurement: Essays in the Scientific Revolution*, edited by Alexandre Koyré (London: Chapman & Hall, 1968). First published in *Pierre Gassendi: Sa vie et son oeuvre* (Paris: Albin Michel, 1955), pp. 118–30.

From the Closed World to the Infinite Universe (Baltimore: Johns Hopkins University Press, 1957).

"Newton and Descartes," in *Newtonian Studies* (Cambridge, Mass.: Harvard University Press, 1965), pp. 53–200.

Kretzmann, Norman, Anthony Kenny, and Jan Pinborg, editors, *The Cambridge History of Later Medieval Philosophy* (Cambridge University Press, 1982).

Kristeller, Paul Oskar, *Renaissance Thought: The Classic, Scholastic, and Humanist Strains* (New York: Harper, 1961; first published 1955).

Eight Philosophers of the Italian Renaissance (Stanford, Calif.: Stanford University Press, 1964).

Renaissance Thought and Its Sources, edited by Michael Mooney (New York: Columbia University Press, 1979).

Kroll, Richard W. F., "The Question of Locke's Relation to Gassendi," *Journal of the History of Ideas,* 1984, *45:* 339–60.

Kubbinga, H. H., "Les premières théories 'moléculaires': Isaac Beeckman (1620) et Sébastien Basson (1621): Le concept d' 'individu substantiel' et d' 'espèce substantielle,' " *Revue d'histoire des sciences,* 1984, *37:* 215–33.

Kuhn, Thomas S., *The Copernican Revolution* (Cambridge, Mass.: Harvard University Press, 1957).

"Mathematical Versus Experimental Traditions in the Development of Physical Science," in *The Essential Tension: Selected Studies in Scientific Tradition and Change,* edited by Thomas S. Kuhn (Chicago: University of Chicago Press, 1977).

Lalande, André, *Vocabulaire technique et critique de la philosophie,* 10th edition (Paris: Presses Universitaires de France, 1968).

Larmore, Charles, "Descartes' Empirical Epistemology," in *Descartes: Philosophy, Mathematics, and Physics,* edited by Stephen Gaukroger (Brighton: Harvester, 1980), pp. 6–22.

Laymon, Ronald, "Transubstantiation: Test Case for Descartes's Theory of Space," in Thomas M. Lennon, John M. Nicholas, and John W. Davis, *Problems of Cartesianism* (Kingston and Montreal: McGill-Queen's University Press, 1982), pp. 149–70.

Lebèque, Raymond, "Une amitié exemplaire, Peiresc et Gassendi," in *Actes du Congrès du Tricentenaire de Pierre Gassendi (1655–1955),* edited by Comité du Tricentenaire du Pierre Gassendi (Digne: CNRS, 1955).

Leclerc, Ivor, "Atomism, Substance, and the Concept of Body in Seventeenth-Century Thought," *Filosophia,* 1967, *18:* 761–78.

The Nature of Physical Existence (London: Allen & Unwin; New York: Humanities Press, 1972).

Leff, Gordon, *Gregory of Rimini: Tradition and Innovation in Fourteenth Century Thought* (Manchester: Manchester University Press, 1961).

Paris and Oxford Universities in the Thirteenth and Fourteenth Centuries: An Institutional and Intellectual History (New York: Wiley, 1968).

William of Ockham: The Metamorphosis of Scholastic Discourse (Manchester: Manchester University Press, 1975).

The Dissolution of the Medieval Outlook: An Essay on the Intellectual and Spiritual Change in the Fourteenth Century (New York: Harper & Row, 1976).

Lennon, Thomas M., "The Epicurean New Way of Ideas: Gassendi, Locke, and Berkeley," in *Atoms, Pneuma, and Tranquillity: Epicurean and Stoic Themes in European Thought,* edited by Margaret J. Osler (Cambridge University Press, 1991).

The Battle of the Gods and Giants: The Legacies of Descartes and Gassendi, 1655–1715 (Princeton, N.J.: Princeton University Press, 1993).

Lenoble, Robert, *Mersenne ou la naissance du mécanisme,* 2d edition (Paris: J. Vrin, 1971; first published 1943).

"A propos des conseils de Mersenne aux historiens et de l'intervention de Jean de Launoy dans la querelle gassendiste," *Revue d'histoire des sciences*, 1953, 6: 112–34.

Lindsay, R. B., "Pierre Gassendi and the Revival of Atomism in the Renaissance," *American Journal of Physics*, 1945, 13: 235–42.

Lloyd, G. E. R., *Aristotle: The Growth and Structure of His Thought* (Cambridge University Press, 1968).

Demystifying Mentalities (Cambridge University Press, 1990).

Lohr, Charles H., "Jesuit Aristotelianism and Sixteenth-Century Metaphysics," in ΠΑΡΑΔΟΣΙΣ *(Paradosis): Studies in Memory of Edwin A. Quain*, edited by Harry George Fletcher III and Mary Beatrice Schulte (New York: Fordham University Press, 1976).

Long, A. A., *Hellenistic Philosophy: Stoics, Epicureans, Sceptics*, 2d edition (Berkeley: University of California Press, 1986; first published 1974).

Lovejoy, Arthur O., *The Great Chain of Being: A Study of the History of an Idea* (Cambridge, Mass.: Harvard University Press, 1936).

Lynch, William T., "Politics in Hobbes' Mechanics: The Social as Enabling," *Studies in History and Philosophy of Science*, 1991, 22: 295–320.

Lynes, John W., "Descartes' Theory of Elements: From *Le Monde* to *The Principles*," *Journal of the History of Ideas*, 1982, 43: 55–72.

Machamer, Peter, "Causality and Explanation in Descartes' Natural Philosophy," in *Motion and Time, Space and Matter: Interrelations in the History and Philosophy of Science*, edited by P. K. Machamer and R. G. Turnbull (Columbus: Ohio State University Press, 1976).

MacIntosh, John J., "Primary and Secondary Qualities," *Studia Leibnitiana*, 1976, 8, (1): 88–104.

"Robert Boyle on Epicurean Atheism and Atomism," in *Atoms, Pneuma, and Tranquillity: Epicurean and Stoic Themes in European Thought*, edited by Margaret J. Osler (Cambridge University Press, 1991).

Maienschein, Jane, "Epistemic Styles in German and American Embryology," *Science in Context*, 1991, 4: 407–27.

Mandelbaum, Maurice, *Philosophy, Science, and Sense Perception: Historical and Critical Studies* (Baltimore: Johns Hopkins University Press, 1964).

Mandon, L., *L'étude sur le Syntagma Philosophicum de Gassendi* (Montpelier: 1858; reprinted, New York: Burt Franklin, 1969).

Mandrou, Robert, *From Humanism to Science, 1480–1700*, translated by Brian Pearce (Harmondsworth: Penguin, 1978; first published 1973).

Marion, Jean-Luc, *Sur la théologie blanche de Descartes: Analogie, création des vérités éternelles et fondement* (Paris: Presses Universitaires de France, 1981).

Martin, Abbé A., *Histoire de la vie et des écrits de Pierre Gassendi* (Paris, 1854).

Massa, Daniel, "Giordano Bruno's Ideas in Seventeenth-Century England," *Journal of the History of Ideas*, 1977, 38: 227–42.

Maull, Nancy L., "Cartesian Optics and the Geometrization of Nature," in *Descartes: Philosophy, Mathematics, and Physics*, edited by Stephen Gaukroger (Brighton: Harvester, 1980). Originally published in *Review of Metaphysics*, 1978, 32: 253–73.

Maurer, Armand, "St. Thomas and Eternal Truths," *Mediaeval Studies,* 1970, 32: 91–107.

"Ockham on the Possibility of A Better World," *Mediaeval Studies,* 1976, 38: 291–312.

Mayo, Thomas Franklin, *Epicurus in England (1650–1725)* (Dallas, Tex.: The Southwest Press, 1934).

Mayr, Ernst, "The Idea of Teleology," *Journal of the History of Ideas,* 1992, 53: 117–35.

McColley, Grant, "Nicholas Hill and the *Philosophia Epicurea,*" *Annals of Science,* 1939, 4: 390–405.

McGuire, J. E., "Boyle's Conception of Nature," *Journal of the History of Ideas,* 1972, 33: 523–43.

McMullin, Ernan, editor, *The Concept of Matter in Greek and Medieval Philosophy,* revised edition (Notre Dame, Ind.: University of Notre Dame Press, 1979; first published in 1963).

Meinel, Christoph, "Early Seventeenth-Century Atomism: Theory, Epistemology, and Insufficiency of Experiment," *Isis,* 1988, 79: 68–103.

"Empirical Support for the Corpuscular Philosophy in the Seventeenth Century," in *Theory and Experience: Recent Insights and New Perspectives on Their Relation,* edited by Diderik Batens and Jean Paul Van Bendegem (Dordrecht: Reidel, 1988).

Messeri, Marco, *Causa e spiegazione: La fisica di Pierre Gassendi* (Milan: Franco Angeli, 1985).

Michael, Emily, and Fred S. Michael, "Gassendi on Sensation and Reflection: A Non-Cartesian Dualism," *History of European Ideas,* 1988, 9: 583–95.

"Corporeal Ideas in Seventeenth-Century Psychology," *Journal of the History of Ideas,* 1989, 50: 31–48.

"Two Early Modern Concepts of Mind: Reflecting Substance vs. Thinking Substance," *Journal of the History of Philosophy,* 1989, 27: 29–48.

Michael, Fred S., and Emily Michael, "The Theory of Ideas in Gassendi and Locke," *Journal of the History of Ideas,* 1990, 51: 379–99.

Miel, Jan, *Pascal and Theology* (Baltimore: Johns Hopkins University Press, 1969).

Miles, Murray Lewis, "Descartes' Mechanicism and the Medieval Doctrine of Causes, Qualities, and Forms," *Modern Schoolman,* 1988, 65: 97–117.

Milhaud, Gaston, *Descartes savant* (Paris: Félix Alcan, 1921).

Millen, Ron, "The Manifestation of Occult Qualities in the Scientific Revolution," in *Religion, Science, and Worldview: Essays in Honor of Richard S. Westfall,* edited by Margaret J. Osler and Paul Lawrence Farber (Cambridge University Press, 1985).

Miller, Leonard G., "Descartes, Mathematics, and God," *Philosophical Review,* 1957, 66: 451–65; reprinted in *Meta-Meditations: Studies in Descartes,* edited by Alexander Sesonske and Noel Fleming (Belmont, Calif.: Wadsworth, 1965).

Milton, John R., "The Origins and Development of the Concept of the 'Laws of Nature,'" *Archives européennes de sociologie,* 1981, 22: 173–95.

Mintz, Samuel I., *The Hunting of Leviathan: Seventeenth-Century Reactions to the Materialism and Moral Philosophy of Thomas Hobbes* (Cambridge University Press, 1962).

Mir, Gabriel Codina, *Aux sources de la pédagogie des Jésuites: Le 'modus Parisiensis'* (Rome: Institutum Historicum S J, 1968).

Mittelstrass, Jürgen, "Remarks on the Nominalistic Roots of Modern Science," *Organon*, 1967, 4: 39–46.

Moody, Ernest A., "Empiricism and Metaphysics in Medieval Philosophy," *Philosophical Review*, 1958, 67: 145–63.

Moore, James R., *The Post-Darwinian Controversies: A Study of the Protestant Struggle to Come to Terms with Darwin in Great Britain and America, 1870–1900* (Cambridge University Press, 1979).

Morris, John, "Descartes' Natural Light," *Journal of the History of Philosophy*, 1973, 11: 169–88.

Morrison, Margaret, "Hypotheses and Certainty in Cartesian Science," in *An Intimate Relation: Studies in History and Philosophy of Science Presented to Robert E. Butts*, edited by James R. Brown and Jürgen Mittelstrass (Dordrecht: Kluwer, 1989).

Murr, Sylvia, "La science de l'homme chez Hobbes et Gassendi," in *Thomas Hobbes: Philosophie première, théorie de la science et politique*, edited by Yves Charles Zarka, with Jean Bernhardt (Paris: Presses Universitaires de France, 1990).

"L'âme des bêtes chez Gassendi," *Corpus: Revue de philosophie*, 1991, 16–17: 41–63.

"Foi religieuse et *libertas philosophandi* chez Gassendi," *Revue des sciences philosophiques et théologiques*, 1992, 76: 85–100.

Nadler, Steven M., "Scientific Certainty and the Creation of the Eternal Truths: A Problem in Descartes," *Southern Journal of Philosophy*, 1987, 25: 175–92.

"Arnauld, Descartes, and Transubstantiation: Reconciling Cartesian Metaphysics and Real Presence," *Journal of the History of Ideas*, 1988, 49: 229–46.

"Deduction, Confirmation, and the Laws of Nature in Descartes' *Principia philosophiae*," *Journal of the History of Philosophy*, 1990, 28: 359–83.

Nielsen, Lauge Olaf, "A Seventeenth-Century Physician on God and Atoms: Sebastian Basso," in *Meaning and Inference in Medieval Philosophy: Studies in Memory of Jan Pinborg*, edited by Norman Kretzmann (Dordrecht: Kluwer, 1988).

Norena, Carlos P., "Ockham and Suárez on the Ontological Status of Universal Concepts," *New Scholasticism*, 1981, 55: 348–62.

Normore, Calvin G., "Divine Omniscience, Omnipotence, and Future Contingents: An Overview," in *Divine Omniscience and Omnipotence in Medieval Philosophy*, edited by Tamar Rudavsky (Dordrecht: Reidel, 1985).

"Meaning and Objective Being: Descartes and His Sources," in *Essays on Descartes' Meditations*, edited by Amélie Oksenberg Rorty (Berkeley: University of California Press, 1986).

"The Tradition of Mediaeval Nominalism," in *Studies in Medieval Philosophy*, edited by John F. Wippel (Washington, D.C.: Catholic University of America Press, 1987).

Norton, David Fate, "The Myth of 'British Empiricism,'" *American Philosophical Quarterly,* 1981, 1: 331–44.

Nussbaum, Martha C., *The Fragility of Goodness: Luck and Ethics in Greek Tragedy and Philosophy* (Cambridge University Press, 1986).

Oakley, Francis, "Christian Theology and Newtonian Science: The Rise of the Concept of the Laws of Nature," *Church History,* 1961, 30: 433–57.

Omnipotence, Covenant, and Order: An Excursion in the History of Ideas from Abelard to Leibniz (Ithaca, N.Y.: Cornell University Press, 1984).

Oberman, Heiko A., "Some Notes on the Theology of Nominalism with Attention to Its Relation to the Renaissance," *Harvard Theological Review,* 1960, 53: 47–76.

The Harvest of Medieval Theology: Gabriel Biel and Late Medieval Nominalism (Cambridge, Mass.: Harvard University Press, 1963).

O'Connor, Daniel, and Francis Oakley, editors, *Creation: The Impact of an Idea* (New York: Scribner, 1969).

Onians, R. B., *The Origins of European Thought About the Body, the Mind, the Soul, the World, Time, and Fate* (Cambridge University Press, 1951).

Osler, Margaret J. "John Locke and the Changing Ideal of Scientific Knowledge," *Journal of the History of Ideas,* 1970, 31: 1–16. Reprinted in *Philosophy, Religion, and Science in the 17th and 18th Centuries,* edited by John W. Yolton (Rochester, N.Y.: University of Rochester Press, 1990).

"Galileo, Motion, and Essences," *Isis,* 1973, 64: 504–9.

"Certainty, Scepticism, and Scientific Optimism: The Roots of Eighteenth-Century Attitudes Towards Scientific Knowledge," in *Probability, Time, and Space in Eighteenth-Century Literature,* edited by Paul R. Backscheider (New York: AMS Press, 1979).

"Descartes and Charleton on Nature and God," *Journal of the History of Ideas,* 1979, 40: 445–56.

"Providence and Divine Will: The Theological Background to Gassendi's Views on Scientific Knowledge," *Journal of the History of Ideas,* 1983, 44: 549–60.

"Baptizing Epicurean Atomism: Pierre Gassendi on the Immortality of the Soul," in *Religion, Science, and Worldview: Essays in Honor of Richard S. Westfall,* edited by Margaret J. Osler and Paul Lawrence Farber (Cambridge University Press, 1985).

"Eternal Truths and the Laws of Nature: The Theological Foundations of Descartes' Philosophy of Nature," *Journal of the History of Ideas,* 1985, 46: 349–62.

"Fortune, Fate, and Divination: Gassendi's Voluntarist Theology and the Baptism of Epicureanism," in *Atoms, Pneuma, and Tranquillity: Epicurean and Stoic Themes in European Thought,* edited by Margaret J. Osler (Cambridge University Press, 1991).

"The Intellectual Sources of Robert Boyle's Philosophy of Nature: Gassendi's Voluntarism and Boyle's Physico-Theological Project," in *Philosophy, Sci-*

ence, and Religion, 1640–1700, edited by Richard Ashcraft, Richard Kroll, and Perez Zagorin (Cambridge University Press, 1991).

"Ancients, Moderns, and the History of Philosophy: Gassendi's Epicurean Project," in *The Rise of Modern Philosophy: The Tension Between the New and Traditional Philosophies from Machiavelli to Leibniz,* edited by Tom Sorell (Oxford University Press, 1993).

Otte, Michael, "Style as a Historical Category," *Science in Context,* 1991, 4: 233–64.

Ozment, Steven, *The Age of Reform, 1250–1550: An Intellectual and Religious History of Late Medieval and Reformation Europe* (New Haven, Conn.: Yale University Press, 1980).

Paganini, Gianni, "Epicurisme et philosophie au XVIIème siècle: Convention, utilité, et droit selon Gassendi," *Studi filosofici,* 1989–90, 12–13: 5–45.

Pagel, Walter, *Paracelsus: An Introduction to Philosophical Medicine of the Renaissance* (Basel: S. Karger, 1958).

Palladino, Paolo, "Stereochemistry and the Nature of Life: Mechanist, Vitalist, and Evolutionary Perspectives," *Isis,* 1990, 81: 44–67.

Panchieri, Lillian Unger, "The Magnet, the Oyster, and the Ape, or Pierre Gassendi and the Principle of Plenitude," *Modern Schoolman,* 1976, 53: 141–50.

"Pierre Gassendi: A Forgotten but Important Man in the History of Physics," *American Journal of Physics,* 1978, 46: 435–63.

Pav, Peter, "Gassendi's Statement of the Principle of Inertia," *Isis,* 1966, 57: 24–34.

Partington, J. R., "The Origins of the Atomic Theory," *Annals of Science,* 1939, 4: 245–82.

Pegis, A. C., *St. Thomas and the Problem of the Soul in the Thirteenth Century* (Toronto: Institute of Medieval Studies, 1934).

Pelikan, Jaroslav, *The Christian Tradition: A History of the Development of Doctrine,* 5 vols. (Chicago: University of Chicago Press, 1978).

Pessel, André, "Mersenne, la pesanteur et Descartes," in *Le discours et sa méthode,* edited by Nicolas Grimaldi and Jean-Luc Marion (Paris: Presses Universitaires de France, 1987).

Peters, R. S., "Hobbes, Thomas," in *The Encyclopedia of Philosophy,* edited by Paul Edwards, 8 vols. (New York: Macmillan and Free Press, 1967).

Pintard, René, "Descartes et Gassendi," in *Travaux du IXe Congrès de Philosophie – Congrès Descartes* (Paris: Études cartésiennes, 1937).

La Mothe le Vayer, Gassendi, Guy Patin: Études de bibliographie et de critique suivies de textes inédits de Guy Patin (Paris: Boivin, 1943).

Le libertinage érudit dans la première moitié du XVIIe siècle, 2 vols. (Paris: Boivin, 1943).

"Modernisme, humanisme, libertinage: Petite suite sur le 'cas' Gassendi," *Revue d'histoire littéraire de la France,* 1948, 48: 1–51.

Popkin, Richard H., "Father Mersenne's War Against Pyrrhonism," *Modern Schoolman,* 1957, 34: 61–78.

The History of Scepticism from Erasmus to Spinoza, revised and expanded edition (Berkeley: University of California Press, 1979; first published 1960).

"Gassendi, Pierre," in *The Encyclopedia of Philosophy*, edited by Paul Edwards, 8 vols. (New York: Macmillan and Free Press, 1967).

"Scepticism, Theology, and the Scientific Revolution in the Seventeenth Century," in *Problems in the Philosophy of Science*, edited by I. Lakatos and A. Musgrave (Amsterdam: North-Holland, 1968).

"Epicureanism and Scepticism in the Early 17th Century," in *Philomathes: Studies and Essays in the Humanities in Memory of Philip Merlan*, edited by R. B. Palmer and R. Hamerton-Kelly (The Hague: Martinus Nijhoff, 1971).

"The Religious Background of Seventeenth-Century Philosophy," *Journal of the History of Philosophy*, 1987, 25: 35–50.

Redondi, Pietro, *Galileo Heretic*, translated by Raymond Rosenthal (Princeton, N.J.: Princeton University Press, 1987).

"Theology and Epistemology in the Scientific Revolution," in *Revolutions in Science: Their Meaning and Relevance*, edited by William R. Shea (Canton, Mass.: Science History, 1988).

Reinbold, Anne, editor, *Peiresc ou la passion de connaître: Colloque de Carpentras, novembre 1987* (Paris: J. Vrin, 1990).

Remsberg, Robert Gotwald, *Wisdom and Science at Port-Royal and the Oratory: A Study of Contrasting Augustinianisms* (Yellow Springs, Ohio: Antioch Press, 1940).

Reymond, A., "Le problème cartésien de vérités éternelles et la situation présente," *Études philosophiques*, 1953, n.s. 8: 155–70.

Reynolds, L. D., editor, *Texts and Transmission: A Survey of the Latin Classics* (Oxford University Press, 1983).

Rist, John M., *Stoic Philosophy* (Cambridge University Press, 1969).

Epicurus: An Introduction (Cambridge University Press, 1972).

"The Stoic Conception of Fate," in *Essays in Ancient Greek Philosophy*, vol. 2, edited by John P. Anton and Anthony Preus (Albany: State University of New York Press, 1983).

Rist, John M., editor, *The Stoics* (Berkeley: University of California Press, 1978).

Rochemonteix, P. Camille de, *Un collège des Jésuites aux XIIe et XVIIIe siècles: Le Collège Henri IV de La Flèche*, 4 vols. (Le Mans: Leguicheux, 1889).

Rochot, Bernard, "Gassendi: Sa place dans la pensée philosophique du XVIIe siècle," *Revue de synthèse: Synthèse historique*, 1940–45, 27–45.

Les travaux de Gassendi sur Épicure et sur l'atomisme, 1619–1658 (Paris: J. Vrin, 1944).

"Le cas Gassendi," *Revue d'histoire littéraire de la France*, 1947, 47 (4): 289–313.

"Les vérités éternelles dans la querelle entre Descartes et Gassendi," *Revue philosophique de la France et de l'étranger*, 1951, 141: 288–98.

"Gassendi et la 'logique' de Descartes," *Revue philosophique de la France et de l'étranger*, 1955, 80: 300–8.

"Comment Gassendi interprétait l'expérience du Puy de Dôme," *Revue d'histoire des sciences*, 1963, 16: 53–76.

"Gassendi et l'expérience," in *Mélanges Alexandre Koyré*, edited by I. Bernard Cohen and René Taton (Paris: Hermann, 1965).

"Gassendi (Gassend), Pierre," in *Dictionary of Scientific Biography*, edited by Charles Coulton Gillispie, 16 vols. (New York: Scribner, 1972), vol. 5, pp. 284–90.

Rodis-Lewis, Geneviève, *L'oeuvre de Descartes*, 2 vols. (Paris: J. Vrin, 1971).

"Polémique sur la création des possibles et sur l'impossible dans l'école carté-sienne," *Studia Cartesiana* 2 (Amsterdam: Quadratures, 1981), pp. 105–23.

"Descartes' Life and the Development of His Philosophy," in *The Cambridge Companion to Descartes*, edited by John Cottingham (Cambridge University Press, 1992).

Rodis-Lewis, Geneviève, editor, *La science chez Descartes: Études en français* (New York: Garland, 1987).

Ronchi, Vasco, *The Nature of Light: An Historical Survey*, translated by V. Barocas (Cambridge, Mass.: Harvard University Press, 1970).

Rorty, Richard, *Philosophy and the Mirror of Nature* (Princeton, N.J.: Princeton University Press, 1979).

Rorty, Richard, J. B. Schneewind, and Quentin Skinner, editors, *Philosophy In History* (Cambridge University Press, 1984).

Rosenfield, Leonora Cohen, *From Beast-Machine to Man-Machine: Animal Soul in French Letters from Descartes to La Mettrie* (New York: Octagon, 1968).

Ross, W. D., *Aristotle: A Complete Exposition of His Works and Thought*, 5th edition (New York: Meridian, 1959; first published 1923).

Royse, James R., "Nominalism and Divine Power in the Chester Cycle," *Journal of the History of Ideas*, 1979, 40: 475–6.

Rubidge, Bradley, "Descartes's *Meditations* and Devotional Meditations," *Journal of the History of Ideas*, 1990, 51: 27–49.

Salmon, J. H. M., "Stoicism and Roman Example: Seneca and Tacitus in Jacobean England," *Journal of the History of Ideas*, 1989, 50: 199–227.

Sambursky, S., *The Physics of the Stoics* (New York: Macmillan, 1959).

Santinello, G., editor, *Storia delle storie generali della filosofia*. Vol. I, *Dalle origini rinscimentali alla "historia philosophica,"* edited by Francesco Bottin, Luciano Malusa, Giuseppe Micheli, Giovanni Santinelllo, and Ilario Tolomio (Brescia: Editrice La Scuola, 1981); Vol. II, *Dall'età cartesiana a Brucker,* edited by Francesco Bottin, Morio Longo, and Gregorio Piaia (Brescia: Editrice La Scuola, 1979).

Sarasohn, Louise [Lisa] Tunick, "The Influence of Epicurean Philosophy on Seventeenth Century Ethical and Political Thought: The Moral Philosophy of Pierre Gassendi," Ph.D. dissertation, University of California, Los Angeles, 1979.

"The Ethical and Political Philosophy of Pierre Gassendi," *Journal of the History of Philosophy*, 1982, 20: 239–60.

"Motion and Morality: Pierre Gassendi, Thomas Hobbes, and the Mechanical World-View," *Journal of the History of Ideas*, 1985, 46: 363–80.

"French Reaction to the Condemnation of Galileo, 1632–1642," *Catholic Historical Review*, 1988, 74: 34–54.

"Epicureanism and the Creation of a Privatist Ethic in Early Seventeenth-Century France," in *Atoms, Pneuma, and Tranquillity: Epicurean and Stoic*

Themes in European Thought, edited by Margaret J. Osler (Cambridge University Press, 1991).

"Nicolas-Claude Fabri de Peiresc and the Patronage of the New Science in the Seventeenth Century," *Isis,* 1993, 84: 70–90.

Freedom in a Deterministic Universe: Gassendi's Ethical Philosophy (Ithaca, N.Y.: Cornell University Press, forthcoming).

Saunders, Jason Lewis, *Justus Lipsius: The Philosophy of Renaissance Stoicism* (New York: Liberal Arts Press, 1955).

Schaffer, Simon, "Godly Men and Mechanical Philosophers: Souls and Spirits in Restoration Natural Philosophy," *Science in Context,* 1987, 1: 55–86.

Schmitt, Charles B., *Gianfrancesco Pico della Mirandola (1469–1533) and His Critique of Aristotle* (The Hague: Martinus Nijhoff, 1967).

Cicero Scepticus: A Study of the Influence of the Academica *in the Renaissance* (The Hague: Martinus Nijhoff, 1972).

Aristotle and the Renaissance (Cambridge, Mass.: Harvard University Press, 1983).

Schmitt, Charles B., Quentin Skinner, and Eckhard Kessler, editors, *The Cambridge History of Renaissance Philosophy* (Cambridge University Press, 1988).

Schuster, John A., "Descartes and the Scientific Revolution, 1618–1634: An Interpretation," Ph.D. dissertation, Princeton University, 1977.

"Descartes' *Mathesis Universalis,* 1619–1628," in *Descartes: Philosophy, Mathematics, and Physics,* edited by Stephen Gaukroger (Brighton: Harvester, 1980).

"Cartesian Method as Mythic Speech: A Diachronic and Structural Analysis," in *The Politics and Rhetoric of Scientific Method: Historical Studies,* edited by John A. Schuster and Richard R. Yeo (Dordrecht: Reidel, 1986).

"Whatever Should We Do with Cartesian Method? Reclaiming Descartes for the History of Science," in *Essays on the Philosophy and Science of René Descartes,* edited by Stephen Voss (New York: Oxford University Press, 1993).

Scott, J. F., *The Scientific Work of René Descartes,* (London: Taylor & Francis, 1952).

Scott, T. K., "Ockham on Evidence, Necessity, and Intuition," *Journal of the History of Philosophy,* 1969, 7: 27–49.

Sebba, Gregor, *Bibliographia Cartesiana: A Critical Guide to Descartes Literature, 1800–1960* (The Hague: Martinus Nijhoff, 1964).

Shanahan, Timothy, "God and Nature in the Thought of Robert Boyle," *Journal of the History of Philosophy,* 1988, 26: 547–69.

Shapin, Steven, "Robert Boyle and Mathematics: Reality, Representation, and Experimental Practice," *Science in Context,* 1988, 2: 23–58.

Shapin, Steven, and Simon Schaffer, *Leviathan and the Air-Pump: Hobbes, Boyle, and the Experimental Life* (Princeton, N.J.: Princeton University Press, 1985).

Shapiro, Barbara J., *Probability and Certainty in Seventeenth-century England: A*

Study of the Relationships Between Natural Science, Religion, History, Law, and Literature (Princeton, N.J.: Princeton University Press, 1983).

Sharp, Lindsay, "Walter Charleton's Early Life, 1620–1659, and the Relationship to Natural Philosophy in Seventeenth-century England," *Annals of Science,* 1973, 30: 311–40.

Shea, William R., *The Magic of Numbers and Motion: René Descartes' Scientific Career* (Canton, Mass.: Science History Publications, 1991).

Simon, Gérard, "Les vérités éternelles de Descartes: Évidences ontologiques," *Studia Cartesiana,* 2 (Amsterdam: Quadratures, 1981), pp. 124–36.

Sleigh, R. C., Jr., *Leibniz and Arnauld: A Commentary on Their Correspondence* (New Haven, Conn.: Yale University Press, 1990).

Snelders, H. A. M., "Christiaan Huygens and the Concept of Matter," in *Studies on Christiaan Huygens: Invited Papers from the Symposium on the Life and Work of Christiaan Huygens, Amsterdam, 22–25 August 1979,* edited by H. J. M. Bos, M. J. S. Rudwick, H. A. M. Snelders, and R. P. W. Visser (Lisse: Swets & Zeitlinger, 1980).

Solmsen, Friedrich, *Aristotle's System of the Physical World* (Ithaca, N.Y.: Cornell University Press, 1960).

Sorabji, Richard, *Necessity, Cause, and Blame: Perspectives on Aristotle's Theory* (Ithaca, N.Y.: Cornell University Press, 1980).

Time, Creation, and the Continuum: Theories in Antiquity and the Early Middle Ages (Ithaca, N.Y.: Cornell University Press, 1983).

Matter, Space, and Motion: Theories in Antiquity and Their Sequel (Ithaca, N.Y.: Cornell University Press, 1988).

Sortais, Gaston, *La philosophie moderne depuis Bacon jusqu'à Leibniz,* 2 vols. (Paris: Paul Lethielleux, 1920–2).

Spink, J. S., *French Free-Thought from Gassendi to Voltaire* (London: Athlone, 1960).

Spragens, Thomas A., Jr., *The Politics of Motion: The World of Thomas Hobbes* (Lexington: University Press of Kentucky, 1973).

Stones, G. B., "The Atomic View of Matter in the Fifteenth, Sixteenth, and Seventeenth Centuries," *Isis,* 1928, 10: 445–65.

Sutton, John, "Religion and the Failures of Determinism," in *The Uses of Antiquity: The Scientific Revolution and the Classical Tradition,* edited by Stephen Gaukroger (Dordrecht: Kluwer, 1991).

Taliaferro, R. Catesby, *The Concept of Matter in Descartes and Leibniz* (Notre Dame, Ind.: Notre Dame University Press, 1964).

Taxil, Nicolas, *Oraison funèbre de P. Gassendi,* edited by M. Tamizey de Larroque (Bordeaux, 1882).

Thirion, Maurice, "La philosophie de Gassendi," *Revue de sythèse,* 1974, 95: 257–70.

Thomas, P.-Félix, *La philosophie de Gassendi* (Paris, 1889; reprinted, New York: Burt Franklin, 1967).

"Descartes et Gassendi," *Annales de philosophie chrétienne,* Sept. 1889, 548–85.

Thorndike, Lynn, *A History of Magic and Experimental Science,* 8 vols. (New York: Columbia University Press, 1923–58).

Tillman, Alexandre, "L'itinéraire du jeune Descartes," Ph.D. dissertation, University of Paris IV, 1975.

Todd, Margo, "Providence, Chance and the New Science in Early Stuart Cambridge," *Historical Journal,* 1986, *29:* 697–711.

Torrance, Thomas F., "The Influence of Reformed Theology on the Development of Scientific Method," in *Theology in Reconstruction,* edited by Thomas F. Torrance (Grand Rapids, Mich.: Eerdmans, 1965).

Trevor-Roper, Hugh, "Nicholas Hill, The English Atomist," in *Catholics, Anglicans and Puritans: Seventeenth Century Essays,* edited by Hugh Trevor-Roper (London: Secker & Warburg, 1987).

Turbayne, Colin Murray, *The Myth of Metaphor* (New Haven, Conn.: Yale University Press, 1962).

Urvoy, Dominique, *Ibn Rushd (Averroes),* translated by Olivia Stewart (London: Routledge, 1991).

van Berkel, "Beeckman, Descartes et 'la philosophie physico-mathématique,'" *Archives de philosophie,* 1983, *46:* 620–6.

van de Pitte, Frederick P., "Some of Descartes' Debts to Eustachius a Sancto Paulo," *Monist,* 1988, *71:* 487–97.

van Gelder, H. A. Enno, *The Two Reformations in the Sixteenth Century: A Study of the Religious Aspects and Consequences of the Renaissance and Humanism* (The Hague: Martinus Nijhoff, 1964).

van Kley, Dale K., *The Jansenists and the Expulsion of the Jesuits from France: 1757–1765* (New Haven, Conn.: Yale University Press, 1975).

van Leeuwen, Henry, *The Problem of Certainty in English Thought, 1630–1680* (The Hague: Martinus Nijhoff, 1963).

Van Melsen, A. G. M., *From Atomos to Atom: The History of the Concept Atom,* translated by Henry J. Koren (Pittsburgh, Pa.: Duquesne University Press, 1952).

van Ruler, J. A., "New Philosophy to Old Standards: Voetius' Vindication of Divine Concurrence and Secondary Causality," *Nederlands archeif voor kerkgeschiedenis/Dutch Review of Church History,* 1991, *71:* 58–91.

Vendler, Zeno, "Descartes' Exercises," *Canadian Journal of Philosophy,* 1989, *19:* 193–224.

Vignaux, Paul, *Nominalisme au XIVe siècle* (Montréal: Institut d'études médiévales, and Paris: J. Vrin, 1948).

von Leyden, W., *Seventeenth-century Metaphysics: An Examination of Some Main Concepts and Theories* (New York: Barnes & Noble, 1968).

Vrooman, Jack R., *René Descartes: A Biography* (New York: Putnam, 1970).

Wade, Ira O., *The Intellectual Origins of the French Enlightenment* (Princeton, N.J.: Princeton University Press, 1971).

Walker, Ralph, "Gassendi and Skepticism," in *The Sceptical Tradition,* edited by Myles Burnyeat (Berkeley: University of California Press, 1983).

Wallace, William A., *Causality and Scientific Explanation,* 2 vols. (Ann Arbor: University of Michigan Press, 1972).

"Galileo and Reasoning *Ex suppositione*," in *Prelude to Galileo: Essays on Medieval and Sixteenth-Century Sources of Galileo's Thought*, edited by William A. Wallace (Dordrecht: Reidel, 1981).

Ward, Benedicta, *Miracles and the Medieval Mind* (Philadelphia: University of Pennsylvania Press, 1987).

Watson, Richard A., *The Downfall of Cartesianism, 1673–1712: A Study of Epistemological Issues in Late 17th Century Cartesianism* (The Hague: Martinus Nijhoff, 1966).

"Transubstantiation Among the Cartesians," in Thomas M. Lennon, John M. Nicholas, and John W. Davis, *Problems of Cartesianism* (Kingston and Montreal: McGill-Queen's University Press, 1982).

Weinberg, Julius R., *A Short History of Medieval Philosophy* (Princeton, N.J.: Princeton University Press, 1964).

Weinstein, Donald, "In Whose Image and Likeness? Interpretations of Renaissance Humanism," *Journal of the History of Ideas*, 1972, 33: 165–76.

Wells, Norman J., "Descartes and the Scholastics Briefly Revisited," *New Scholasticism*, 1961, 35: 172–90.

"Objective Being: Descartes and His Sources," *Modern Schoolman*, Nov. 1967, 45: 49–61.

"Old Bottles and New Wine: A Rejoinder to J. C. Doig," *New Scholasticism*, 1979, 53: 515–23.

"Suarez on the Eternal Truths," *Modern Schoolman*, 1981, 58: 73–104, 159–74.

"Descartes' Uncreated Eternal Truths," *New Scholasticism*, 1982, 56: 185–99.

"Objective Reality of Ideas in Descartes, Caterus, and Suárez," *Journal of the History of Philosophy*, 1990, 28: 33–61.

Wessely, Anna, "Transposing 'Style' from the History of Art to the History of Science," *Science in Context*, 1991, 4: 265–78.

Westfall, Richard S., *Science and Religion in Seventeenth-Century England* (New Haven, Conn.: Yale University Press, 1958).

"The Foundation of Newton's Philosophy of Nature," *British Journal for the History of Science*, 1962, 1: 171–82.

The Construction of Modern Science: Mechanisms and Mechanics (New York: Wiley, 1971).

Force in Newton's Physics: The Science of Dynamics in the Seventeenth Century (New York: American Elsevier, 1971).

Never at Rest: A Biography of Isaac Newton (Cambridge University Press, 1980).

Westman, Robert S., "The Melancthon Circle, Rheticus, and the Wittenberg Interpretation of the Copernican Theory," *Isis*, 1975, 66: 165–93.

"Three Responses to the Copernican Theory: Johannes Praetorius, Tycho Brahe, and Michael Maestlin, in *The Copernican Achievement*, edited by Robert S. Westman (Berkeley: University of California Press, 1975).

"Huygens and the Problem of Cartesianism," in *Studies on Christiaan Huygens: Invited Papers from the Symposium on the Life and Work of Christiaan Huygens, Amsterdam, 22–25 August 1979*, edited by H. J. M. Bos, M. J. S.

Rudwick, H. A. M. Snelders, and R. P. W. Visser (Lisse: Swets & Zeitlinger, 1980).

Whyte, Lancelot Law, *Essay on Atomism: From Democritus to 1600* (Middletown, Conn.: Wesleyan University Press, 1961).

Wildiers, N. Max, *The Theologian and His Universe: Theology and Cosmology from the Middle Ages to the Present* (New York: Seabury, 1982).

Williams, Bernard, *Descartes: The Project of Pure Inquiry* (Harmondsworth: Penguin Books, 1978).

Wilson, Catherine, "Visual Surface and Visual Symbol: The Microscope and the Occult in Early Modern Science," *Journal of the History of Ideas*, 1988, 49: 85–108.

Wilson, Margaret Dauler, *Descartes* (London: Routledge & Kegan Paul, 1978).
"Superadded Properties: The Limits of Mechanism in Locke," *American Philosophical Quarterly*, 1979, 16: 143–50.

Wippel, John F., "The Condemnations of 1270 and 1277 at Paris," *Journal of Medieval and Renaissance Studies*, 1977, 7: 169–201.
"Divine Knowledge, Divine Power, and Human Freedom in Thomas Aquinas and Henry of Ghent," in *Divine Omniscience and Omnipotence in Medieval Philosophy*, edited by Tamar Rudavsky (Dordrecht: Reidel, 1985).

Wisan, W. L., "Galileo and the Emergence of A New Scientific Style," in *Pisa Conference Proceedings*, edited by J. Hintikka, D. Gruender, and E. Agazzi (Dordrecht: Reidel, 1980).

Wojcik, Jan W., "Robert Boyle and the Limits of Reason: A Study in the Relationship Between Science and Religion in Seventeenth-Century England," Ph.D. dissertation, University of Kentucky, Lexington, 1992.

Yates, Frances, *Giordano Bruno and the Hermetic Tradition* (New York: Vintage, 1969; first published 1964).

Yoder, Joella G., *Unrolling Time: Christiaan Huygens and the Mathematization of Nature* (Cambridge University Press, 1988).

Index

Abelard, Peter, 18
abortion, Gassendi on, 67
abstraction
    Aquinas on, 25–6
    Gassendi on, 114
    Ockham on, 32
active principles, 214n
Adams, Marilyn McCord, 21n, 26n,
    29n, 30n, 32n, 33n, 91n, 136n
Alberti, Antonina, 36n, 108n
Ames, William, 93n, 126n
Amicus, Bartholomeus, 184n
angels, 98
annihilation, imaginary, 183–4
    Newton on, 184n
anti-essentialism
    in Galileo, 216n
    in Gassendi, 9, 11, 53, 102, 110,
    113, 115, 117, 158–9, 163–5
Aphrodisias, Alexander of, 62
Aquinas, Thomas
    Christian Aristotelianism of, 20, 45
    and Descartes, 128, 128n, 129n,
    133–4
    on epistemology, 25–7
    on essences, 22, 25, 55
    on forms, 25–6
    on God's freedom, 134
    on God's intellect, 22
    on God's power, 21–5, 21n, 54–5,
    134, 147
    on God's will, 22–3
    intellectualism of, 19, 20–7, 55,
    133–4
    in Jesuit curriculum, 123–4

on necessity, 22–5, 22n
on the soul, 70n
on universals, 22, 26–7, 114n, 115
Ariew, Roger, 122n, 124n
Aristotelianism
    decline of, 3, 4n
    in Descartes, 122, 123, 211–12
    essentialism of, 3, 4, 25
    on form, 172–4
    on matter, 172–4
    in Mersenne, 122
    rejection of, 38–9, 104, 205, 214–
    16
    in seventeenth-century, 3n, 4n
    on the soul, 179
    theological problems of, 20
Aristotle
    on certainty as goal of science, 110
    on chance, 86
    on final causes, 162
    on fortune, 86–7
    on future contingents, 90
    in Jesuit curriculum, 123–4
    on prime matter, 173
    on scientific method, 173
    on the soul, 64
    translation of works into Latin, 3,
    16
    on universals, 116
Armogathe, J.-R., 119n, 123n
Armstrong, D. M., 113n
Arnauld, Antoine, 125, 125n
Arthur, Richard T. W., 149n
Asmis, Elizabeth, 108n
astrology

Gassendi on, 55, 99–100
Hermetic revival of, 82
Mersenne on, 97
objections to, 82
Pomponazzi on, 82–3, 96
atheism, 58n
atomism, 38, 176, 181
Eucharist explained by, 5n
Gassendi on, 9, 180, 188–94
theological problems of, 41
atoms
activity of, 188n, 191, 192
Gassendi on, 182, 189–90, 192–4
swerve of (*clinamen*), 91–2, 91n, 188n
Augustine, Saint, 81–2
Augustinianism, 16, 139n
in Bérulle, 129
in Descartes, 125
Austin, William H., 225n
Auzout, Adrien, 183
Averroës (Ibn Rushd), 20, 21, 27, 28, 61
Averroism, in thirteenth-century Paris, 27, 28

Bacon, Francis, 110, 110n
Barker, Peter, 42n
barometric experiments, 186–8
Barry, Sister M. Inviolata, 21n
Basso, Sebastian, 180
Beeckman, Isaac, 7, 37, 119
and Descartes, 118
and Gassendi, 40
on the mechanical philosophy, 5, 5n, 6
Belenky, Mary Field, 224n
Bernier, François, 47n, 80n
Berr, Henri, 40n, 106n
Bérulle, Pierre
influence on Descartes, 124–6, 129
bête-machine doctrine, in Descartes, 220n
Beyssade, Jean-Marie, 121n
Blackwell, Richard J., 213n
Bloch, Olivier René, 36n, 44n, 45n,

47n, 63n–64n, 65n, 96n, 102, 102n, 106n, 113n, 155, 155n, 157n, 191n–192n, 192, 192n
on Gassendi as materialist, 43–4, 60n, 192
on Gassendi's theology, 46, 47
Boas, Marie, 41n, 181n
Bobik, Joseph, 172n
Boethius, 82, 165
Boutroux, Émile, 121n
Boylan, Michael, 225n
Boyle, Robert, 150n, 174n, 224, 224n
on atomism, 76
criticism of Descartes, 149–51
on final causes, 162n–163n
on God's omnipotence, 147
influenced by Gassendi, 229
on mathematics, 230
on matter, 174–5
on the mechanical philosophy, 9–10, 174–5
on providence, 59
on revelation, 160n
on scientific method, 227–30
on the void, 225, 228
voluntarism of, 227, 229
Bramhall, John, 89
Brandt, Fritiof, 37n, 71n
Bréhier, Émile, 121n
Brockliss, L. W. B., 48n, 124n
Brown, Gregory, 206n–207n
Brundell, Barry, 36n, 37n, 40, 40n, 45n, 46n, 49n, 50n, 51n, 182n, 183n, 194n
Bruno, Giordano, 181
Buckley, Michael J., 58n
Burns, Norman T., 60n
Busson, Henri, 46, 46n, 129n

Cajetan, Thomas, 128, 129
Calvin, John
on free will, 83–4
on predestination, 83–4, 88, 93
Campanella, Tommaso, 65n
Capreolus, John, 128
Cariou, Marie, 36n

Carlton, Thomas Compton, 119
causality, 81, 96, 137*n*, 138*n*, 171,
  172, 193–4
causes, final
  Aristotle on, 162
  Boyle on, 162*n*–163*n*
  Descartes on, 161–2, 212–13, 213*n*
  Gassendi on, 49, 161–2, 193
  Newton on, 163*n*
Cavendish Circle, 7*n*
Cavendish, William, 119
  and the mechanical philosophy, 7
Céard, Jean, 96*n*
Centore, F. F., 162*n*
certainty
  degrees of, 107, 107*n*, 218
  goal of science, 110, 120
chance, 15
  Aristotle on, 86–7
  in Epicureanism, 51–2, 81
  Gassendi on, 52, 87–8, 92–6
Charleton, Walter, 6, 7, 41*n*, 184, 195
Charron, Pierre, 103
Chrysippus, on fate, 88
Cicero, Marcus Tullius, 15*n*, 74, 91*n*,
  106*n*
Cioffari, Vincenzo, 92*n*
Clark, Stuart, 96*n*
Clarke, Desmond, 140*n*
Clarke, Samuel, 232
Clericuzio, Antonio, 228*n*
Clerselier, Claude, 125
clinamen (swerve of atoms), 91–2,
  91*n*, 188*n*
Clinchy, Blythe McVicker, 224*n*
Cochois, Paul, 125*n*
Coimbrian commentators, 21*n*, 174*n*
Coirault, Gaston, 111*n*
Colish, Marcia L., 81*n*
Collier, Raymond, 46*n*
Collins, James, 140*n*
conceptualism, 20, 113
Condemnations of 1277, 27, 28
  possible influence on Descartes, 126
contingency, 1, 2, 11, 235–6
  Aquinas on, 26

Descartes on, 140
  in evolutionary theory, 236
  Gassendi on, 53–7, 80–1, 87–8,
    92, 115
  Ockham on, 30
  voluntarists on, 19
contingents, future, *see* future
  contingents
Copenhaver, Brian P., 96*n*, 178*n*,
  199*n*
Copernicanism, 5
  realism of, 3
Copleston, Frederick C., 21*n*, 25*n*,
  26*n*, 28*n*, 74*n*, 114*n*, 129*n*
Cordemoy, Gérauld de, 125
Cottingham, John, 120*n*, 130*n*, 134*n*,
  220*n*
Council of Trent, 84
Courtenay, William J., 18*n*, 19*n*, 27*n*,
  29*n*, 31*n*, 34*n*
covenant, Ockham on, 33, 34
Craig, William Lane, 22*n*, 84*n*, 94*n*
Cranefield, Paul T., 108*n*
creation, Gassendi on, 53, 57
Cronin, T. J., 128*n*
Cunningham, Andrew, 10*n*
Curley, E. M., 120*n*, 121*n*

Dalbiez, R., 123*n*
Damian, Peter, 19, 19*n*
Damrosch, Leopold Jr., 8*n*, 229*n*
Dante Alighieri, 60–1, 61*n*
Darwin, Charles, on contingency, 235
Daston, Lorraine, 107*n*
Davis, Edward B., 126*n*, 226*n*
Dear, Peter, 118*n*, 122*n*, 124*n*, 128*n*,
  227*n*
Defarrari, Roy J., 21*n*
Degrood, David H., 113*n*, 116*n*
Democritus, 61, 89–90, 92
demons
  Gassendi on, 97–9
  mortality of, 73
  Pomponazzi on, 96
density and rarity
  in Descartes, 209–10

in Gassendi, 185
Descartes, René, 157, 207–8
 on animal spirits, 219–20, 220*n*
 and Aquinas, 128, 128*n*
 anti-Aristotelianism of, 203, 214–16
 and Beeckman, 118
 on *bête-machine* doctrine, 220*n*
 on certainty as goal for science, 110, 111*n*, 120
 on clear and distinct ideas, 136
 on contingency, 140
 on continuous creation, 148–9
 on the creation of mathematical truths, 129, 132
 denial of miracles, 147–8
 education and biography, 118–19, 123–4
 on elements, 210–12
 essentialism in, 1, 153, 157–9, 216
 on eternal truths, 1, 119–35, 163–7
 on the Eucharist, 119
 on experiment, 140, 140*n*, 142
 on final causes, 161–2, 212–13, 213*n*
 on force, 214*n*
 and Galileo, 122, 127, 127*n*, 207–8
 on God's freedom, 1, 134, 135
 on God's existence, 159–62
 on God's immutability, 134–5, 138–9
 on God's power, 129, 134, 143–4, 145, 149
 on God's unity, 133, 146
 on impact, 138, 213, 213*n*–214*n*
 on inertia, 138
 on innate ideas, 130, 158
 intellectualism of, 11, 120–1, 130–1, 134, 143–4, 146–52, 165, 167, 225–6
 and Kepler, 127, 127*n*
 on the laws of nature, 137–40
 Le monde, suppression of, 122*n*
 on matter, 9, 11, 176, 201–21

on mechanical models, 141–2, 218
 on the mechanical philosophy, 5, 8, 201–21
 on mechanics, 138, 139*n*, 213, 213*n*–214*n*
 on method, 140–6, 143*n*, 218
 on method *ex suppositione*, 144
 on mind–body problem, 70–1, 219–20
 on natural light, 139, 139*n*
 on necessity, 134
 on necessity of mathematical truths, 126–7, 132
 on primary and secondary qualities, 215
 Principia philosophiae, placed on Index, 119
 on qualities, 195
 rationalism of, 11, 137–46, 147–50, 152, 157–8, 205
 religious beliefs of, 7, 122
 on scepticism, 106, 120, 156
 and Scholasticism, 123–4, 127
 search for new foundations, 119–20, 135, 203
 on sense perception, 156
 on the soul, 66*n*, 220–1
 on space and place, 207
 on substance, 159
 theory of matter, 204–19
 on universals, 163
 on the void, 183, 186, 202, 208–10, 224*n*
Desharnais, Richard P., 18*n*
design, argument from
 in Gassendi, 51–2, 51*n*, 56–8, 161
 Newton on, 59, 163
determinism, 2, 90, 92, 234–5
 Calvin on, 83–4, 93
 in Democritus, 89–90
 Dominicans on, 93
 Gassendi on, 88
 in Hobbes, 229–30
 in Laplace, 234–5
 in Pomponazzi, 82–3
 in Stoicism, 81

Digby, Kenelm, 6, 7, 41*n*, 60*n*, 72, 72*n*, 195
Dijksterhuis, E. J., 5*n*, 40*n*–41*n*, 45*n*, 187*n*
Diogenes Laertius, 43
divination
    Gassendi on, 96–101
    in Stoicism, 96, 99, 99*n*
Dobbs, Betty Jo Teeter, 42*n*, 59*n*, 151, 151*n*, 225*n*, 232*n*
Dod, Bernard G., 16*n*
Doig, James C., 128*n*
Dominicans, on predestination, 84–5, 88, 93, 133
Doyle, John P., 128*n*
Drake, Stillman, 37*n*
Dugas, René, 187*n*
Duhem, Pierre, 3*n*, 126*n*
Duns Scotus, 30*n*, 85*n*
    on uncreated essences, 128
DuPuy, Jacques, 119
DuPuy, Pierre, 119

elements, in Descartes, 210–12
Elzinga, Aant, 224*n*, 231*n*
Emerton, Norma E., 172*n*, 174*n*, 196*n*
empiricism
    Gassendi on, 102–17
    and voluntarism, 20, 102
Empiricism, British
    origins of in Gassendi, 111
Epicurus, 89–90
    atomism of, 38
    on *clinamen* (swerve of atoms), 91–2, 91*n*, 188*n*
    on ethics, 39, 104
    on free will, 91–2
    on future contingents, 90–1, 91*n*, 97
    Gassendi's restoration of, 9, 36, 39–45, 50, 87, 91–2, 102, 104–5, 188*n*
    heterodoxy of, 38, 45
    materialism of, 38, 58
    on the soul, 59–60, 63

religion of, 38, 58
on the void, 183, 184–5
Erasmus, Desiderius, on free will, 83
essences, 22
    Aquinas on, 25
    Aristotelian, 22
    Descartes on, 216
    Gassendi on, 53, 113, 117, 163–5
    uncreated, 18
essentialism
    in Aquinas, 25
    in Aristotelianism, 3, 4, 25
    in Descartes, 163, 216
    in Duns Scotus, 128
eternity, 165–6
ethics, Epicurean, 39, 104
    Gassendi on, 39, 42, 104
Eucharist, 173
    atomistic explanation of, 5*n*
    Descartes on, 119
    Paschasius Radbertus on, 16
Euclid, 167
Eustachius a Sancto Paulo, 159*n*
evil, problem of
    Epicurus on, 58
    Gassendi on, 58, 75
evolutionary theory, contingency in, 236
exorcism, Gassendi on, 98
experiment
    Descartes on, 140, 140*n*, 142
    Hobbes on, 228
explanation
    Aristotelian, 115
    Descartes on, 140–1, 218
    Gassendi on, 115

fate, 15, 81, 88
    Gassendi on, 92–6
    in Stoicism, 81, 81*n*, 88
Feldhay, Rivka, 85*n*, 133*n*, 145*n*
Ferguson, Wallace K., 39*n*
Ferrier, Francis, 133*n*
Ficino, Marsilio, 82
force, concept of, in Descartes, 214*n*
Force, James E., 161*n*

forms, 172
  Aquinas on, 25–6
  Gassendi on, 114, 196*n*
  substantial, 6, 173–4, 215–16
fortune, 15, 86–8, 92–6
Foster, M. B., 17*n*, 19*n*
Foster, Michael, 6*n*, 7*n*
Frankfurt, Harry, 121*n*, 132*n*, 146
Franklin, James, 107*n*
Freddoso, Alfred J., 83*n*, 137*n*
free will, 8, 80–1
  Calvin on, 83, 88
  Epicurus on, 91–2
  Erasmus on, 83
  Gassendi on, 56, 80–1, 85–6, 92–7
  Luther on, 83
  Molina on, 86*n*
Funkenstein, Amos, 11*n*, 121*n*, 123*n*,
  160*n*
Furley, David, 38*n*
future contingents
  Aristotle on, 90
  Epicurus on, 90–1, 91*n*, 97
  Gassendi on, 94–5, 97
  Ockham on, 91*n*

Gabbey, Alan, 138*n*, 175*n*, 213*n*,
  217*n*, 224*n*, 226*n*, 232*n*
Galileo Galilei, 37, 207
  anti-essentialism of, 216*n*
  on barometric experiments, 186
  and Descartes, 122, 127, 127*n*,
    207–8
  Gassendi on, 37–8
  on mathematical physics, 208*n*
  on mathematical truths, 127
  and the mechanical philosophy, 5, 7
  on primary and secondary qualities,
    196*n*, 215, 215*n*
  reactions to trial of, 122, 122*n*
Garber, Daniel, 9*n*, 66*n*, 119*n*, 122*n*,
  123*n*, 126*n*, 137*n*, 138*n*, 140*n*,
  142*n*, 143*n*, 148*n*, 149*n*, 172*n*,
  173*n*, 202*n*, 203*n*, 207*n*, 212*n*,
  213*n*, 218*n*

Garin, Pierre, 121*n*, 124*n*
Gassendi, Pierre
  on the animal soul, 64, 64*n*, 65, 66,
    220
  anti-Aristotelianism of, 38, 39, 49*n*,
    103–5
  anti-essentialism of, 9, 11, 53, 102,
    110, 113, 115, 117, 153, 158–9,
    163–5
  argument from design in, 51–2,
    51*n*, 56–8, 161
  on astrology, 55, 99–100
  atomism of, 9, 49, 91*n*, 180–200
  on barometric experiments, 186–8
  and Beeckman, 40
  on causality, 49, 49*n*, 54, 161–2,
    182, 193–4
  on chance, 87–8, 92–6
  on the *clinamen* (swerve of atoms),
    91*n*
  on demons, 97–9
  on Descartes, 153–7
  empiricism of, 11, 50*n*, 51, 102,
    105, 107–8, 111–13, 157–8,
    229
  Epicurean project, 9, 39–45, 50,
    87, 91–2, 102, 104–5, 188*n*
  epistemology, 75, 155, 157
  on ethics, 39, 42, 104
  on fate, 92–6
  on final causes, 49, 162, 193
  flos materiae, 64, 64*n*–65*n*
  on fortune, 86–8, 92–6
  on free will, 56, 80–1, 85–6, 92–7
  on Galileo, 37–8
  on God's existence, 50–1
  on God as first cause, 52, 192
  on God's intellect, 53–6
  on God's power, 1, 53–7, 69, 147,
    164
  on Hobbes, 92, 100–1
  humanism of, 38–9, 56, 86
  on immortality of the soul, 58, 59–
    77
  on innate ideas, 50*n*, 158, 160
  intellectual relations, 36–8

on mathematical truths, 94*n*, 163–7
on matter, 9, 11, 176, 182–200
on mechanical explanations, 199–200
on mind–body problem, 70–1
on miracles, 55–6
Molina in, 100, 184*n*
on necessity, 55, 80–1
nominalism of, 9, 56, 68, 102, 105, 113–17, 157–8
on the phantasy, 68
on Pomponazzi, 101
on predestination, 93–6
probabilism of, 110, 112
on providence, 54, 56–9, 73, 80–101
on Puy de Dôme experiment, 186, 186*n*
on qualities, 194–200
on the rational soul, 66–71
on the regularity of the world, 55–7, 73*n*
religious convictions of, 7, 45–7
on revelation, 160
on scepticism, 105, 109, 117, 197
science of appearances, 110, 116, 158–9
"science," redefinition of, 105–7, 110
on scientific method, 104, 111–13
on sense perception, 63, 194–200
on the soul, 62–77
on space, 182–3, 193*n*–194*n*
on Stoicism, 75, 89
his theory of signs, 111–13, 185
on time, 182
on transdiction, 185, 192–3
on universals, 163–5
on the void, 63, 182–8
voluntarism of, 11, 47, 48–56, 57, 63, 69, 85, 98, 102, 105, 115, 164–5, 167
on Zeno's paradoxes, 190, 194
Gaukroger, Stephen, 202*n*, 208*n*
Gibieuf, Guillaume, 133

Gilligan, Carol, 224*n*
Gilson, Étienne, 20*n*, 21*n*, 26*n*, 28*n*, 61*n*, 62*n*, 66*n*, 70*n*, 96*n*, 121*n*, 125, 125*n*, 128*n*, 131*n*, 133*n*, 159*n*, 174*n*
God
 as cause of motion, 191, 192, 212
 as first cause, 49, 52, 178, 192, 193, 212
 concurrence of, 138*n*
 creation by, 24
 existence of, proved by Descartes, 159–62
 freedom of, 1, 10, 19, 80, 129, 133–4, 135, 138
 intellect of, 10, 15, 53–6, 131
 omnipotence of, 1, 2, 15, 16, 31, 53–7, 69, 147, 164
 omniscience of, 15, 93–5
 ordinary concourse of, 137, 137*n*
 power of, 1, 10, 13–14, 15, 16, 18–19, 21–5, 21*n*, 28–9, 31, 34, 53–7, 69, 127, 129, 130, 134, 137, 143–4, 145, 147–9, 154, 164–5
 relationship to the creation, 10–11, 16–18, 57, 130, 146–8, 164
 unity of, 125, 131, 146
 will of, 23, 125, 131
 wisdom of, 57–8
God, immutability of
 in Aquinas, 22
 Boyle on, 150
 in Descartes, 131, 134, 135, 138–9
Goddu, André, 33*n*
Goldberger, Nancy Rule, 224*n*
Goldstein, Bernard R., 42*n*
Golinski, Jan V., 41*n*
Gouhier, Henri, 119*n*, 125, 125*n*, 126*n*, 129*n*
Gould, Stephen Jay, 235*n*, 236*n*
Grafton, Anthony, 4*n*
Grant, Edward, 3*n*, 27*n*, 28*n*, 56*n*, 126*n*, 176*n*, 183*n*, 184*n*, 185*n*
Greene, Robert A., 226*n*
Gregory, Tullio, 103*n*, 106*n*, 181*n*

Gregory of Rimini, 113
Grene, Marjorie, 155, 155n, 158n
Gueroult, Martial, 120n

Hacking, Ian, 107n, 223n
Hahn, Roger, 235n
Halbronn, Jacques E., 100n
Hall, A. Rupert, 224n
Hall, Marie Boas, 224n
Hariot, Thomas, 6n, 181
Harré, R., 171n
Hartlib, Samuel, 150, 150n
Harvey, William
    Descartes on, 5n
    Hobbes on, 6
Hatfield, Gary, 121n, 122n, 214n,
    221n
Henry, John, 178n, 183n, 214n,
    225n, 226n, 227n
Henry of Ghent, 128
Hermetic Corpus, 4, 82
Hero of Alexandria, 183, 185, 185n
Hevelius, Johannes, 37
Hill, Nicholas, 181
Hillerbrand, Hans J., 84n
Hine, William, 97n
Hobbes, Thomas, 5, 17, 37, 41n, 119,
    195, 224, 224n
    and Descartes, 153
    determinism of, 89-90, 229-30
    Gassendi on, 100-1
    on God, 8n, 229
    materialism of, 8, 71, 179
    on mathematics, 230-1
    on scientific method, 227-31
    on the void, 225, 228
Hooykaas, R., 5n, 7n, 27n
Huet, Pierre-Daniel, 125
humanism, Renaissance, 3, 4
    definition of, 38n-39n
    of Gassendi, 38-9, 56, 86
Hutchison, Keith, 77n, 176n, 177n,
    178n, 179n, 216n
Hutton, Sarah, 224n

Huygens, Christiaan, 119, 207, 224,
    224n
    on impact, 231
    on mathematics, 231-2
    on matter, 232n
    rationalism of, 232
Huygens, Constantijn, 119

ideas, innate
    Descartes on, 158
    Gassendi on, 50n, 158
    of God, 160
impact
    Descartes on, 138, 213, 213n-214n
    Huygens on, 231
inertia, 192n
    Descartes on, 138
intellectualism, 19n, 135
    in Abelard, 18
    in Aquinas, 19, 20-7, 55, 133-4
    definition of, 11, 17, 17n
    in Descartes, 120-1, 130, 134,
        143-4, 146-52, 225-6
    in More, 227
    and scientific method, 143-4
    in Suárez, 18
    varieties of, 18-19, 130
Isidore of Seville, 99n

Jacquot, Jean, 6n, 181n
Javellus, Chrysostomus, 128
Jeannel, Charles, 46n
Jedin, Hubert, 84n
Jenkins, Jane Elizabeth, 176n-177n,
    187n, 226n
Jerome, Saint, 16
Jesuits
    and education, 122-4
    on predestination, 84-5, 133
Job, 2
Jobe, Thomas Harmon, 179n
Jones, Howard, 36n, 40n, 42n, 45n
Joy, Lynn Sumida, 37n, 39n, 40, 40n,
    45n, 63n, 103n, 182n, 189n,
    190n

Kant, Immanuel, 17n
Kargon, Robert Hugh, 6n, 8n, 46n, 181n
Kepler, Johannes, influence on Descartes, 127, 127n
Kessler, Eckhard, 61n, 96n
Klaaren, Eugene M., 226n
Koyré, Alexandre, 6, 6n
Kretzmann, Norman, 91n
Kristeller, Paul Oskar, 38n, 96n
Kroll, Richard W. F., 77n, 111n
Kubbinga, H. H., 181n
Kuhn, Thomas S., 230n

La Forge, Louis de, 125
La Mothe le Vayer, François de, 96
Lamy, Bernard, 125
Laplace, Pierre Simon Marquis de, determinism in, 234–5
laws of nature, 16, 18
  contingency of, 20
  in Descartes, 137, 138–40, 213–14
  Gassendi on, 53
Leclerc, Ivor, 172n, 175n
Leff, Gordon, 27n, 32n
Leibniz–Clarke correspondence, 59, 154, 232
Lennon, Thomas M., 9n, 108n, 111n
Lenoble, Robert, 6n, 97n
Lewis, Charleton T., 50n, 166n
libertins érudits, 46, 154
Lipsius, Justus, 41, 41n
Lloyd, G. E. R., 116n, 173n
Locke, John, 108n
logic, 19, 23, 24, 29, 53
Lohr, Charles H., 16n, 62n, 124n
Long, A. A., 38n, 45n, 51n, 60n, 81n, 91n
Lovejoy, Arthur O., 15, 15n, 145n
Loyola, Ignatius, 124n
Lucretius, Titus Carus, 4, 191n
  Gassendi's knowledge of, 9, 39
  on the soul, 63n
  on the void, 183

Luther, Martin, 83
Lynes, John W., 210n

McColley, Grant M., 181n
McGuire, J. E., 9n, 225n
Machamer, Peter, 203n, 212, 212n
MacIntosh, J. J., 76n, 176n
McMullin, Ernan, 173n
Malebranche, Nicholas, 125
Mandelbaum, Maurice, 177n
Mandon, L., 46n
Marion, Jean-Luc, 120n, 121n, 122n, 124n, 127n, 128n, 129n, 136n
Markus, R. A., 129n
Martin, Abbé A., 46n
Massa, Daniel, 181n
materialism, 179, 192
  in Democritus, 89–90
  Epicurean, 38, 60
  in Hobbes, 8, 179
mathematical truths, 163–7
  Descartes on, 126–7, 132, 163–7
  Kepler on, 127
matter, 16
  activity of, 172, 175, 177–9, 178n, 191, 191n, 191n–192n
  Aquinas on, 26, 172
  Aristotelian theory of, 171n, 172, 173, 205
  Boyle on, 174–5
  Descartes on, 176, 204–19
  Gassendi on, 9, 11, 176, 182–200
  infinite divisibility of, 176, 190, 202, 209
  in the mechanical philosophy, 172, 174–5, 178
  in Platonism, 172
Maull, Nancy L., 208n
Mayo, Thomas Franklin, 181n
Mayr, Ernst, 162n
mechanical philosophy, the
  appeal of, 4–6
  Descartes on, 9, 10, 201–21
  Gassendi on, 9, 10, 180, 182–200

mechanics, Descartes' contribution to, 138, 139n, 213, 213n–214n
mechanization, limits of, 60, 71, 96, 219
Meinel, Christoph, 189n, 193n
Mersenne, Marin, 6, 7, 37, 40, 58, 96–7, 119, 122, 125, 125n, 126, 153, 186
Messeri, Marco, 45n, 191n
method *ex suppositione*, 144
method, scientific
  Boyle on, 226–30
  Descartes on, 140–6, 143n, 218
  Gassendi on, 104, 111–13
  Hobbes on, 227–31
Metzger, Hélène, 41n
Michael, Emily, 61n, 62n, 66n, 68n, 72n
Michael, Fred S., 61n, 62n, 66n, 68n, 72n
microscope, 112, 190
Miel, Jan, 83n, 85n
Millen, Ron, 179n, 216n
Miller, Leonard, 121n
mind–body problem, 70–1, 219–20
Mintz, Samuel I., 6n, 7n, 37n, 89n
miracles
  Aquinas on, 31n
  Boyle on, 31n
  and contingency, 20
  Descartes on, 147–8
  Gassendi on, 55–6
  Ockham on, 31
Mirandola, Gianfrancesco Pico della, 103, 103n
Mirandola, Giovanni Pico della, 103n, 167, 167n
  on astrology, 82
Mitchell, Stephen, 2n
models, mechanical, 199, 218
  in Descartes, 141–2
Molina, Luis de, 83n, 84, 85n
  on freedom, 86n
  influence on Gassendi, 94
  on predestination, 133

scientia media, 85, 94
Montaigne, Michel, 103
Moore, James R., 236n
More, Henry, 119, 224, 224n
  on Boyle, 226
  on the immortality of the soul, 72–3, 73n
  intellectualism of, 227, 227n
  on the Spirit of Nature, 226
  on the void, 225
Morris, John, 139n
Morrison, Margaret, 111n, 143n
mortalism, 60n
Murr, Sylvia, 64n, 71n, 72n

Nadler, Steven, 130n, 138n, 139n, 143n, 214n
  on Descartes' method, 143
natural light, in Descartes, 139, 139n
nature, uniformity of, in Descartes, 210, 217
natures
  Aquinas on, 25
  Aristotelian, 22, 177
  individual, in Gassendi, 114–15
Naudé, Gabriel, 96
necessity, 24, 91n, 234–5
  in Aquinas, 23, 132
  in Averroës, 28
  in Descartes, 132, 139, 139n, 148
  epistemological, 10, 11
  Gassendi on, 55, 80–1, 94
  metaphysical, 10–11
  Ockham on, 30, 32
  in voluntarism, 148
necessity of supposition
  Aquinas on, 22, 22n, 24, 133–5
  Descartes on, 134–5
  Gassendi on, 94–5
Neoplatonism, 129, 225
Newton, Isaac, 9, 207, 224
  criticism of Descartes, 140n, 151–2
  on final causes, 163n
  influence of Gassendi on, 184n
  scientific style of, 231–4

theology of, 59, 76, 151–2, 232
Nielsen, Lauge Olaf, 181*n*
nominalism
  in Gassendi, 9, 56, 68, 102, 105,
    113–17, 157–8
  in Ockham, 30–4
  and voluntarism, 20, 115, 135
noncontradiction, principle of, 23, 24,
    29, 53
Normore, Calvin, 84*n*, 91*n*
Northumberland Circle, 181
Norton, David Fate, 40*n*, 111*n*
Nussbaum, Martha C., 2*n*, 15*n*

Oakley, Francis, 16*n*, 18*n*, 19*n*, 29*n*,
    34*n*, 73*n*, 148*n*
Oberman, Heiko Augustinus, 30*n*,
    34*n*
occasionalism, in Descartes, 149
Ockham, William of, 136*n*, 206*n*
  epistemology of, 31–4
  on future contingents, 91*n*
  on God's power, 29, 30, 31, 54,
    147
  on necessity, 30
  on nominalism, 30–4
  on noncontradiction, 29
  on universals, 30–4
  voluntarism of, 20, 28–34, 134
Oldenburg, Henry, 150, 150*n*
Olscamp, Paul J., 140*n*
Oratory, The, 125, 133
Osler, Margaret J., 4*n*, 39*n*, 76*n*,
    111*n*, 216*n*, 227*n*, 229*n*
Otte, Michael, 223*n*
Owens, Joseph, 139*n*
Ozment, Steven, 84*n*, 124*n*

Palladino, Paolo, 223*n*
parhelia, 202, 202*n*
Pascal, Blaise, 183, 186, 186*n*
Patrizi, Francesco, on space and time,
    183
Pegis, Anton Charles, 70*n*

Peiresc, Nicolas-Claude Fabri de, 37,
    37*n*, 43
Pelikan, Jaroslav, 17*n*
Percy, Henry, 181
Peters, R. S., 224*n*
Petit, Pierre, 186
phantasy
  definition of, 68*n*
  Gassendi on, 68
Pierre d'Ailly, 113
Pintard, René, 44*n*, 46, 46*n*, 154,
    154*n*, 157*n*
Plato, 4, 21
Platonism, 2, 88, 139*n*, 225
plenitude, principle of, 145*n*
Pomponazzi, Pietro, 83*n*, 96
  on astrology, 82–3, 96
  determinism of, 82–3
  on the soul, 61–2, 61*n*
Pontano, Giovanni, on astrology, 82
Popkin, Richard H., 4*n*, 8*n*, 36*n*, 40*n*,
    46, 46*n*, 75*n*, 105, 106*n*, 200*n*
Poppi, Antonio, 82*n*, 83*n*, 84*n*, 92*n*
prayer
  Descartes on, 131
  Gassendi on, 54
predestination
  Calvin on, 83, 88, 93
  Dominicans on, 84–5, 88, 93, 133
  Gassendi on, 93–6
  Jesuits on, 84–5, 133
  Luther on, 83
  Molina on, 84–5, 133
  Suárez on, 84–5
probabilism
  in Boyle, 228
  in Gassendi, 75, 112
  and voluntarism, 20
probability, 107
prophecy, Gassendi on, 98, 99
providence, 8
  Boyle on, 59
  Gassendi on, 54, 56–9, 73, 80–101
  Newton on, 59
  in Stoicism, 41, 81

Puy de Dôme experiment, 186, 186*n*
Pythagoreanism, 2

qualities, 176*n*, 195
   Aristotelian, 172–3, 175–6, 195,
      205
   Descartes on, 195, 215–19
   Gassendi on, 194–200
   primary and secondary, 8, 175,
      176*n*, 195–6, 196*n*, 215, 215*n*
qualities, occult, 6, 178, 186, 198–9,
   216*n*, 226
   Descartes on, 216–17
   in the mechanical philosophy, 178–
     9
   in Renaissance naturalism, 177–8
qualities, real, 205
   rejected by Descartes, 215–16

Radbertus, Paschasius, 16, 17*n*
Ramus, Peter, 103
rationalism
   in Descartes, 205, 216
   epistemological, 19*n*
   in Hobbes, 228
   in Huygens, 232
   theological, 17, 18, 19*n*
realism
   metaphysical, 114
   scientific, 3
Redondi, Pietro, 5*n*, 119*n*
Reformation, 3–4
Reinbold, Anne, 37*n*
Resurrection, The, 188*n*
resurrection
   atomic explanation of, 60*n*
   in Christianity, 60
Reymond, Arnold, 121*n*
Rist, J. M., 38*n*, 81*n*, 108*n*
Roberval, Gilles Personne de, 186
Rochemonteix, P. Camille De, 123*n*,
   124*n*
Rochot, Bernard, 36*n*, 39*n*, 40, 40*n*,
   42*n*, 44*n*, 46, 46*n*, 103*n*, 154,
   155*n*, 167*n*, 186*n*
Rodis-Lewis, Geneviève, 118*n*, 121*n*

Rohault, Jacques, 125
Ross, W. D., 64*n*, 116*n*, 173*n*, 216*n*
Rubidge, Bradley, 124*n*

Sambursky, S., 81*n*, 96*n*, 99*n*, 100*n*
Sarasohn, Lisa T., 37*n*, 40, 40*n*, 42*n*
   44*n*, 45*n*, 71*n*, 80*n*, 84*n*, 85*n*,
   86*n*, 89, 89*n*, 90*n*, 92*n*, 94*n*–
   95*n*, 100*n*, 133*n*, 155, 155*n*
Saunders, Jason Lewis, 41*n*
Schaffer, Simon, 37*n*, 177*n*, 179*n*,
   187*n*, 225*n*, 226*n*, 228*n*
Scheiner, Christopher, 202
Schmitt, Charles B., 3*n*, 4*n*, 103*n*,
   106*n*
Schuster, John Andrew, 147, 147*n*,
   201*n*
"*scientia*," 8, 107
scientific instruments, 112, 190
scientific method, 16
   Aristotle on, 173
   Descartes on, 140–6, 143*n*, 218
   Gassendi on, 104, 111–13
Scotists, 113
Sedley, D. N., 38*n*, 51*n*, 60*n*
sempiternity, 165–6
Seneca, 89
sense perception
   Aristotle on, 216, 216*n*
   Epicurus on, 108*n*
   Gassendi on, 63, 108–10, 194–7
   the mechanical philosophy on, 105
Sextus Empiricus, 106*n*, 197
Shanahan, Timothy, 59*n*
Shapin, Steven, 37*n*, 177*n*, 187*n*,
   225*n*, 226*n*, 228*n*, 230*n*
Shapiro, Barbara J., 76*n*, 107*n*
Sharp, Lindsay, 6*n*, 7*n*
Shea, William R., 118*n*, 202*n*
ship experiment, 37
Shirley, John W., 6*n*, 181*n*
Short, Charles, 50*n*, 166*n*
Siger of Brabant, 28*n*
Simon, Gérard, 121*n*
skepticism, 3–4, 4*n*, 40, 46, 106*n*
   Descartes on, 8, 106, 120, 136, 156

Gassendi on, 8, 105–6, 106, 109, 136, 156, 197
Mersenne on, 106, 136
Soncinas, Paulus Barbus, 128, 129, 129n
Sorabji, Richard, 2n, 87n, 165n, 173n
Sortais, Gaston, 36n
soul
　Alexander of Aphrodisias on, 62
　Aquinas on, 61, 70n
　Aristotle on, 64
　Averroës on, 61
　corporeality of, 74
　Descartes on, 66n, 74, 220–1
　Epicurus on, 59–60, 63
　Gassendi on, 58, 62–77
　Hobbes on, 71
　immortality of, 8, 59–77, 220–1
　More on, 72–3, 73n
　Pomponazzi on, 61–2, 61n
soul, animal, Gassendi on, 64, 64n, 65, 66, 220
soul, rational, Gassendi on, 66, 68–70
space
　Charleton on, 184
　Descartes on, 207
　Gassendi on, 182–3, 193n–194n
　Hobbes on, 184
Spink, J. S., 46, 46n
Spirit of Nature, 226–7
spirits, animal, Descartes on, 219–20, 220n
Spragens, Thomas A. Jr., 172n, 224n
Stanley, Thomas, 77, 77n
Stoicism
　on causality, 81
　on divination, 96, 99, 99n
　on fate, 81, 81n, 88, 89
　on life after death, 75
　on Logos, 81, 81n
　on providence, 41, 81
　and the new science, 41, 42
style of science, 10, 12, 223n, 224n
　definition of, 223–4
　Hacking on, 223n

Kuhn on, 230
Newtonian, 233–4
Suárez, Francisco, 21n, 84, 128n, 129n
　intellectualism of, 18–19
Sutton, John, 81n, 83n, 90n, 97n
swerve of atoms, *see clinamen*
Sylvester of Ferrara, 128, 129n

Tamny, Martin, 9n, 225n
Tarule, Jill Mattuck, 224n
Taylor, Richard, 17n, 19n
teleology, 162, 162n
　in Descartes, 213
telescope, 112
Telesio, Bernardino, 65n
Tempier, Étienne, 27
Tertullian, 15
　on the soul, 74
Thorndike, Lynn, 181n, 199n
Todd, Margo, 93n
Torricelli, Evangelista, 183, 186
transdiction
　definition of, 177
　in Descartes, 217
　in Gassendi, 189
Trentman, John A., 128n
Trevor-Roper, Hugh, 182n

universals, 20, 22
　Aquinas on, 22, 26–7, 114n, 115
　in Aristotelianism, 116
　Descartes on, 163
　Gassendi on, 9, 56, 68, 102, 105, 113–17, 157–8, 163–5
　Ockham on, 30–4
　Plato on, 115–16
universities
　curriculum, 48
Urvoy, Dominique, 28n

vacui, horror, 186
vacuum, *see* void
Valois, Louis-Emmanuel de, 37n
van Goorle, David, 180
Vanini, Lucilio, 58

van Kley, Dale, 83*n*
van Leeuwen, Henry, 75*n*, 196*n*
van Ruler, J. A., 138*n*
Vasquez, Gabriel, 128
Vendler, Zeno, 124*n*
Victor, Ambrosius, 125
Vignaux, Paul, 30*n*, 32*n*
Vives, Juan Luis, 103
void, the
 coacervatum, 183, 186–8
 Descartes on, 183, 186, ʼ.02, 208–
  10, 224*n*
 Epicurus on, 59, 184–6
 extracosmic, 176*n*, 183–4
 Gassendi on, 63, 182–8
 interstitial, 176, 183, 184–6
voluntarism, 18, 19, 230
 in Boyle, 227, 229
 in Damian, 19
 definition of, 11, 17
 in Gassendi, 11, 47, 48–56, 57, 63,
  69, 98, 102, 105, 115, 135, 164–
  5, 167
 in Newton, 151–2, 232
 and nominalism, 20, 115, 135
 in Ockham, 20, 28–34, 134
 and regularity of nature, 55–7,
  73*n*, 148

Walker, Robert, 106*n*
Wallace, William A., 144*n*
Warner, Walter, 181
Weinberg, Julius R., 25*n*, 28*n*, 30*n*
Weinstein, Donald, 39*n*
Wells, Norman J., 19*n*, 121*n*, 123*n*,
  127*n*, 128*n*, 129*n*, 132*n*, 139*n*
Wesseley, Anna, 223*n*
Westfall, Richard S., 9*n*, 31*n*, 58*n*,
  139*n*, 152*n*, 161*n*, 171*n*, 187*n*,
  207, 207*n*, 213*n*, 217*n*, 225*n*,
  231*n*
Westman, Robert S., 3*n*, 224*n*
Williams, Bernard, 17*n*, 141*n*
Wilson, Catherine, 193*n*
Wilson, Margaret Dauler, 120*n*
Wisan, W. L., 223*n*
witchcraft, Gassendi on, 98
Wojcik, Jan W., 76*n*
Woozley, A. D., 113*n*

Yates, Frances A., 82*n*
Yoder, Joella G., 224*n*, 232*n*

Zeno of Citium, on fate, 88
Zeno's paradoxes, Gassendi on, 190,
  194